SCIENCE AND TECHNOLOGY IN THE ISLAMIC WORLD

DE DIVERSIS ARTIBUS

COLLECTION DE TRAVAUX
DE L'ACADÉMIE INTERNATIONALE
D'HISTOIRE DES SCIENCES

COLLECTION OF STUDIES
FROM THE INTERNATIONAL ACADEMY
OF THE HISTORY OF SCIENCE

DIRECTION
EDITORS

EMMANUEL POULLE

ROBERT HALLEUX

TOME 64 (N.S. 27)

BREPOLS

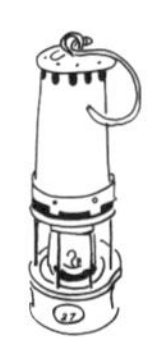

PROCEEDINGS OF THE XXth INTERNATIONAL CONGRESS OF HISTORY OF SCIENCE (Liège, 20-26 July 1997)

VOLUME XXI

SCIENCE AND TECHNOLOGY IN THE ISLAMIC WORLD

Edited by

S.M. Razaullah ANSARI

with a Proceedings' author index by

Daniela BERARIU

BREPOLS

The XXth International Congress of History of Science was organized by the Belgian National Committee for Logic, History and Philosophy of Science with the support of :

ICSU
Ministère de la Politique scientifique
Académie Royale de Belgique
Koninklijke Academie van België
FNRS
FWO
Communauté française de Belgique
Région Wallonne
Service des Affaires culturelles de la Ville de Liège
Service de l'Enseignement de la Ville de Liège
Université de Liège
Comité Sluse asbl
Fédération du Tourisme de la Province de Liège
Collège Saint-Louis
Institut d'Enseignement supérieur "Les Rivageois"

Academic Press
Agora-Béranger
APRIL
Banque Nationale de Belgique
Carlson Wagonlit Travel - Incentive Travel House
Chambre de Commerce et d'Industrie de la Ville de Liège
Club liégeois des Exportateurs
Cockerill Sambre Group
Crédit Communal
Derouaux Ordina sprl
Disteel Cold s.a.
Etilux s.a.
Fabrimétal Liège - Luxembourg
Generale Bank n.v. - Générale de Banque s.a.
Interbrew
L'Espérance Commerciale
Maison de la Métallurgie et de l'Industrie de Liège
Office des Produits wallons
Peeters
Peket dè Houyeu
Petrofina
Rescolié
Sabena
SNCB
Société chimique Prayon Rupel
SPE Zone Sud
TEC Liège - Verviers
Vulcain Industries

D/2002/0095/90
ISBN 2-503-51415-4
Printed in the E.U. on acid-free paper

TABLE OF CONTENTS

The emergence of scientific tradition in Islam 7
Alparslan AÇIKGENÇ

Al-Ghazālī and the emergence of modern science 23
Cemil AKDOGAN

Los itinerarios de aprendizaje exterior de los intelectuales Hispano-Musulmanes : estudio estadístico 43
Carmen ESCRIBANO RÓDENAS y Juan MARTOS QUESADA

Computer applications to the history of the medieval exact sciences : suggestions for future research 65
E.S. KENNEDY

Un complément arabe aux *Données* d'Euclide : Le *Kitāb al-mafrūḍāt* de Ṯābit Ibn Qurra 71
Hélène BELLOSTA

Ḳāḍīzāde al-Rūmī on Samarḳandī's Aṣẖ kāl al-Ta'Sīs : a mathematical commentary 83
Gregg DE YOUNG

The solution of Apollonius' problem in the Medieval Arab East 91
Irina LUTHER

Transmission et innovation : l'exemple du miroir parabolique 101
Roshdi RASHED

Ibn al-Kammād's astronomical work in Ibn al-Hā'im's *al-Zīj al-Kāmil fi-l-Taᶜālīm* 109
Emilia CALVO

Some new Maghribī sources dealing with trepidation 121
Mercè COMES

Eclipses and comets in the Rawḍ al-Qirṭās of Ibn Abī Zarᶜ 143
Mònica RIUS

Contradictions in Taghwim : recent past and present 155
Mashallah ALI-AHYAIE

Quelques aspects du problème de la transmission des connaissances scientifiques en mécanique 177
Mariam M. ROZHANSKAYA

The problems of physics in the works of ar-Razi and Ibn-Sina 183
Abdulhai KOMILOV

Aristotle's *Meteorology* in the Arabic World 189
Paul LETTINCK

Medieval Arabic sources on the magnetic compass 195
Petra G. SCHMIDL

Pionnier dans la recherche du degré de qualité des médicaments composés : le médecin et philosophe arabe al-Kindī (IXe siècle) 209
Nouha STÉPHAN

Séparation de la médecine et la pharmacie : plaidoyer d'al-Rāzī 217
Mehrnaz KATOUZIAN-SAFADI

Arabic *Materia Medica* in Byzantium during the 11th century A.D. and the problems of transfer of knowledge in Medieval science 223
Alain TOUWAIDE

Contributors 249

Proceedings of the XXth International Congress : volumes' list 251

Proceedings of the XXth International Congress : author index 255

THE EMERGENCE OF SCIENTIFIC TRADITION IN ISLAM

Alparslan AÇIKGENÇ

The epistemological ground of science can be deduced from primarily its cognitive nature. A tradition, on the other hand, is a social phenomenon which springs from the social constitution of our nature and as such cannot be deduced from the cognitive aspect of science. This shall lead us to distinguish the cognitive, or rather the epistemic ground of science from its social aspect. In fact these two aspects of science spring from two aspects of man which must be somehow reflected in all human activities as well ; epistemological and sociological. We do not mean, however, that all aspects of man are reducible to these two alone ; on the contrary, our aim, being rather pragmatic, is to show that science as a human activity must manifest such characteristics of man which will be examined here as the social and epistemological foundation of science. This is also the case with the concept of science in Islam. Without developing these two grounds of scientific activities we cannot investigate how a scientific tradition emerged in Islam[1].

A scientific tradition is actually the foundation upon which sciences are built within a certain civilization (or society). But this proposition leaves us with a dilemma which is theoretically circular. This is because before there is any science in a civilization, no learning activity can be characterized as

1. Crucially related to this discussion is the history of Islamic philosophy, which has not yet been questioned with regard to its framework. The classical framework has been formulated and critically evaluated perhaps for the first time by H. Corbin as " Arabic Philosophy began by al-Kindi, reached its height with al-Fārābī and Avicenna, suffered the disastrous shock of the criticism of al-Ghazālī and made heroic effort to rise again with Averroës. That is all. ". See his *Avicenna and the Visionary Recital*, in Willard R. Trask (transl.), Irving, Texas, 1980, 13. Parviz Morewedge also rightly cites the names of original great Philosophers all of whom lived after al-Ghazālī ; " Suhrawardī, Kāshānī, Ibn Khaldun, Dawwānī, Mīr Dāmād, Mullā, Ṣadrā and Sabzawārī, as well as Iqbal ". It is possible to give more examples for the defenders of this false assumption but I shall suffice with referring the reader to Morewedge's article. " Contemporary Scholarship on Near Eastern Philosophy ", *Philosophical Forum*, 2 (1970-1971), 122-140. For a theoretical framework of the Islamic philosophy see the present author's " The Framework for a History of Islamic Philosophy ", *Al-Shajarah : Journal of The International Institute of Islamic Thought and Civilization* (ISTAC), 1 (1996).

scientific ; this being the case, since any tradition of learning or an intellectual tradition can be described as " scientific " only after the existence of sciences, scientific tradition is required for the emergence of sciences, but sciences are required in turn for the emergence of a scientific tradition. Our disapproval of the use of the adjective " scientific " for the intellectual activities prior to the emergence of sciences is defended on the basis of a totally new concept which we would like to introduce here as *scientific consciousness* that is required by the systematic nature of our mind. We shall try to expose this in order to resolve the apparent vicious circularity in our theoretical foundation.

If a specific subject of inquiry is investigated for a long period of time with an uninterrupted chain of investigators, which will be called here " scientific community " (or the *ulamā* within the Islamic civilization), the knowledge accumulated therein will be perceived gradually within a disciplinary unity. When this awareness emerges in the minds of the scholars involved in that activity they become conscious of the fact that subjects or any problems of learning they have been investigating constitute a specific discipline, which is then given a certain name designating thus a particular science. It is such an awareness which we call " scientific consciousness ". If scientific consciousness belongs thus to our mind *naturally,* then it cannot be conventional[2]. But a tradition is almost totally conventional ; hence, if there is such a thing as scientific tradition, then we may infer from this analysis that science is at once conventional and universal. The former ensuing from the ways and manners adopted by the scientific community in question, and the latter from the epistemological character of our mind. We thus conclude that sciences emerge through a process, identified here as " scientific process ". We may distinguish three stages in the history of a scientific process :

1. The stage of problems, where scattered and discrete studies of various problems are carried out for a period of time ;

2. The stage of disciplinary tradition, where a tradition arises as a result of conventional consensus among the scholars ; general subject master and method are determined ;

3. The stage of naming this scientific enterprise.

Accordingly, we need to show in the first place that the Islamic worldview as it emerged out of the Revelation was suitable for the emergence and development of sciences in Islam. It is this worldview that eventually led to scientific progress right from the first century of Islam. We may argue in this vein that there must be some conditions at the social level with all its aspects for

2. The term natural is used here in relation to the Kantian term *a priori*, though as it is clear we would like to emphasize the biological aspect rather than the *categorical* as done by Kant.

the rise of learning in a given society. Since these conditions are the causes for the rise of learning within a certain social and cultural context, they can be named " contextual causes " for the rise of sciences. These contextual causes act as dynamic forces that lead to a pre-scientific tradition of learning and intellectualism and if the society is able to provide some suitable ground for the development of a worldview which acts as the conceptual ground for the emergence of sciences.

In order to be able to enumerate these causes which are necessary for the rise of a scientific tradition we may distinguish certain contextual causes as more rudimentary than others, and hence, prior to all other contextual causes for the emergence of any kind of scientific activity, called " nucleus contextual causes ". Other peripheral elements that help the nucleus contextual causes lead to the emergence of a scientific tradition can be termed " marginal contextual causes "[3]. We distinguish two phenomena as corresponding to the nucleus contextual causes : the first is moral dynamism ; and the second is intellectual dynamism[4].

It is possible to elucidate how moral and intellectual dynamism may take place as social phenomena. The moral unrest within a particular society demonstrates a struggle mainly between two classes of people ; the morally sensitive and the selfish class. The masses remain as the middle class between the two. When the struggle is taking place, although it is only between the morally sensitive and the selfish, it is immediately passed on to the masses, which become the battle ground of the good and evil forces. Some of the masses are thus won to the moral side, and yet others to the selfish front.

There is no human society in which this struggle as *sunnatullah* cannot be found in one form or another. When the morally sensitive people have the suf-

3. A nucleus contextual cause is a dynamism which manifests itself at two levels : first is at the social level, which causes certain unrest and stirring within the society as if the whole structure of the society is reshaping itself and thus every social institution is affected by this dynamism, but most importantly, the political and educational institutions are re-organized as a result of this unrest ; second is at the level of learning and it is this dynamism which causes a lively exchange of ideas on scientific and intellectual subjects among the learned of the community. For instance, in case of Islam we may explain how it was internally generated by the thought of the Qur'an through its dissemination within the first Muslim community. But it is possible to find a universal rule (or rules) governing the generation of that dynamism.

4. Both of these causes fall within the domain of *sunnatullah* because the " nucleus contextual causes " of intellectual progress is a *natural* phenomenon, and therefore, it is deeply rooted within the human dispositions. In fact, for that reason it must be included in the meaning of the Qur'anic concept, *sunnatullah*. Since we claim that the nucleus contextual causes of any intellectual progress are natural, *i.e. sunnatullah*, we thereby accept that it will be the same universally in every society. But the way they are manifested in a given society will definitely vary from society to society, due to the fact that cultures, temperaments and inclinations of societies are different. Moreover, by " natural " we mean " an essentially inert characteristic or a trait *given* or *activated* by God ", referring thereby to nature as " something that is created ". It is in this sense that we must use the Qur'anic term *sunnatullah* to express any contextual cause of intellectual progress that is natural.

ficient vigor, dynamism and energy, they win to their side an adequate number of the masses and thereby produce intellectual and social dynamism.

When the moral struggle between the two groups continue with a victory of the moral class (for this struggle never ends with a victory, but always continues in different forms as long as the society exists), the morally sensitive individuals either produce intellectuals or are themselves intellectuals who formulate original ideas, doctrines and systems by introducing fresh and novel definitions of key concepts that are moral and intellectual or otherwise. This way a lively exchange of ideas and alternative views come into existence within the society ; a phenomenon which is necessary to produce intellectual dynamism.

The moral struggle, which is essentially a strife between the good and evil, may either directly give rise to social dynamism, or to intellectual dynamism first, which, then, in turn produce social dynamism. Hence, although in certain cases social dynamism may precede the intellectual one, it does not mean that social dynamism is a nucleus contextual cause. Therefore, it is still a secondary contextual cause with regard to the nature of the activity in question. For the activity in question is of a cognitive nature, viz., science. But it is this social dynamism that usually leads to an overall activity within the society, that we call " institutional dynamism ", which are indeed many, but we may mention primarily three in order to show their significance in the emergence of sciences and scientific progress : educational, political and economic dynamisms. The society as a whole enters into a process of reshaping its institutions, according to the worldview constructed and fundamental concepts developed therein.

When the new Muslim community faced certain serious challenges, it was set on a tremendous dynamism with all respects. Both the speculative challenge of previous civilizations, more particularly the *Jāhiliyyah* culture, and the Qur'anic encouragement for reflection on the nature of man, his moral and religious responsibility as the *khalīfah*[5] on earth and on the universe, must have led the early generations of Muslims to speculate upon certain problems. As they dealt with these questions, the Prophet enlightened them under the guidance of the Revelation. This is the unfolding process of the construction of the Islamic worldview.

It is plausible, therefore, to infer that, from the very beginning, Islamic civilization was based on *rational* thinking[6] which was guided by the Qur'anic

5. There is no need here to concentrate on the moral struggle of the early Muslim community in this context, as it is quite obvious and irrelevant for our purpose, which is specifically to show the emergence of a scientific tradition.

6. For a detailed discussion of the concept of rationality in the Islamic perspective see the present author's " Transcendent Rationality, Ibn Rushd and Kant : A Critical Synthesis ", *Alif*, 16 (1996).

teaching ; and in this vein the very early Muslim generations began to explain, supplement and rethink the speculative allusions of both the Qur'an and *ḥadīth*. As we argued elsewhere[7] scientific activities are carried out within three mental frameworks : first is the worldview (of the scientist), which is the conceptual environment within which scientific activities are cultivated ; second is the network of a well-defined body of scientific concepts, entitled " context " (of sciences), or more properly called " scientific conceptual scheme " ; and the third is the network of technical vocabulary and the outlook resulting from such a network of concepts within a specific science which I call " specific scientific conceptual scheme ".

It is possible to identify the general scientific conceptual scheme as the " scientific tradition " *if* it is manifested within a certain civilization and thus takes the name of that civilization. This is because a tradition by its very nature requires, besides the conceptual scheme, a community (of scholars).

Therefore, the Islamic scientific tradition is the manifestation of the Islamic scientific conceptual scheme in the Islamic civilization. As such it is primarily the general scientific conceptual scheme with its community, but since this scheme cannot be without its environment, the Islamic scientific tradition necessarily assumes the Islamic worldview. Hence, the Islamic scientific tradition is the Islamic environmental context handed down from one generation of scientists, *i.e.*, the '*ulamā*', to the next[8]. Let us now try to provide the historical content of our theoretical framework.

After the Prophet moved to Medina, he began setting up certain institutions that became the model of education in later Islamic history. The School of the Bench, known as Aṣḥāb al-Ṣuffah, or Ahl al-Ṣuffah (*i.e.*, the People of the Bench) is only one of these educational establishments which was founded by the Prophet himself in Medina[9]. Ṣuffah was originally set apart for the lodging of newcomers and those of the local people who were too poor to have a house of their own. But soon it acquired the character of a regular residential school where reading, writing, Muslim law, the memorizing of chapters of the Qur'an, *tajwīd* (how to recite the Qur'an correctly), and other Islamic sciences were taught under the direct supervision of the Prophet[10]. 'Ubādah ibn al-Ṣāmit says

7. *Islamic Science : Towards A Definition*, Kuala Lumpur, 1996.

8. Obviously, the community of scientists involved in the Islamic scientific tradition is Muslim scientists, but it must also be pointed out that non-Muslim scientists are also included in this community if they accept and work within the same tradition, which was the case, for example, with Zakariyyā al-Rāzī, Ḥunayn ibn Isḥāq, and Maimonides.

9. Ibn Sa'd gives the following names as members of this School : Abū Hurayrah, Abū Dharr al-Ghifārī, Wāthilah ibn al-Asqā', Qays ibn Tihfah al-Ghifārī, 'Abd al-Raḥmān ibn Ka'b al-Asamm, etc. Al-Ḥujwirī mentions 34 names in his *Kashf al-Maḥjūb*, in R. Nicholson (transl.), Leyden, London, 1911, 81.

10. The Prophet was so much concerned with the education of Muslims that when some Meccans were taken prisoners by him after the victory of Badr, he asked those among them who were literate to teach ten children of Medina how to write in exchange for their freedom See Ibn Ḥanbal, *Musnad*, 1 (21), Istanbul, 1992, 247 ; Ibn Sa'd, *Ṭabaqāt al-Kubrā*, 2 (1), 14.

that the Prophet appointed him a teacher in the school of Ṣuffah for classes in writing and in Qur'anic studies[11].

Moreover, there is sufficient evidence that Ṣuffah was not the only school in Medina. Ibn Ḥanbal, for example, records that at a certain time, " a batch of 70 students attended the lectures of a certain teacher in Medina, and worked there till morning "[12]. In fact, there were at least nine mosques in Medina at the time of the Prophet that served as schools : " The people inhabiting the locality sent their children to these local mosques. Quba is not far from Medina. The Prophet sometimes went there and personally supervised the school in the mosque of that place. There are general dicta of the Prophet regarding those who studied in the mosque-schools. He also enjoined upon people to learn from their neighbors[13] ".

In order to understand this tremendous social phenomenon, we need to elaborate only one aspect of this early Islamic worldview which provided the adequate mental environment for the subsequent flourishing of sciences. First of all, the concept of *'ilm* was introduced as a fundamental element :

" Amongst His servants, only the scholars [*'ulamā'*] are God-fearing "[14].

" Are those who know, to be considered equal to those who do not know ? Only prudent men reflect [on this] "[15].

" God will raise in rank those of you who believe and those who are given knowledge "[16].

Many more verses in the Qur'ān can be given to this effect ; the Prophet was even asked to supplicate " O my Lord ! increase my knowledge "[17]. The scholars are honored by being mentioned in rank next to the angels : " God is the witness that there is no deity except Himself, and so are the angels and those endued with knowledge, standing firm on justice "[18]. Same approach was reflected in the mentality of the Prophet :

" Among the signs of the Hour (*ashrāṭ al-sā'ah*-Doomsday) are the decreasing of knowledge and the appearance of ignorance "[19].

" God does not take away knowledge by wresting it from the people, but takes it away by the death of the scholars (*'ulamā'*) until no scholar is left. People begin to accept the ignorant as leaders. When they are asked, they fur-

11. *Sunan Abū Dāwūd*, 2, in Ahmad Hasan's (transl.), Lahore, 1984, 972.

12. *Sunan Abū Dāwūd*, 2, *op. cit.*

13. M. Hamidullah. " Educational System in the Time of the Prophet ", *Islamic Culture*, 13 (1939), 53-55.

14. 35/*al-Fāṭir*, 28.

15. 39/*al-Zumar*, 9.

16. 58/*Al-Mujādalah*, 11.

17. 20/*Ṭā Hā*, 114.

18. 3/Āli 'Imrān, 18.

19. Al-Bukhāri, " Kitāb al-'Ilm ", 71.

nish information without knowledge. They thus go astray and lead the people astray "[20].

" He who is asked about something that he knows but conceals it will have a bridle of fire put on him on the Day of Resurrection "[21].

" If anyone travels on a road in search of knowledge, God will cause him to travel on one of the roads of Paradise, the angels will lower their wings from good pleasure with one who seeks knowledge, and the inhabitants of the heavens and the earth and the fish in the depth of water will ask forgiveness for the scholar (*'ālim*). The superiority of a scholar over a pious (*zāhid*) is like that of the moon on the night when it is full over the rest of the stars. The scholars are the heirs of the Prophets who leave neither money nor property behind, but only knowledge. He who takes it, takes an abundant portion "[22].

" An intellectual (*faqīh*) is more vehement to the Satan than one thousand devout persons (*'ābid*) "[23].

" If God wants to do good to a person, He makes him an intellectual (*faqīh*) in religion "[24].

It is clear that Islamic worldview begins with an immense emphasis on the concept of knowledge. At this time the terms *'ilm* and *fiqh* are used in a peculiar way. Both terms refer to knowledge : the former expresses exact, precise and definite knowledge, while the latter signifies, as we shall show below, *scientific*, and hence knowledge of the rational kind. That is why *'ilm* is used by both the Qur'ān and *ḥadīth* to refer to revealed knowledge which is definite and absolute[25]. The Prophet's prayer for Ibn 'Abbās uses both terms in this contrast : " O God, grant him the understanding of religion and instruct him in interpretation "[26]. As it is clear *'ilm* is used to refer to knowledge which is either revealed or related to that which is revealed[27].

'Ilm was conceived as absolute, and as such it was identified with the Revelation ; but we also find the literal usage of *'ilm* by both the Qur'ān and

20. Al-Bukhāri, " Kitāb al-'Ilm ", 86.

21. Abū Dāwūd, " Bāb al-'Ilm ", *Sunan*, 3650. Translations of the *aḥādīth* from the *Sunan* of Abū Dāwūd are adopted from Ahmad Hasan's translation *Sunan Abū Dāwūd*, Lahore, 1984.

22. Abū Dāwūd, " Bāb al-'ilm ", *Sunan*, 3634 ; also in al-Tirmidhī, " 'Ilm ", 19, al-Nasa'ī, *Ṭahārah*, 112, Ibn Mājah, *Muqaddimah*, 17, Aḥmad ibn Ḥanbal, *Musnad*, IV, 239.

23. Ibn Mājah, " Muqaddimah ", 222.

24. Al-Bukhāri, " Kitāb al-'Ilm ", chapter 14.

25. For this usage, see the following verses : 2/*Al-Baqarah*, 120 ; 3/*Āli 'Imrān*, 61 ; 6/*al-An'ām*, 119, 140, 143 ; 11/*Hūd*, 14, 49 ; 13/*Al-Ra'd*, 37, 43 ; 19/*Maryam*, 43.

26. *Allāhumma faqqihhu fi'l-dīn wa 'allimhu al-ta'wīl* ; see *Kashf al-Khafā' wa Muzīl al-Ilbās*, 1, Ismā'īl ibn Muḥammad al-'Ajlūnī, Beyrut, 1985, 220-221.

27. We are not, however eliminating its usage in the literal sense. See, for example, " We have given them a book (*i.e.*, Revelation) and explained it with a knowledge as a guidance and mercy for people who believe ". (7/*al-A'rāf*, 52 ; also see 4/*al-Nisā'*, 157 ; 6/*al-An'ām*, 119 ; 27/*al-Naml*, 15-16 ; 31/*Luqmān*, 20). " If anyone acquires knowledge of things by which God's own pleasure is sought, yet acquires it only to get some worldly advantage, he will not reach the smell of Paradise ". (Abū Dāwūd, " Bāb al-'Ilm ", *Sunan*, 3656).

ḥadīth. The usage *bi ghayri 'ilm* — without having any knowledge[28], then, means " *'ilm* devoid of revelational content when it should not be so devoid ". Hence, the general meaning of *'ilm* is intimately linked in the knowledge-structure of the Islamic worldview with its usage in the sense of Revelation[29]. The Qur'ān is trying to impart a moral dimension into its knowledge mentality without which disastrous consequences may follow[30]. Therefore, the Qur'anic approach qualifies knowledge with a moral dimension which is provided again by the revelation ; the attitude of indifference to knowledge is thereby excluded from the Islamic worldview. Hence, knowledge is not conceived to be neutral to values, it is inherently linked with values[31]. Once general knowledge is thus invested with values, it becomes *illumined*, which is no longer knowledge that is harmful[32].

As all these conceptions concerning *'ilm, fiqh* and other knowledge-related terms were developed, a doctrinal understanding gradually began to emerge within the Islamic worldview ; it is this comprehensive doctrinal understanding that we call the " knowledge-structure " of the Islamic worldview. As we have seen, this conception emphasizes knowledge with an utmost care, without even leaving it with a mere emphasis, for it also states that " seeking knowledge is an obligation for every Muslim "[33]. Moreover, besides this emphasis, a framework is also given together with the doctrinal understanding of knowledge.

Considering also the Qur'anic encouragement to examine and understand the universe and the nature of certain related problems, it becomes inevitable that as a result of all these comprehensive knowledge-seeking activities, a network of concepts emerges, the Islamic *conceptual scheme*.

The educational activities in early Islam led to the emergence of a group of scholars (a pre-scientific community) who handed down the Prophetic tradition

28. 6/*al-An'ām*, 119 ; 31/*Luqmān*, 20, and so on.

29. E.g. " The ones who do wrong follow their own whims *without having any knowledge*. Who will guide someone whom God has let go astray ? They will have no supporters. So keep your face set straight to the true religion, God's natural handwork along which He has patterned mankind. There is no way to alter God's creation. That is the correct religion, though most men do not *know* " (30/*al-Ra'd*, 29-30).

30. See, for example, " Those who have stupidly killed their own children *without having any knowledge* and forbidden something God has provided them with, have lost out through inventing things about God ; they have gone astray and not been guided...Who can be more harmful than the one who invents a lie about God to mislead people *without having any knowledge* " (6/*al-An'ām*, 140, 144).

31. As this can be observed in the Prophet's pray : " O God, I seek refuge with You from the knowledge which is not useful ". Muslim, " Kitāb al-Dhikr ", 73 ; Abū Dāwūd, " Witr " 32, and so on. See also the verse " They learn what is harmful and not useful to them " (2/*al-Baqarah*, 102).

32. For example, first it states that " the true knowledge is with God alone " (46/*al-Aḥkāf*, 23), then it points out : " above all those who possess knowledge is an All-knowing " (12/*Yūsuf*, 76). Moreover, it categorically declares that " God knows you do not know " (2/*al-Baqarah*, 216 ; see also 3/*Āli 'Imrān*, 65-66).

33. Ibn Mājah, " Muqaddimah ", 17, 224.

of teaching and searching for knowledge to the next generation of scholars who became their students. Soon, as a new generation of scholars began to take over this scholarly tradition, the desire for learning increased ; as a result, a group of scholars with a sophisticated scientific mentality emerged[34]. As a result of the learning activities of these scholars soon various schools of thought emerged, such as the Madenese School, the School of Kūfa, the School of Baṣrah, and also such schools as the Khārijiyyah, Qadariyyah, Murji'a, Shi'ah, Jabriyyah and Ash'ariyyah. Some of these schools emerged as a result of the socio-political upheavals within the Muslim community. It is exactly such events which change the course of contextual causes in a given society. We must, then, acknowledge such social forces that may affect the course of scientific process.

It is through the worldview of the group of scholars working under one scientific tradition that gives science its social character ; we shall define such a group of scholars constituting a unity in outlook and scientific conceptual scheme as " scientific community ", or *'ulamā'* in the Islamic scientific tradition. Our definition of scientific community, or *'ulamā'* leads us to ascribe all social aspects found in scientific activities to this community rather than directly to science. We may, therefore, lay down the following characteristics of a scientific community in general : 1. methodological aim ; 2. scientific ideals ; 3. formal linkage ; 4. marginal ideals[35]. Of course particular scientific communities will have more characteristics than what we have enumerated here. For instance, in the Islamic case, the scientific community, called *'ulamā'* or formerly *fuqahā*, has characteristics which other scientific communities do not have, because of the Islamic worldview and the worldview of the other scientists.

A scientific community is necessary for the emergence of a scientific tradition, and hence, prior to it. In fact, for the existence of a scientific tradition a scientific community is required with a long history. When the initial group of scholars begin to work on certain issues, they attract students who are interested in their knowledge seeking activity in the way they carry it out. In this way a *group* is formed as a result of their knowledge-seeking activity. It is possible to cite two characteristics that belong primarily to the group of scholars

34. Among them, we can give the following names : 'Abdullāh ibn 'Umar (d. 692), Ḥasan ibn Muḥammad ibn al-Ḥanafiyyah (d. 700), Ma'bad al-Juhāni (d. 703), Sa'īd ibn al-Musayyab (d. c. 709), 'Urwah ibn al-Zubayr ibn al-'Awwām (d. 712) Ibrāhīm Nakha'ī (d. c. 717), Ābān ibn 'Uthmān (d. 718), Mujāhid ibn Jabr (d. 718), 'Umar ibn 'Abd al-'Azīz (d. 720), Wahb ibn Munabbih (d. 110, 114/719, 723), Ḥasan al-Baṣrī (d. 728), 'Aṭā' ibn Abī Rabāh (d. 732), Ḥammād ibn Abū Sulaymān (d. 737), Ghaylān al-Dimashqī (d. c. 740), al-Zuhrī (d. 742), Jahm ibn Ṣafwān (d. 746), Wāsil ibn 'Aṭā' (d. 748), 'Amr ibn 'Ubayd (d. 760-1), Ibn Isḥāq (d. 768), Ja'far al-Ṣādiq (d. 765), Abū Ḥanīfah (d. 767), Zurārah ibn A'yān (d. 767), al-Awzā'ī (d. 774), Sulaymān ibn Jarīr (d. c. 793), Hishām ibn al-Ḥakam (d. 795-6), Mālik ibn Anas (d. 796), Abū Yūsuf (d. 799), Sufyān al-Thawrī (d. 778), al-Shāfi'ī (d. 819), and so on.

35. A detailed exposition of these characteristics of scientific communities is dealt with in my unpublished work : *Scientific Thought in Islam : An Essay in the History and Philosophy of Islamic Science.*

which their fellows of the same society do not have : first is that the group of scholars are those who are interested in knowledge-seeking ; second but more importantly is that their interest in knowledge-seeking is in a way that is more systematic and methodical, which distinguishes them from the same activity that may be manifest by everyday people ; therefore, that which brings scholars together as a group is the methodological aim of their activity, not the daily needs of life. In fact, the daily needs of life bring them together with their other fellow beings into the same society, but not into the scientific community.

This aim to pursue knowledge systematically and having the objective of searching for truth leads the group of scholars to organize their community in accordance with the needs and requirements of their activity. Once such an attempt is made a cognitive organization is usually achieved in almost all scientific communities. For the *cognitive organization* is required by our epistemological nature ; if all humans acquire knowledge in the same way then there will necessarily be similarities as a result of their epistemological nature. *Cognitive organization* means setting up the necessary means and the tools needed not only for executing their search for knowledge and truth, but also teaching the knowledge they acquired and the ways in which they thrived to search for that knowledge. In this way an educational initiation prepares and thus passes on the scientific tradition developed by the earlier members of the scientific community.

We, therefore, distinguish the scientific community from its society with respect to their aims and organization ; all characteristics that distinguish both groups of people from each other are expressed here as " methodological aim ", because, as we have shown, they are primarily related to the cognitive aspects of the activities of the members of the scientific community which involves their method. But the scientific community usually idealizes these aims, which do not belong as characteristics to the society in general. There are, therefore, certain scientific objectives which may change from one scientific tradition to another, such as the fact that there are impersonal criteria, impartiality and even certain moral ends that are attached to scientific inquiries. All such objectives that are idealized in a scientific tradition can be referred by a general name as " scientific ideals ". Since the term *science* is strictly applied to the product of the activities of a scientific community in the sense of discipline, scientific ideals cannot be applied to science, but only to the scientific community and their usual practices, the product of which is science[36].

36. R.K. Merton applies the scientific ideals to science as a social institution but identifies them as " disinterestedness ". First of all, there is no institution called science, however, there may be in a society an institution that is governed by the scientific activities and thus can be called " scientific institution ". Science has only four characteristic elements ; subject matter, method, a body of theories and accumulated knowledge. Secondly, it is clear that all these elements are intimately related to our epistemological constitution and thus are cognitive, not social. Therefore, these ideals cannot belong to science, but rather to the scientific community as defined here. See R.K. Merton, *The sociology of Knowledge : Theoretical and Empirical Investigations*, in N.W. Storer (ed.), Chicago, London, 1973, 275.

Science, as we see it, is the product of also a master-student relationship which is linked in an unbroken chain of successors and followers to produce a tradition. It is this self-maintained continuity that we call " formal linkage " as a characteristic of the scientific community. It is indeed the scientific community that prepares the ground for such a scientific continuity which thus enables the establishment of a scientific tradition at the same time. The formal linkage as a characteristic of scientific communities is based like the others upon the epistemological make up of our faculties of knowledge. For instance, we do science in the way we *learn* from our instructors, just as we live in the way as we learn from our environment including our parents and social surrounding[37]. Since formal linkage is also a necessary element in the rise of a scientific tradition, no scientific community can avoid to dispense with it. The establishment of such a link requires a well organized teaching system and an educational institution.

There is also a set of rudimentary characteristics which appears peripheral to scientific activities, such as scientific career and education should be open to talents, scientific activities must be supported not only financially, but socially and politically as well. All such idealized principles of a scientific community we term " marginal ideals ". Just like the scientific ideals, marginal ideals also vary from one scientific tradition to another. But since they are not based on the epistemological nature of our faculties, they are not necessary requirements for the emergence of a scientific tradition, but rather they are complimentary to the necessary ones. They may as such speed up the process of the emergence of such a tradition.

Islamic scientific tradition is, then, the manifestation of the *Islamic scientific conceptual scheme* within the Islamic milieu. *Islamic scientific conceptual scheme* is, on the other hand, the early Islamic conceptual scheme as elaborated after the emergence of special sciences. We thus argue that sciences in Islam first emerged out of a conceptual scheme[38]. The purpose of our exposi-

37. This learning cannot be transcended totally, but only minimally which is what we call " originality ". Therefore, *originality* is a break from the tradition and it cuts off the usual continuity of a scientific tradition. On the other hand, since originality itself is the product of the continuity implanted within the formal linkage, there is a superimposed formal continuity that governs the very process of originality itself. Therefore, when such originalities are continually attached through the formal linkage, a new scientific scheme is produced in individual sciences ; a process that may take hundreds of years. For example, the Ptolemaic and the Copernican models in astronomy ; Aristotelian dynamics and the Newtonian mechanics in physics ; the Ash'arite atomism and the existentialist theory of creation by the Ṣūfīs in Kalām.

38. As an example of this scheme the following concepts can be cited : *' ilm, uṣūl, ra'y, ijtihād, qiyās, fiqh, 'aql, qalb, idrāk, wahm, tadabbur, fikr, naẓar, ḥikmah, yaqīn, waḥy, tafsīr, ta'wīl, 'ālam, kalām, nuṭq, ẓann, ḥaqq, bāṭil, ṣidq, kidhb, wujūd, 'adam, dahr, ṣamad, sarmad, azal, abad, khalq, khulq, firāsah, fiṭrah, ṭabī'ah, ikhtiyār, kisb, khayr, sharr, ḥalāl, ḥarām, wājib, mumkin, amr, īmān, irādah.*

tion is to demonstrate that all these technical terms formed a sophisticated web of concepts until the end of the second century of Islam which eventually led to the rise of individual sciences within this pre scientific tradition (around 800s). Then out of these scientific activities emerged the Islamic scientific tradition. As such it includes not only the conceptual network utilized in scientific activities, but also the norms, practices, educational instruments as well as " set of cultural values and mores "[39] adopted by the group of scholars involved in these activities. We can show this by examining the scientific meanings gradually attached to the concepts located in the early Islamic conceptual scheme. In order to do this we shall select only the most fundamental concepts in Islamic science, as they are situated within the Islamic scientific conceptual scheme. As we still need to do more historical research to bring out materials, we cannot go into a detailed exposition of this. We shall rather concentrate on certain key terms only, which will sufficiently prove our case in this context, in order to exhibit the emergence of the early Islamic scientific tradition.

Al-Zuhrī, for example, says that a sound theory (*al-ra'y al-ḥasan*) is a good piece of knowledge[40]. Ibn 'Abbās's usage " this (Qadariyyah) is the first polytheism of this community. By God, their wrong theory (*sū' ra'yihim*) shall eventually lead them to exclude God from predetermining good, just as they had already excluded Him from predetermining evil "[41]. Of course it may not always be possible to find an equivalent translation of a scientific term of a scientific tradition, coined for a specific meaning within a certain worldview in another scientific tradition. This is the case with the concept of *ra'y*, which does not have an exact corresponding term in the Western scientific vocabulary ; except that the term " theory " is used very much in a meaning close to the term *ra'y*. This is attested also in the report of Ibn Sa'd who states that when 'Aṭā' ibn Abī Rabāḥ was asked concerning his judgment whether it was *'ilm* or *ra'y*, he replied that it was *'ilm*, if his judgment is derived from a precedent, *i.e.*, *athar* ; otherwise, it was implied that the judgment in question was grounded upon *ra'y*[42]. This means that *'ilm* is understood as a definite piece of knowledge which is either directly taken from a revealed source, or derived from it on the basis of a precedent practice of the Prophet. But *ra'y* cannot be *'ilm* in this sense because it is the view of an individual on a certain problem. Hence, *ra'y* actually means " theory " in the Western scientific terminology. Not only does a theory, *i.e.*, *ra'y*, mean " provisional opinion ", it also expresses a rational argumentation because a scientific theory is based on reasoning. This understanding of *ra'y* is also clear from the following usage ;

39. I borrow this phrase from R.K. Merton, *The sociology of Knowledge : Theoretical and Empirical Investigations*, in N.W. Storer (ed.), Chicago, London, 1973, 268.

40. See Ahmad Hasan, *Analogical Reasoning in Islamic Jurisprudence*, Islamabad, 1986, 8.

41. Aḥmad ibn Ḥanbal, *Musnad*, 1 (21), *op. cit.*, 330.

42. *Kitāb Ṭabaqāt al-Kubrā*, 5, in Iḥsān 'Abbās (ed.), Beyrut, 1968, 469.

ni'ma wazīr al-'ilm al-ra'y al-ḥasan (what a good minister of knowledge is the correct theory)[43].

Moreover, since reason is not authoritative in the absolute transcendent realm, the Prophet says that " if one interprets the Qur'ān on the basis of his theory, he has committed an error even if he is correct in his interpretation " (*man qāla fi'l-Qur'ān bi ra'yihi fa aṣāba, fa qad akhṭa'*)[44], since no knowledge can be based on a theory. It is also reported that " sometimes Ibn 'Abbās held a theory which later he abandoned "[45].

The knowledge based on a rational argumentation is reached as a result of *ra'y*, and such a knowledge was actually defined as *fiqh* in the early scholarly tradition to mean *science*, such as al-Tha'ālibī's *Fiqhu'l-Lughah*, *i.e.*, the science of lexicography. Later developments, however, diverted this usage, and perhaps as an influence of the Greek scientific tradition, this usage was dropped and thus replaced by the term *'ilm*. Moreover, according to Abū Ḥanīfah, *fiqh* meant " speculative thinking "[46]. Al-Dhahabī says of 'Abdullāh ibn al-Mubārak that he " recorded knowledge, *i.e.*, *ḥadīth*, in chapters and concerning *fiqh* "[47].This usage of the term has a basis in the Qur'ān[48], as well as in the *ḥadīth*[49].

Ijtihād is also a closely related term in the network of concepts of the Islamic scientific tradition ; it means the *effort to search for knowledge through ra'y*. Hence, *ijtihād* is also a scientific effort which is theoretical. For this reason it is not a definite knowledge, but it must, of course, be based on revealed knowledge. It must be for this reason that the Prophet says : *faḍlu'l-'ālim 'ala'l-mujtahid mi'atu darajah* namely, the scholar who bases himself on true knowledge is a hundred times higher in rank than the theoretical scholar[50]. If we want to show the relation of *ra'y* to *ijtihād*, we can say that *ra'y* is the theory which is produced in an *ijtihād*. This is clear in Mu'ādh ibn Jabal's interesting usage of *ijtihād* and *ra'y* together in the famous *ḥadīth* of *ijtihād* : *ajtahidu ra'yī lā ālū* ; *i.e.,* I shall make my best effort to come up with a theory[51]. But *ijtihād* is necessarily based on the Qur'ān and *ḥadīth*, as understood from this *ḥadīth* as well. It is, therefore, the theoretical knowledge based on the Qur'ān and the *ḥadīth*.

43. Al-Dārimī, *Sunan*, " Muqaddimah ", 30.
44. *Sunan Abū Dāwūd, op. cit.*, 1036.
45. Al-Dārimī, *Sunan*, " Muqaddimah ", 52.
46. L. Gardet, " 'Ilm al-Kalām ", *EI*[2].
47. *Tadhkirat al-Ḥuffāẓ,* 1, Hyderabad, 1955, 275.
48. e.g., 9/al-Tawbah, 122 ; *li yatafaqqahu fī al-dīn*.
49. See the *ḥadīth* quoted above in relation to the Prophet's prayer for Ibn 'Abbās.
50. Al-Dārimī, *Sunan*, " Muqaddimah ", 32.
51. Aḥmad ibn Ḥanbal, *Musnad*, 5, 230.

What about the theoretical knowledge which is primarily derived from discursive thinking ? The early Islamic scientific tradition used the term *kalām* to refer to this kind of knowledge. As such *kalām* meant " speculative knowledge ". The earliest reference in this regard can be taken from Ḥasan al-Baṣrī's (d. 728) letter in which he states that " we initiated the speculative study of *qadar* ; just as people initiated the denial of it " (*aḥdathnā al-kalām fīhi*)[52]. It is also reported that once our Prophet's wife, 'Ā'ishah, heard Ḥasan al-Baṣrī speaking, and asked : " who is this discoursing with the word of the veracious "[53]. Moreover, Muslim reports that " the first to open [speculative] discussions (*kalām*) at Baṣra on free will was Ma'bad al-Juhanī "[54]. Ibn Qutaybah confirms this report on the authority of al-Awzā'ī[55]. In this sense, *kalām* comes very close to the term " philosophy " as it is used today ; *i.e.*, speculative thinking. It is clear why Muslims chose the word '*kalām*' for this kind of knowledge, for *kalām* means " language " or " speech " but not in the ordinary sense. It rather refers to the kind of human language which is discursive. In this sense, it comes close to the term *logos* in the Greek scientific conceptual scheme. It may be translated into English as " discourse ", but in the technical sense of today's Western scientific terminology, it means precisely " philosophy ".

It is clear that all these usages determined the scientific vocabulary of the early Muslims. Not only is the meaning of each term clarified, but so is its relation to other terms and the way, viz., method, they ought to be used is also given. For instance, it is possible to think that since *fiqh* is a rational understanding, it may be a kind of knowledge that is to be avoided by Muslims, as indicated in the above *ḥadīth* that the scholar who bases himself on true knowledge is a hundred times higher in rank than the theoretical scholar, *i.e.*, the rationalist. But another *ḥadīth* clarifies that rational understanding may be decisive in certain cases (e.g., *faqīh wāḥid ashadd 'alā al-shayṭān min alf 'ābid*)[56]. Of course, the Islamic worldview also clarifies in which cases which is to be preferred.

To the concept of knowledge in the Islamic scientific tradition of the Islamic science, the term *ḥikmah* also proved indispensable. Mujāhid, for exemple, explains the term *ḥikmah* in the verse *wa man yu'ta al-ḥikmah fa qad ūtiya*

52. J. Obermann " Political Theology in Early Islam ", *Journal of the American Oriental Society*, 55 (1935), 145 ; Arabic text, H. Ritter, " Studien zur Islamischen Fromigkeit I : Hasan al-Basri ", *Der Islam*, 21 (1933), 68, lines 10-11.

53. *Man hādhā alladhī yatakallam bi kalām al-ṣiddiqīn* ; see Ibn al-Jawzī, *Al-Ḥasan al-Baṣrī*, 8-9 ; quoted by Muḥammad 'Abd al-Raḥīm, *Tafsīr al-Ḥasan al-Baṣrī*, 1, Al-Qāhirah, 1992, 21.

54. Ṣaḥīḥ, " Īmān ", 1.

55. See Ibn Qutayabah, " Awwal man takallama bil'qadar ... ", *Ma'ārif*, in Tharwat 'Ukkashah (ed.), Cairo, 1981, 484.

56. Al-Tirmidhī " 'Ilm ", 13 ; Ibn Mājah " Muqaddimah ".

khayran kathīrā[57] as comprising three things : 1. Al-Qur'ān ; 2. *al-'ilm* ; 3. *al-fiqh*[58].

Here *'ilm* refers to the knowledge of Islamic tradition and the Sunnah, *fiqh* was held as a rational understanding on the basis of the revealed sources. *Ḥikmah*, on the other hand, was understood as knowledge derived rationally from a revealed source, as such it is both *'ilm* and *fiqh* at once, but different from independent speculation, which can be understood as *kalām*. That is why al-Ṭabarī reports that *ḥikmah* was defined by his predecessors as the Qur'ān and its (rational) understanding[59].

In this way every term which was gradually given a specific place in the Islamic conceptual scheme acquired a technical *scientific* meaning, but always in relation to each other. We see, therefore, already towards the end of the first century, these learning activities gradually enter into a disciplinary stage. At this stage, it is possible to observe the scientific consciousness, as a result of which individual sciences are specifically named and referred to by these names, hence particular sciences begin to emerge after the second century. It is in these sciences that the conceptual scheme is elaborated into a *scientific conceptual scheme*. Therefore, the concepts in this scheme are so related to each other that when they are thus held together, they yield a vision, an insight, and an outlook in the mind of the Muslim scientists ; and as a framework in the mind of the Muslim scientists it constitutes what we call here " Islamic scientific conceptual scheme ", whose manifestation through the Islamic scientific community is the Islamic scientific tradition. No matter in what field the scientist is working, by the very epistemological constitution of his mind, he will necessarily reflect this tradition ; and it is this outlook that characterizes a scientific activity as Islamic, since it springs from the Islamic worldview.

And again it is in this sense that a scientific activity will render itself as a part of the Islamic science. We may show the emergence of this tradition in a chronological order on the following table.

57. 2/*al-Baqarah*, 269.

58. Al-Ṭabarī, *Jāmi 'al-Bayān fi Tafsīr al-Qur'ān*, 3, Beyrut, 1980, 60.

59. Al-Ṭabarī, *Al-ḥikmah hiya al-Qur'ān wa'l-fiqh bihi.*

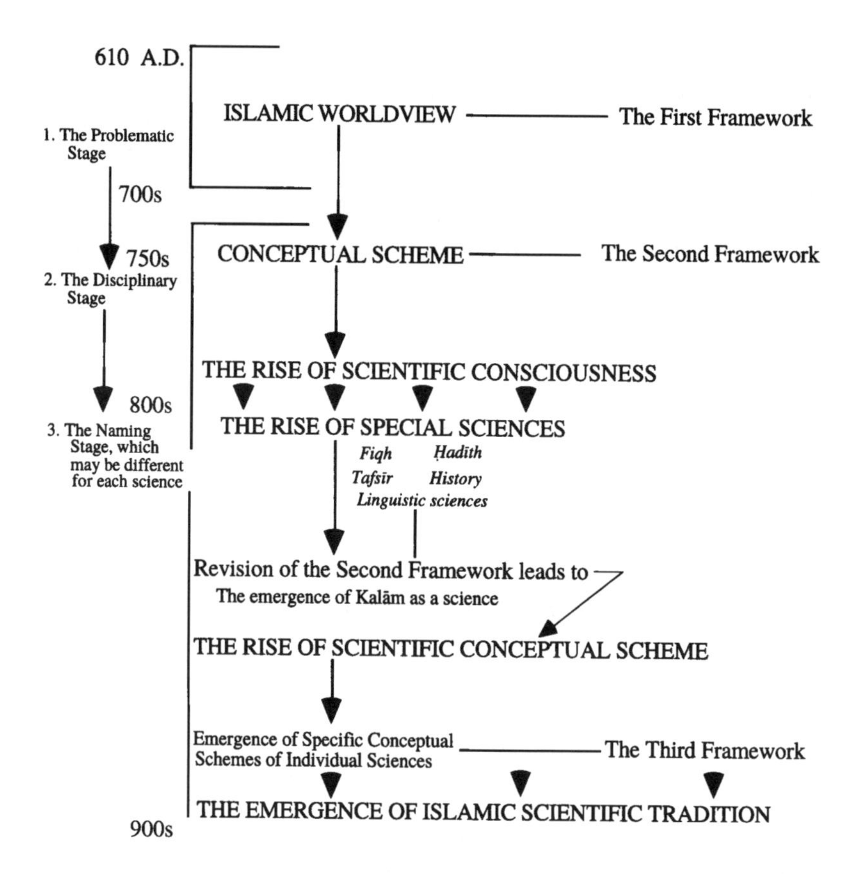
610 A.D.
1. The Problematic Stage
700s
750s
2. The Disciplinary Stage
800s
3. The Naming Stage, which may be different for each science
900s
ISLAMIC WORLDVIEW
The First Framework
CONCEPTUAL SCHEME
The Second Framework
THE RISE OF SCIENTIFIC CONSCIOUSNESS
THE RISE OF SPECIAL SCIENCES
Fiqh
Ḥadīth
Tafsīr
History
Linguistic sciences
Revision of the Second Framework leads to
The emergence of Kalām as a science
THE RISE OF SCIENTIFIC CONCEPTUAL SCHEME
Emergence of Specific Conceptual Schemes of Individual Sciences
The Third Framework
THE EMERGENCE OF ISLAMIC SCIENTIFIC TRADITION

AL-GHAZĀLĪ AND THE EMERGENCE OF MODERN SCIENCE

Cemil AKDOGAN

Almost everybody including many Muslim historians of science such as Aydın Sayılı and Seyyed Hossein Nasr as well as many Western historians of science such as Giorgio de Santillana, G.E. Grunebaum, Edward Grant and J.J. Saunder accuses Ghazali of destroying science or natural philosophy because of his attacks on al Farabi's and Ibn Sina's heretical ideas. More specifically, they all claim that the main cause of scientific decline in Islam is none other than Ghazali's attacks on philosophers[1] but no claim could be further from truth than this one.

How could al Ghazali, a religious thinker with no political power, stop the flourishing of science all by himself ! Although Ghazali attacked the Aristotelian metaphysics developed by al Farabi and Ibn Sina, he supported natural sciences and placed them within the new metaphysical framework which is the mechanical philosophy with its theological clothing. According to him the subject matters of mathematics, logic and natural science are neutral, *i.e.*, they are not against the spirit of Islam even within the context of Greek framework, if they are utilized properly. He writes :

1. H. Floris Cohen writes as follows with regard to Islamic science : " Von Grunebaum, Sayili, and Saunders are all agreed that the root cause of its decline is to be found in the faith, and in the ability of its orthodox upholders to stifle once-flowering science ", *The Scientific Revolution,* Chicago, 1994, 389. Concerning Seyyed Hossein Nasr's stand on al Ghazali Cohen asserts that, " in al-Ghazali's fierce attack against science in the name of the Islamic faith is seen the fundamental cause of the decline of Islamic science by Seyyed Hossein Nasr in his fascinating book *Science and Civilization in Islam.* Only, for Nasr, a Sufi philosopher, the inward turn that ensued does not signify a matter of decline at all ". *Idem*, 590 n. Giorgio de Santillana in the preface he wrote for Seyyed Hossein Nasr's book, *Science and Civilization in Islam*, XII, finds al-Ghazali the main culprit in the decline of Islamic science : " The decline of science inside a great culture is in itself a fascinating study and a terrible object lesson. We can find here the key, in the documents that allow us to judge for ourselves, of the showdown between Averroes and al-Ghazzali. Averroes speaks with the clarity and passionate honesty that we would expect of him, for here was the great Greek tradition at bay, whereas al-Ghazzali's famous eloquence, undistinguished intellectually as it is, and to us ethically uninspiring, went to building up the whirlwind of intolerance and blind fanaticism which tore down not only science, but the very School system and the glorious *ijtihad,* the interpretation of the Quran ".

" MATHEMATICS. This embraces arithmetic, geometry and astronomy. None of its results are connected with religious matters, either to deny or to affirm them. They are matters of demonstration which it is impossible to deny once they have been understood and apprehended… ".

" LOGIC. Nothing in logic is relevant to religion by way of denial or affirmation… ".

" NATURAL SCIENCE OR PHYSICS. This is the investigation of the sphere of the heavens together with the heavenly bodies, and of what is beneath the heavens, both simple bodies like water, air, earth, fire, and composite bodies like animals, plants and minerals, and also of the causes of their changes, transformations and combinations. This is similar to the investigation by medicine of the human body with its principal and subordinate organs, and of the causes of the changes of temperament. Just as it is not a condition of religion to reject medical science, so likewise the rejection of natural science is not one of its conditions "[2].

He not only supported mathematics, astronomy and natural science and but also wrote books on logic for theologians. Moreover, he sternly warned Muslims not to attack science, by presenting an example from astronomy : " There are those things in which the philosophers believe, and which do not come into conflict with any religious principle… An example is their theory that the lunar eclipse occurs when the light of the Moon disappears as a consequence of the interposition of the Earth between the Moon and the Sun. For the Moon derives its light from the Sun, and the Earth is a round body surrounded by Heaven on all the sides. Therefore, when the Moon falls under the shadow of the Earth, the light of the Sun is cut off from it. Another example is their theory that the solar eclipse means the interposition of the body of the Moon between the Sun and the observer, which occurs when the Sun and the Moon are stationed at the intersection of their nodes at the same degree. We are not interested in refuting such theories either ; for the refutation will serve no purpose. He who thinks that it is his religious duty to disbelieve such things is really unjust to religion, and weakens its cause. For these things have been established by astronomical and mathematical evidence which leaves no room for doubt. If you tell a man, who has studied these things — so that he has sifted all the data relating to them, and is, therefore, in a position to forecast when a lunar or solar eclipse will take place ; whether it will be total or partial ; and how long it will last — that these things are contrary to religion, your assertion will shake his faith in religion, not in these things. Greater harm will be done to religion by an immethodical helper than by an enemy whose actions, how-

2. W. Montgomery Watt (transl.), *The Faith and Practice of al-Ghazali,* Oxford (UK), 1994, 33-37.

ever hostile, are yet regular. For as the proverb goes, a wise enemy is better than an ignorant friend "[3].

Moreover, for Ghazali to ideally understand religious knowledge the essential requirement or rather the first step was to study and master natural sciences which encompass God's works[4].

We know that by 1500 natural sciences " in Islam had reached lofty heights, greater than they achieved in medieval Western Europe "[5]. Thus science never came to a stop with Ghazali or after Ghazali, as claimed. Rather than declining it continued to flourish particularly in arithmetic and astronomy long after Ghazali[6].

Although the exact sciences had reached their pinnacles and the technical basis for heliocentricism was ready due to the planetary theories of Ibn al-Shatir, the Scientific Revolution did not take place in Islam. It is my humble opinion that if the external conditions were right Muslims might have been able to bring the Scientific Revolution to completion most probably in the 15th century, thus gaining some precious time for the further development of science.

By inheriting the achievements of Islamic civilization in all areas through translations and personal contacts and also by borrowing and utilizing the magnetic compass, gunpowder, paper making and printing techniques from Muslims the Westerners set out to build the edifice of modern science and technology. But surely they first became the students of Muslims. In the words of Edward Grant, " Latin scholars in the 12th century recognized that all civilizations were not equal. They were painfully aware that with respect to science and natural philosophy their civilization was manifestly inferior to that of

3. Sabih Ahmad Kamali (transl.), *Al-Ghazali's Tahafatul al-Falasifah,* Lahore, 1974, 6.

4. A.H. Al-Ghazali, " Al-Risalat Al-Laduniyya ", Part III, in Margeret Smith (transl.), *Journal of the Royal Asiatic Society* (1938), 373.

5. E. Grant, *The Foundations of Modern Science in the Middle Ages,* Cambridge (UK), 1996, 185.

6. For instance, approximately twenty astronomers worked together at the Maragha observatory in the east in the 13th century and collected data for twenty years. As far as we know this was the first organized observatory in which there were constructed instruments and also a library. Although they worked within the framework of Ptolemaic astronomy, they were also critical of it. That is why the head of this observatory, Nasir al-Din al-Tusi and his student, Qutb al-Din al-Shirazi worked together on a model which was more conservative in accepting the uniform motion than the Ptolemaic system. Later in the Copernican system we also find the same conservative attitude. In the 14th century Ibn al-Shatir, a Damascene astronomer, brought the model of Tusi and Shirazi to completion by developing non-Ptolemaic planetary models and his lunar theory. There are striking similarities between planetary theories of Ibn al-Shatir and Copernicus, although one theory is geostatic and the other theory heliostatic. In the words of E.S. Kennedy and V. Roberts : " The planetary machinery of Ibn al-Shatir " is quite different from that of Copernicus ... to the extent that the universes of the two individuals are geostatic and heliostatic respectively. In all other respects, particularly in the case of Mercury and Venus, the solutions worked out in Copernicus' *De Revolutionibus* for corresponding planets show a remarkable similarity to those of our source ". See E.S. Kennedy, V. Roberts, " The Planetary theory of Ibn al-Shatir ", *Isis*, 50 (1959), 227.

Islam. They faced an obvious choice : learn from their superiors or remain forever inferior. They chose to learn and launched a massive effort to translate as many Arabic texts into Latin as was feasible. Had they assumed that all cultures were equal, or that theirs was superior, they would have had no reason to seek out Arabic learning, and the glorious scientific history that followed might not have occurred "[7].

Because of the process of learning and assimilation of Islamic heritage by the Westerners modern science lost at least several centuries. If the Scientific Revolution had occurred in Islam, today the contemporary science would have been far more advanced.

According to Edward Grant science declined in Islam mostly because of the diminishing role of natural philosophy : " After 1500, Islamic science effectively ceased to advance, but Western science entered upon a revolution that would culminate in the 17th century. What can we learn from this state of affairs ? Let me propose the following : that the exact sciences are unlikely to flourish in isolation from a well-developed natural philosophy, whereas natural philosophy is apparently sustainable at a high level even in the absence of significant achievement in the exact sciences. One or more of the exact sciences, especially mathematics, was practiced in a number of societies that never had a fully developed, broadly disseminated natural philosophy. In none of these societies had scientists attained as high a level of competence and achievement as they had in Islam. Was then the subsequent decline of science in Islam perhaps connected with the relatively diminished role of natural philosophy in that society and to the fact that it was never institutionalized in higher education ? This is a distinct possibility, if natural philosophy played as important a role as I attribute to it throughout this study. Thus in Islamic society, where religion was so fundamental, the absence of support for natural philosophy from theologians, and more often, their open hostility toward the discipline, might have proved fatal to it and eventually, to the exact sciences as well "[8].

If natural or exact sciences cannot develop without a developed natural philosophy, how come " before 1500, the exact sciences [in Islam] had reached lofty heights … without a vibrant natural philosophy "[9]. This is definitely a contradiction, but Edward Grant cannot find a way out of this contradiction. Why ?

Although Grant is sympathetic to Ghazali, he still views Ghazali's achievement in the light of positivism and secularism. That is why he accuses Ghazali and Islamic theologians of being hostile toward natural philosophy and also claims that the exact sciences developed in Islam without having a metaphysical foundation. Due to his bias he did not even ponder the possibility that al

7. E. Grant, *The Foundations of Modern Science in the Middle Ages, op. cit.*, 206.

8. *Idem,* 185-186.

9. *Idem,* 185.

Ghazali or other Muslims might have developed an alternative natural philosophy. If he had realized such a possibility, he would not have contradicted himself.

If we evaluate Ghazali's achievement independently of the simple and wrong prejudice of revelation-reason conflict, we will see that he destroyed not the science itself, but rather the obstacles (particularly Aristotelianism) inhibiting the progress of science, and supplied science with its new metaphysical foundation, namely the mechanical philosophy by making the laws of nature perfect, regular and universal. Today we know for sure that without a metaphysical foundation science cannot develop. Since Ghazali's phenomenal and staggering achievement in the metaphysical arena preceded both Descartes and Gassendi, we must accept him not as a destroyer of science, but as the real and unique pioneer of modern science. Of course, with this statement I do not claim that the modern science actually began with Ghazali or that he directly influenced Descartes[10], Hume and others. Notice further that Ghazali is speaking from the standpoint of Islamic theology whereas the above mentioned thinkers, if they ever profited from the ideas of Ghazali, took them out of context and used them for their own purposes.

My main goal in writing this paper is to point out some close parallels or similarities that existed between what Ghazali achieved concerning natural philosophy and what happened in the West later.

As we will see, Ghazali first pointed out the anomalies or inconsistencies of Aristotelianism in view of Islamic theology and then established the mechanical philosophy, the metaphysical foundation of modern science. In this way Al Ghazali achieved roughly the same things with regard to natural philosophy between 1094 and 1108 what Europeans as peoples would achieve in five centuries, from the 13th century until the eighteenth century.

AL-GHAZALI'S ATTACKS ON THE ARISTOTELIAN NATURAL PHILOSOPHY

By rejecting authority or by symbolizing individualism, a cherished notion in the Renaissance, in the best possible way, Ghazali attacked and destroyed the heretical ideas of Aristotle and Aristotelianism in the three years, most probably from 1092 until 1095. In other words, instead of following the authority in philosophy, al Ghazali broke away from the shackles of ancient or Aristotelian natural philosophy.

From the beginning he wanted to reject the metaphysics of Aristotle developed by al Farabi and Ibn Sina. But to do so he had to learn philosophy by

10. I claim no direct influence despite the fact that according to Mustafa Abu-Sway " Uthman Ka'ak related that he found a translated copy of *al-Munqidh (Deliverance from Error)* in Descartes' library in Paris with his comments in the margin ". See Mustafa Abu-Sway, *Al-Ghazzaliyy : A Study in Islamic Epistemology,* Kuala Lumpur (Malaysia), 1966, 142.

himself, since in Islam nobody could learn philosophy at schools or *madrasas*, which came too late and did include neither philosophy nor natural sciences in their curriculums. That is why Ibn Sina and Ghazali learned philosophy by themselves. Interestingly enough, Ghazali was professor of Islamic sciences at the Nizamiyah Madrasa in Baghdad. When he wanted to criticize philosophy, he studied it in his spare time and learned it in two years. In the third year he assimilated and evaluated what he had learned before. Thus he was ready to write his famous book entitled *Tahafut al-Falasifah (The Incoherence of the Philosophers)* after three years. As he says : " I realized that to refute a system before understanding it and becoming acquainted with its depths is to act blindly. I therefore set out in all earnestness to acquire a knowledge of philosophy from books, by private study without the help of an instructor. I made progress towards this aim during my hours of free time after teaching in the religious sciences and writing, for at this period I was burdened with the teaching and instruction of three hundred students in Baghdad. By my solitary reading during the hours thus snatched God brought me in less than two years to a complete understanding of the sciences of the philosophers. Thereafter I continued to reflect assiduously for nearly a year on what I had assimilated, going over it in my mind again and again and probing its tangled depths, until I comprehended surely and certainly how far it was deceiptful and confusing and how far true and a representation of reality "[11].

But *Tahafut al-Falasifah* was not translated into Latin before the 14th century and therefore was not directly influential in the process of condemnations.

Contrary to the West, where from the 13th century onward theologians first became philosophers by attending the faculty of arts to learn the Aristotelian natural philosophy and even to teach it, before going to the advanced faculty of theology ; in Islam theologians (*mutakallimun*) and philosophers were separate and independent. *Mutakallimun* developed the metaphysics of Islam, which also included a belief in atomism, without coming under the influence of Greek philosophers. On the other hand, philosophers such as al Farabi and Ibn Sina defended Neo-Platonized Aristotelianism. So in Islam there existed two rival systems of metaphysics, one belonging to *mutakallimun* and the other to the followers of Aristotle, namely al Farabi and Ibn Sina. Of course, Ghazali could not let this rivalry and contradictory situation go unchallenged. To that extent he effectively destroyed twenty ideas of Aristotelianism. He found al Farabi and Ibn Sina infidels on three points and heretical on seventeen points. He writes : " The views of Aristotle, as expounded by al-Farabi and Ibn Sina, are close to those of Islamic writers. All their errors are comprised under twenty heads, on three of which they must be reckoned infidels and on seventeen heretics. It was to show the falsity of their views on these twenty points

11. W. Montgomery Watt (transl.), *The Faith and Practice of al-Ghazali, op. cit.*, 29-30.

that I composed *The Incoherence of the Philosophers.* The three points in which they differ from all the Muslims are as follows :

(a) They say that for bodies there is no resurrection ; it is bare spirits which are rewarded or punished ; and the rewards and punishments are spiritual, not bodily. They certainly speak truth in affirming the spiritual ones, since these do exist as well ; but they speak falsely in denying the bodily ones and in their pronouncements disbelieve the Divine revelation ;

(b) They say that God knows universals but not particulars. This too is plain unbelief. The truth is that 'there does not escape Him the weight of an atom in the heavens or in the earth' (Q. 34, 3) ;

(c) They say that the world is from eternity, without beginning. But no Muslim has adopted any such view on this question "[12].

To enter upon a new period such as the Renaissance and then to achieve the Scientific Revolution the first prerequisite was to destroy the hold of Aristotle, which Ghazali had accomplished wonderfully in his *Tahafut al-Falasifah* towards the end of the 11th century. As we will see, from 1210 until 1277 and also in the Renaissance period the West also tried to destroy the authority of Aristotle. Particularly in 1277 the Westerners repeated in essence what Ghazali had achieved with *Tahafut al-Falasifah,* but in a wider scope. Although Ghazali had rejected twenty heretical ideas of Aristotle advocated by al Farabi and Ibn Sina, the Westerners rejected 219 ideas of Aristotle, Ibn Rushd and Thomas Aquinas.

AL-GHAZALI'S NATURAL PHILOSOPHY

After showing the inconsistencies of the Aristotelian natural philosophy, Ghazali laid down the groundwork for the modern metaphysics.

He made soul pivotal to natural philosophy and distinguished it from matter. In his own words, matter or the body is " evil, gross, subject to generation and corruption, composite, made up of parts, earthy "[13] whereas the soul is " substantial, perfect, simple, enlightened, comprehending, acting, moving "[14]. Without soul body is not complete and cannot function. As Ghazali states : " The accident [the body] does not subsist after the substance has passed away, because it does not subsist in itself. For the body is subject to dissolution as it was subject to being compounded of matter and form, which is set forth in the books. And from ... verses and traditions and intellectual proofs, we have come to know that the spirit [the soul] is a simple substance, perfect, having

12. W. Montgomery Watt (transl.), *The Faith and Practice of al-Ghazali, op. cit.,* 37-38.

13. A.H. Al-Ghazali, " Al-Risalat Al-Laduniyya ", Part II, in Margeret Smith (transl.), *Journal of the Royal Asiatic Society* (1938), 193.

14. *Ibidem.*

life in itself, and from it is derived what makes the body sound or what corrupts it "[15].

God created man from two different substances, namely the body and the soul[16]. Although the body is passive and mortal, the soul is alive, immortal and perfect[17]. In the words of Ghazali, " The scholastic theologians, who are skilled in discussion, reckon the soul to be a body and state that it is a subtle body, corresponding to this gross body, and they hold that there is no difference between the spirit and the flesh except in respect of subtlety and grossness. Then certain of them reckon the spirit as an accident ... Now this spirit ... is only a servant, a captive which dies with the death of the body "[18].

Ghazali ended the hold of ancient metaphysics by freeing human soul from its material attachments and making it part of God[19]. Thus for the first time and long before Descartes he revolutionized metaphysics on the basis of the Qur'an : " God related the spirit sometimes to Himself and sometimes to His command and sometimes to His glory for He said : " I breathed into him of My Spirit ", and He said also : " Say, the Spirit (proceedeth) at the command of my Lord ". ...Now God Most High is too glorious to attach unto Himself a body or an accident, because of their lowliness and their liability to change and their swift dissolution and corruption "[20].

Ghazali defined the soul as " perfect, simple substance which is concerned solely with remembering and studying and reflection and discrimination and careful consideration "[21]. and made it an independent and different substance. It is no longer a refined matter, a collection of atoms or accident[22]. Furthermore, the souls, created by God in a special way, are the only direct links between God and men.

After proving the distinction between mind and body to his satisfaction, Ghazali wanted to find out the absolute truth. To that extent he went through a period of personal scepticism during which he even doubted the truth of sense

15. A.H. Al-Ghazali, " Al-Risalat Al-Laduniyya ", Part II, *op. cit.,* 197-198.

16. *Idem,* 193.

17. *Idem,* 197.

18. *Idem,* 195.

19. As we know, soul was part of cosmos in the Greek philosophy. It was made up of delicate materials such as pure water, pure air, pure fire or light atoms and was like an inner engine which made things move whether they were objects or human beings. Even Plato placed soul between the perfect ideas and the sensory world and made it part of the world. As for Aristotle it is the form which is not separable from body or matter. In other words soul cannot exist independently of body and thus in the Islamic world Farabi, Ibn Sina and Ibn Rushd who were Aristotelians could not satisfactorily explain the immortality of human souls. Furthermore, they all accepted that the active intellect which is immortal did not belong to particular individuals, but rather to the whole human species. Thus according to Greek and Islamic philosophers soul either originated from matter or was part of the world or body.

20. A.H. Al-Ghazali, " Al-Risalat Al-Laduniyya ", Part II, *op. cit.,* 197.

21. *Idem,* 194.

22. *Idem,* 196-197.

perception and mathematics. He compared consciousness to the state of dreaming. Later, he once again began to believe logic, mathematics and self evident ideas. As we know, these features also exist in Descartes's philosophy. The main difference between them is that Ghazali works within the framework of Islamic theology whereas Descartes places his emphasis upon human mind or philosophy.

To distinguish between true and false Ghazali doubted everything and even gave up the authority of Islam : " To thirst after a comprehension of things as they really are was my habit and custom from a very early age ... Consequently as I drew near the age of adolescence the bonds of mere authority *(taqlid)* ceased to hold me and inherited beliefs lost their grip upon me, for I saw that Christian youths always grew up to be Christians, Jewish youths to be Jews and Muslim youths to be Muslims ... My inmost being was moved to discover what this original nature really was and what the beliefs derived from the authority of parents and teachers really were, and also to make distinctions among the authority-based opinions, for their bases are oral communications and in distinguishing between the true and the false in them, there are differences of view. I therefore said within myself : To begin with, what I am looking for is knowledge of what things really are, so I must undoubtedly try to find what knowledge really is "[23].

As it is clear from this quotation, he is a rationalist *par excellence,* since even in theology, instead of endorsing authority passively, he sought knowledge by subjecting every belief to the scrutiny of reason[24].

If authority is not trustable, then is it possible to believe in sense perception and necessary truths which seem to be self-evident ? The senses deceive us. For instance, when we look at the sun we see it as big as a coin. But actually it is bigger than even the whole earth itself. How do we know this ? Through our intellect, *i.e.*, geometrical calculations[25]. Then how about mathematical and logical truths ? Can we trust them ? Ghazali responded to this inquiry as follows : " Perhaps only those intellectual truths which are first principles (or derived from first principles) are to be relied upon, such as the assertion that ten are more than three, that the same thing cannot be both affirmed and denied at one time, that one thing is not both generated in time and eternal, not both existent and non existent, not both necessary and impossible ".

23. W. Montgomery Watt (transl.), *The Faith and Practice of al-Ghazali, op. cit.,* 19.

24. A.I. Sabra makes the same claim for all *kalam* : " All kalam, whether that of the Mu'tazila or of the later, " orthodox ", Ash'arites, declares itself against the passive acceptance of authority in matters of faith, an attitude which it calls by the name of *taqlid* (the imitation or unquestioning following of authority), and which it seeks, expressly and as a matter of principle, to replace by a state of knowledge *('ilm)* rooted in reason *('akl)* ". See his article, " Science and Philosophy in Medieval Islamic Theology ", *Zeitschrift fur Geschichte der Arabisch-Islamischen Wissenschaften,* Band 9 (1994), 1-42.

25. A.I. Sabra, " Science and Philosophy in Medieval Islamic Theology ", *op. cit.*, 21-22.

Sense-perception replied : " Do you not expect that your reliance on intellectual truths will fare like your reliance on sense-perception ? You used to trust in me ; then along came the intellect-judge and proved me wrong ; if it were not for the intellect-judge you would have continued to regard me as true. Perhaps behind intellectual apprehension there is another judge who, if he manifests himself, will show the falsity of intellect in its judging, just as, when intellect manifested itself, it showed the falsity of sense in its judging. The fact that such a supra-intellectual apprehension has not manifested itself is no proof that it is impossible "[26].

According to Ghazali we cannot even trust intellect, since it may also deceive us. In order to prove the last point Ghazali offers the example of dreams : " sense-perception heightened the difficulty [of reliance on intellectual truths] by referring to dreams. 'Do you not see', it said, 'how, when you are asleep, you believe things and imagine circumstances, holding them to be stable and enduring, and, so long as you are in that dream-condition, have no doubts about them ? And is it not the case that when you are awake you know that all you have imagined and believed is unfounded and ineffectual ? Why then are you confident that all your waking beliefs, whether from sense or intellect, are genuine ? They are true in respect of your present state ; but it is possible that a state will come upon you whose relation to your waking consciousness is analogous to the relation of the latter to dreaming. In comparison with this state your waking consciousness would be like dreaming ! When you have entered into this state, you will be certain that all the suppositions of your intellect are empty imaginings' "[27].

After God cured Ghazali of scepticism, he once again began to believe in necessary truths. As he says, " The disease [scepticism] was baffling, and lasted almost two months, during which I was a sceptic in fact though not in theory nor in outward expression. At length God cured me of the malady ; my being was restored to health and an even balance ; the necessary truths of the intellect became once more accepted, as I regained confidence in their certain and trustworthy character "[28].

Since Ghazali's belief in *tawhid*, the unity of God was fully restored, he established his metaphysics or natural philosophy. The metaphysical concepts developed first by Ghazali and supported by modern science later include uniformity of laws which govern all motions in the universe, the possibility of an essentially wider universe than that of the traditional (Aristotelian) cosmology and stable causality.

Ghazali established the mechanical philosophy that solely concerns the macroscopic world, by making the laws of nature perfect, regular and univer-

26. A.I. Sabra, " Science and Philosophy in Medieval Islamic Theology ", *op. cit.*, 22.
27. *Idem*, 23.
28. *Idem*, 24.

sal : " If the meaning of ruling is to arrange the causes and apply them to their effects, He [God] will be an absolute arbitrator, because He is the one who causes all the causes, in general and in detail. Branching out from the arbitrator are the divine decree and predestination. His planning the principles positing the causes is so that this ruling may be applied to the effects. His appointing the universal causes — original, fixed and stable, like the earth, the seven heavens, the stars and celestial bodies, with their harmonious and constant movements which neither change nor corrupt — which remain without change until what is written be fulfilled : this is his decree "[29].

According to Ghazali the whole universe works like a machine or a mechanical device such as the horologe (water-clock) which will stop working only when " what is written " is " fulfilled ", but until then nothing will change, *i.e.*, the horologe or the universe will run in accordance with the unchanging laws of God.

In other words, Ghazali's universe is a perfect clock which does not require any interference or winding until the Last Day. In the words of Ghazali : " His applying these causes with their harmonious, defined, planned, and tangible movements to the effects resulting from them, from moment to moment, is His predestination. The ruling is the initial planning of the whole, together with the initial command ... The decree is the positing of universal and constant causes. Predestination applies universal causes with their ordained and measured movements to their effects, numbered and defined, according to a determined measure which neither increases nor decreases. And for that reason nothing escapes His decree and His predestination. This cannot be understood without an example. Perhaps you have seen the horologe by which the times of prayer are announced "[30].

After offering a detailed description of the horologe Ghazali described the universe as " an integrated system of entities and events bound together in an interlocking order of causes and intermediaries "[31].

According to Ghazali terrestrial events result from celestial causes which are permanent. In his own words, " The cause that moves the spheres and the stars and the sun and the moon in a predetermined measure is like the aperture that causes the water to descend necessarily in a predetermined measure. The way in which the motion of the sun and the moon and the stars results in the occurrence of events on the earth is analogous to the way the motion of the

29. Al-Ghazali, *The Ninety-Nine Beautiful Names of God,* in David B. Burrell, Nazih Daher (transl.), Cambridge (UK), 1995, 86.

30. Al-Ghazali, *The Ninety-Nine Beautiful Names of God, op. cit.,* 86-87.

31. R.M. Frank, *Creation and the Cosmic System : Al-Ghazali & Avicenna,* Heidelberg, 1992, 18.

water results in those motions which terminate in the falling of the ball that makes it known that the hour has elapsed "[32].

Thus Ghazali makes the existence of every subsequent event a necessity, by linking all causes and their effects[33].

Amazingly enough, even the prophetic miracles are natural happenings, since as Ghazali says, " the natural processes [in the atomic world] can be accelerated to produce what is called a prophet's miracle "[34].

Moreover, there are no differences between the terrestrial and the celestial regions in terms of diversity and corruption. According to Ghazali there are 1200 different and individual stars in the stellar sphere, differing from one another in sizes, shapes, locations, colors and influences. Since stars are as varied as terrestrial objects, the universe becomes uniform and significantly wider than the Aristotelian one[35]. He also anticipated the work of Galileo in astronomy by theoretically suggesting that the sun decays[36]. So heaven is not perfect, but it is as corrupt and varied as the earth itself.

To annul the distinction between these two regions further Ghazali correlated the movement of a stone with that of celestial objects. On the ground that both movements are constrained and caused by God there exists no difference between them[37].

In this way he described the universe in mechanical terms and also anticipated the developments of the Scientific Revolution. Indeed, the destruction of the medieval cosmos and the establishment of the uniformity of physical laws are essentially what happened in the 16th and 17th centuries.

Now we can ask what makes Ghazali's universe causally stable and constant, since matter decays due to the will of God or the second law of thermodynamics. According to his metaphysics the stability of nature is dependent upon energy, but not on atoms. In the words of Ghazali " bodies are formed by nature, the world-spirit which through energy is the direct cause of shape, bulk quantity and quality of matter. Nature is the sum-total of all energies in the world and assumes various forms made active by the reflection of the Real "[38]. Thus he escaped the chaos of occasionalism with regard to the macroscopic world.

32. I got this quotation from Frank's book, *Creation and the Cosmic System : Al-Ghazali & Avicenna*, 42. A different translation of the same passage can also be found in al-Ghazali's book, *The Ninety-Nine Beautiful Names of God,* in David B. Burrell, Nazih Daher (transl.), 88-89.

33. R.M. Frank, *Creation and the Cosmic System : Al-Ghazali & Avicenna, op. cit.*, 45.

34. Sabih Ahmad Kamali (transl.), *Al-Ghazali's Tahafatul al-Falasifah, op. cit.*, 191.

35. *Idem,* 86.

36. *Idem,* 55-56.

37. *Idem,* 165-166.

38. *Letters of Al-Ghazzali,* in Abdul Qayyum (transl.), Lahore, 1976, 33.

It is interesting to see that Ghazali identified matter or properties of matter with energy ! Even Isaac Newton did not have such a notion. Only with Einstein's famous formula ($E = mc^2$) that we identify matter with energy. Ghazali was also aware of the notion of the conservation of energy, because he suggested that nature consists of " the sum-total of all energies in the world ".

As for the microscopic world, he supported the atomistic belief in a theological clothing long before Pierre Gassendi christianized atomism, and denied the necessity of causality by preceding David Hume.

Mutakallimun believed that God continually destroys the universe in one moment and recreates it in the next one and that " there does not escape Him [even] the weight of an atom in the heavens or in the earth "[39].

This doctrine called occasionalism depends on an atomistic structure of the universe. Being a brilliant *mutakallim* and As'harite, Ghazali also believed that the physical reality, *i.e.* the light of the sun, heavenly bodies, the earth and whatever exists between them consists of atoms[40], but he probably realized that what happens in the world of atoms is beyond human comprehension. That is why he did not discuss what is going on in the microscopic world.

In fact, quantum mechanics confirms the stand of Ghazali on this issue, since we know that the sub-atomic region is weird and acausal from the standpoint of human beings.

In Ghazali's view, God causes everything, therefore there is no room for secondary causation. Cause and effect relation is just a construct of our mind and has no basis in nature. For instance, fire does not actually burn a piece of cotton, but God does.

Ghazali asserted that " the connection between what are believed to be the cause and the effect is not necessary "[41] despite the fact that " the Norm in the past is indelibly impressed upon our minds "[42]. So there are striking similarities between Ghazali's and David Hume's views on causality.

As Eugene A. Myers says : " Al Ghazali held that events are brought about by the will of God rather than by external causes. He therefore denied the principle of causality. This view was adopted by the English thinker David Hume (1711-1776), who defined the relationship of cause and effect as the result of recollections rather than of principle, emphasizing that even though one event follows another the first is not a priori the cause of the second. While al Ghazali referred the ultimate ground to God, Hume referred the ultimate ground to recollections.

39. W. Montgomery Watt (transl.), *The Faith and Practice of al-Ghazali, op. cit.,* 38.
40. Al-Ghazali, *The Ninety-Nine Beautiful Names of God, op. cit.,* 145.
41. Sabih Ahmad Kamali (transl.), *Al-Ghazali's Tahafatul al-Falasifah, op. cit.*, 185.
42. *Idem,* 189.

The similarity of al Ghazali's and Hume's thinking on this subject prompted Ernest Renan the eminent French historian to remind his readers, " Hume has said [about the causal nexus] nothing more than al Ghazzali had already said "[43].

WHAT HAPPENED IN THE WEST ?

At the beginning of the 13th century the Westerners had already translated almost all the books of Aristotle, of other Greek philosophers, and of Islamic philosophers, theologians and scientists from Arabic to Latin. They received technical treatises on astronomy, optics, astrology, mathematics, alchemy and medicine avidly and with great exhilaration, since they filled an existing gap. But Aristotle's books were different because of the heretical ideas they carried. Although Ghazali had written *Tahafatul al-Falasifah* in order to expose those ideas in three years, the Westerners spent at least ten years to understand some of the implications of Aristotle's ideas interpreted by Ibn Sina. The first reaction against Aristotelianism came in 1210. To mollify the charge that students were learning pantheism at the faculty of arts in Paris the Provincial Synod of Sens ordered the faculty members and students of that faculty to read neither Aristotle's books nor Ibn Sina's commentaries under the penalty of excommunication[44]. Five years later the same decree was renewed. About 1230 the Westerners began to read Ibn Rushd's commentaries to learn the genuine ideas of Aristotle instead of the Neoplatonised version offered by Ibn Sina.

Following the spirit of the decrees of 1210 and 1215, Pope Gregory IX ordered that a committee be established in order to weed out the heretical ideas from Aristotle's books and from the commentaries in 1231. As far as we know, the committee never got together and carried out the order. As a result in 1240s, particularly after the death of Gregory IX, the censure on Aristotle and the commentaries lost its power, and Roger Bacon began to teach Aristotle's ideas to the students at the faculty of arts in Paris[45].

In 1255 because of a new statute it became compulsory for the students of faculty of arts to study Aristotle's books to receive their Master's degree. In the period between 1255 and 1270 Albert the Great and Thomas Aquinas compromised Aristotelian natural philosophy and Christianity. But at the same period the followers of Ibn Rushd such as Siger de Brabant and Boethius of Dacia were publicly defending the eternity of the world, the unity of intellect (Ibn Rushd's doctrine of Monopsychism), and the impossibility of the resurrection of the dead body at the faculty of arts in Paris[46].

43. E.A. Myers, *Arabic Thought and the Western World,* New York, 1964, 40.
44. D.C. Lindberg, *The Beginnings of Western Science,* Chicago, London, 1992, 218.
45. *Idem,* 217.
46. *Idem,* 217-218.

In 1270 Etienne Tempier, the Bishop of Paris, condemned thirteen propositions of Aristotle including the above mentioned ideas. More specifically, five of these propositions are related to the doctrine of monopsychism, three to determinism, two to the eternity of the world and three to the denial of divine providence[47]. When this condemnation did not fully eradicate the radical Aristotelianism, in 1277 Etienne Tempier reacted against Aristotelianism more harshly by condemning 219 propositions which belonged to Aristotle, Ibn Rushd, and Thomas Aquinas.

With the condemnations of 1270 and 1277 the Westerners followed the suit of Ghazali who had already criticized Aristotelianism on twenty points towards the end of the 11th century. Although the Westerners or Etienne Tempier listed the heretical ideas one by one, Ghazali had written an impressive and sophisticated book arguing against the main heretical ideas of Aristotelianism. Therefore, Ghazali's achievement is more meaningful, more scholarly, and more conscious. Ralph Lerner, Muhsin Mahdi, and Ernest L. Fortin readily point out the unsystematic character of the condemnation of 1277 : " One cannot say whether the propositions were extracted textually from the writings of the Aristotelians. No ostensible effort was made to introduce a logical order among them. Moreover, their precise meaning remains in some instances obscure, due to the lack of any immediate context. Although visibly influenced by Averroism, many of the propositions represent at best a crude version of the genuine Averroistic teaching "[48].

Condemnations or refusal of Aristotle's heretical ideas weakened the hold of Aristotelianism and encouraged the alternatives to the Aristotelian world view. Later, in the Renaissance period, this anti-Aristotelian movement gave rise to mysticism and animism.

Ghazali was especially influential upon the 14th century and I will dare to say that this century was the Ghazalian century *par excellence,* since there are striking similarities between the ideas of Duns Scotus, Nicholas of Cusa, William of Ockham, Nicolaus of Autrecourt, John Buridan, Nicole Oresme and those of Ghazali concerning causality, the general attitude towards philosophy and the theme of divine omnipotence[49].

47. Abdelali Elamrani-Jamal, " La Philosophie Arabe à l'Université de Paris ", in Ch.E. Butterworth, Blake Andree Kessel (eds), *The Introduction of Arabic Philosophy into Europe,* Leiden, 1994, 37.

48. Ralph Lerner, Muhsin Mahdi (eds), *Medieval Political Philosophy,* Ithaca, New York, 1963, 336.

49. In this paper I am arguing that there were parallel developments, but some believe that al-Ghazali directly influenced the 14th century thinkers. For instance, Harry Wolfson argued that Ghazali influenced Nicolaus of Autrecourt on the problem of causality. See his paper, " Nicolaus of Autrecourt and Ghazali's Argument against Causality ", *Speculum* (1969), 234-238.

THE RENAISSANCE PERIOD

As a reaction against the vain speculation of Scholasticism and Christianity the Westerners focused their attention on the ideas of ancient figures in literature, architecture and the arts. But their real purpose was to create a new age.

In science they rebelled against Aristotelianism and weakened its hold by the recovery of ancient philosophies such as Platonism, Pythagoreanism, scepticism, and animism (Hermetism). Thus, like Ghazali they did not follow authority, namely Aristotle.

In this period of crisis thinkers, scientists and philosophers believed that the universe is an organism and that there are occult forces everywhere. Superior forces which belong to the heavenly bodies can influence inferior forces. With these ideas the Renaissance became a period of magic *par excellence,* since magic or anti-rationalism for the first time gained an intellectual status. Thus, this period, in which witch hunting took place, was not as rational as generally thought, and was anti-scientific.

THE MECHANICAL PHILOSOPHY

The mechanical philosophy that became the foundation of modern science was diametrically opposed to the Renaissance animism and as an alternative both to the Aristotelianism and animism it gained more and more ground with the passage of time. Particularly after the great successes of Isaac Newton it became the dominant paradigm and reigned supreme from the 17th century until the present time.

Descartes (1596-1650), the main architect of the mechanical philosophy, followed the method of Ghazali, but he placed the emphasis on philosophy, rather than theology : " I have always considered that the two questions respecting God and the Soul were the chief of those that ought to be demonstrated by philosophical rather than theological argument "[50].

His main motive in doing philosophy was to refute that soul perishes when body dies so that he could revolutionize philosophy and begin with a new slate. As he says : " And as regards the soul, ... some have even dared to say that human reasons have convinced us that it would perish with the body, and that faith alone could believe the contrary... I have ventured in this treatise to [disprove these tenets] "[51].

So, like Ghazali, Descartes was also after the " real and true distinction between the human soul and the body ". According to him soul is alive, active, and immaterial whereas body is passive. Soul is not body and body is not soul.

50. *The Philosophical Works of Descartes,* in E.S. Haldane, G.R.T. Ross (transl.), Cambridge (UK, 1973, 133.

51. *The Philosophical Works of Descartes, op. cit.,* 134.

Moreover, soul does not occupy a space, but body does. In this way Descartes divorced soul from body by making it a thinking substance that does not perish even after the death of the body : " I knew that I was a substance the whole essence or nature of which is to think, and that for its existence has no need of any place, nor does it depend on any material thing ; so that this 'me', ...is entirely distinct from body, ... and even if body were not, the soul would not cease to be what it is "[52].

As we all know, Descartes rejected all the authority in philosophy and began to doubt everything. He first doubted senses on the ground that they sometimes deceive us. For instance, to misjudge things is possible if they are far away or not clearly perceptible. But how about the things that we clearly perceive such as our wakefulness or the fire in front of which we are sitting ? Descartes argues that we can even be wrong about them. To prove this point he uses dreams as Ghazali had done so a long time ago : " How often has it happened to me that in the night I dreamt that I found myself in this particular place, that I was dressed and seated near the fire, whilst in reality I was lying undressed in bed ! At this moment it does indeed seem to me that it is with eyes awake that I am looking at this paper ; that this head which I move is not asleep, that it is deliberately ... that I extend my hand and perceive it ; what happens in sleep does not appear so clear nor so distinct as does all this. But in thinking over this I remind myself that on many occasions I have in sleep been deceived by similar illusions, and in dwelling carefully on this reflection I see so manifestly that there are no certain indications by which we may clearly distinguish wakefulness from sleep that I am lost in astonishment. And my astonishment is such that it is almost capable of persuading me that I now dream "[53].

Then he doubted necessary truths by raising the following question : " How do I know that I am not deceived every time that I add two and three, or count the sides of a square, or judge of things yet simpler, if anything simpler can be imagined ?[54].

Although he could doubt everything, he could not doubt that he was doubting. Thus he got his first certain proposition : *Cogito, ergo sum.* He later proved the existence of God by depending on the idea of a perfect being in his mind. Thus making theology secondary or God's power over nature ineffective by accepting that the universe is a perfect clock which does not need any winding, he opened the way for secular science.

Descartes, Gassendi, Kepler, Galileo, Boyle and other mechanical philosophers distinguished between primary and secondary qualities to strip animism from nature. According to them, primary qualities (geometrical properties) be-

52. *The Philosophical Works of Descartes, op. cit.,* 101.
53. *Idem,* 145-146.
54. *Idem,* 147.

long to the nature, but secondary qualities such as color, pain, hardness, titillation, and bitter taste arise only in the human mind.

Of course, the separation of mind and nature resulted in the dichotomy between subjectivity and objectivity. In order to become objective and value-free science had to deal with primary qualities which belong to the nature and ignore secondary qualities which are subjective. As a result of this approach mechanical philosophers generally viewed nature as a machine working according to mechanical principles and claimed that the qualities we observe in nature do not actually exist, but arise out of the process of perception.

In this way the Westerners separated nature from both mind and God in order to understand and manipulate it in the sense of Francis Bacon. Thus, if we use the term of Alexander Koyre, nature became " devalorized ". As Prof. Syed Muhammad Naquib al-Attas puts it aptly : " The Cartesian revolution in the 17th century effected a final dualism between matter and spirit in a way which left nature open to the scrutiny and service of secular science, and which set the stage for man being left only with the world on his hands "[55].

However, Ghazali, following the principle of *tawhid*, did not accept the dualism or the distinction between primary and secondary qualities.

As indicated before, Descartes was by no means the only architect of the mechanical philosophy. For instance, another mechanical philosopher, Gassendi revived and Christianized the ancient atomistic school and accepted the conclusions of scepticism underscoring the complexity of nature. Contrary to Descartes he claimed that to understand the essence of anything is impossible and that the only thing to do is to describe the appearances.

As we pointed out before, *mutakallimun* believed atomism, and Ghazali also supported *kalam* atomism. Probably following the example of *mutakallimun* and Ghazali, Pierre Gassendi compromised atomism with Christianity. Although Gassendi revived Epicurean atomism, *kalam* atomism was unique and very different. For instance, according to Gassendi secondary qualities exist only in human mind, but according to *kalam* atomism secondary qualities are part of or accidents of individual atoms, *i.e.*, they reside in nature.

CONCLUSION

As we have seen, there are indeed striking similarities between what al Ghazali achieved with regard to natural philosophy and what happened in the West later. From Kuhnian and progressive standpoint what Ghazali accomplished was actually to destroy the obstacles impeding the progress of science and to supply science with its new metaphysical foundation, namely the mechanical philosophy without which modern science could not develop.

55. Syed Muhammad Naquib Al-Attas, *Islam and Secularism*, Kuala Lumpur, 1993, 36.

Roughly speaking, the Westerners also achieved the same things from the thirteenth century until the 18th century. Since Ghazali's achievements anticipated the later developments in Europe, his contribution to natural philosophy is surely more revolutionary and more phenomenal.

Los itinerarios de aprendizaje exterior de los intelectuales Hispano-Musulmanes : estudio estadístico

Carmen Escribano Ródenas y Juan Martos Quesada

Los contactos de al-Andalus con Oriente y el Magreb, en particular los culturales, han sido analizados desde muy diversos puntos de vista y sus conclusiones utilizadas frecuentemente para dar vida o sostener teorías acerca de la independencia del Islam andalusí con respecto a Oriente, o bien han dado paso a teorías de signo contrario[1].

Desde esta perspectiva, una visión global de todos los datos acerca de viajes al exterior de las biografías de los intelectuales hispano-musulmanes es necesariamente la piedra de toque indispensable que nos puede permitir o no hablar de la posibilidad de una existencia de autonomía del desarrollo de la ciencia y la culture andalusí con respecto a Oriente[2].

Afortunadamente, la literatura árabe medieval acoge en su seno a un género muy específico basado en la relación de biografías de personajes importantes culturalmente : el género *ṭabaqāt* o de " repertorios biográficos ".

Con la denominación general de " repertorios biográficos " queremos referirnos al conjunto de obras cuyo principal motivo es ocuparse de las biografías de los principales maestros de una época o de una clase determinada[3]. La abundancia y proliferación de estos diccionarios biográficos es notable y, desde

1. *Cf.* M.ᶜA. Makkī, " Ensayo sobre las aportaciones orientales en la España musulmana ", *Revista del Instituto de Estudios Islámicos*, XI-XII (1963-1964), 7-140.

2. Efectivamente, así ha sido estudiado ya por algunos investigadores como, por ejemplo, L. Molina, " Lugares de destino de los viajeros andalusíes en el *Ta'rīj* de Ibn al-Faraḍī ", *Estudios Onomásticos-biográficos de Al-Andalus*, I (1988), 585-610.

3. *Cf.* González Palencia, *Historia de la literatura arábigo-española*, Barcelona, 1928, 188 ; J. Mª Fórneas Besteiro, *Elencos biobibliográficos arábigo-andaluces. Estudio especial de la Fahrasa de Ibn ᶜAṭiyya al-Garnāṭī (481-541/1088-1147)*, (extracto de Tesis Doctoral), Madrid, 1971.

luego, los escritores musulmanes eran muy aficionados a la producción de este género historiográfico[4].

Esta literatura de repertorios biográficos, tan típicamente islámica, encuentra su explicación en el concepto musulmán de autoridad y el papel central que ocupó el *ḥadīṯ*[5] en la cultura islámica : para poder transmitir un *ḥadīṯ* era necesario, en principio, haberlo escuchado de viva voz, así como determinadas circunstancias y condiciones de honestidad y honradez del individuo que formaba parte del *isnād*[6], la cadena de transmisores. De esta manera, el conocimiento biográfico de los transmisores se hacía imprescindible[7].

Autores como Heffening[8] se inclinan por considerar este género biográfico, no como una consecuencia de la necesidad de justificar la autoridad de los transmisores, sino un interés primario por la biografía genealógica entre los árabes que, posteriormente, encontró una utilización particular en la necesidad de crítica de las tradiciones y sus transmisores.

En realidad, el género *ṭabaqāt* es muy antiguo en la historiografía musulmana ; su inicio ya comenzó con las biografías de Mahoma, de sus compañeros y de sus discípulos[9]. El género se extendió pronto y en al-Andalus, la predilección por esta forma de literatura fue manifiesta, hasta el punto que este género, en opinión de Pons Boigues, ocupa las dos terceras partes de la historiografía andalusí[10].

Ni qué decir tiene que este género ha sido muy utilizado por los estudiosos[11] como una fuente de información privilegiada para investigaciones de muy diverso signo : profesiones, esperanza de vida, demografía, relaciones entre intelectuales, ciencias más cultivadas, etc. Nosotros los utilizaremos para

4. *Cf.* H.A.R. Gibb, *Islamic biographical literature*, Londres, 1963 ; P. Chalmeta, " Le barnāmaŷ d'Ibn Abī-l-Rabī[c] ", *Arabica*, XV (1968) ; P. Chalmeta, " De historiografía hispanomusulmana ", *Revista de la Universidad de Madrid,* XX (1972), 144 y, especialmente, la bibliografía que da en nota.

5. Se refiere este término a la Tradición respecto a los actos o palabras del Profeta, o bien a su aprobación tácita de palabras o actes efectuados en su presencia ; *cf.* M.Z. Siddiqi, " The importance of Hadith as a source of Islamic law ", *Studies in Islam*, I (1964), 19-25 ; A.J. Wensinck, " The importance of tradition for the study of Islam ", *The Muslim World*, XI (1921), 239-245.

6. Relación de personajes que se habían transmitido el *ḥadīṯ* de unes a otros, formando una cadena de transmisores.

7. *Cf.* J. M[a] Fórneas Besteiro, *Elencos biobibliográficos arábigo-andaluces. Estudio especial de la Fahrasa de Ibn [c]Aṭiyya al-Garnāṭī (481-541/1088-1147), op. cit.*, 3 ; J. Vernet, *La cultura hispanoárabe en Oriente y Occidente*, Barcelona, 1978, 18.

8. Heffening, *Encyclopédie de l'Islam*, 1[a] ed., Leiden, París, 1913-1934, 229-230, s.v. *ṭabaqa*.

9. *Cf.* Pons Boigues, *Ensayo bio-bibliográfico sobre los historiadores y geógrafos arabigo-españoles*, Madrid, 1918, 572.

10. Pons Boigues, *Ensayo bio-bibliográfico sobre los historiadores y geógrafos arabigo-españoles, op. cit.*

11. Véase, por ejemplo, los *Cahiers d'onomastique Arabe*, publicados, desde 1979, por el Centre National de la Recherche Scientifique de Paris ; o bien, los *Estudios Onomástico-biográficos de Al-Andalus*, publicados, desde 1988, por el Consejo Superior de Investigaciones Científicas de Madrid.

analizar de forma cuantitativa los viajes llevados a cabo fuera de los limites de al-Andalus y extraer, en la medida de lo posible, algunas conclusiones.

Como prototipo de intelectual hispano-musulmán, hemos elegido al muftí, que podemos definir, en un principio, como un especialista en el *fiqh,* en el Derecho práctico, y en la aplicación de éste, que ayuda al juez a resolver las dudas planteadas ante casos concretos y que está al servicio de la comunidad como hombre de consulta de todo tipo de temas[12].

El muftí está considerado como una persona culta, cuyo saber desborda en numerosas ocasiones — casi siempre — el del mero campo de la jurisprudencia para sobresalir en disciplinas como la lengua, la poesía, la ciencia, la astronomía, la medicina, la filosofía o la especulación racional. Los muftíes son definidos en numerosas ocasiones por Ibn al-Faraḍī como personas " preocupadas por la ciencia "[13]. Casi todos ellos han viajado al Oriente para ampliar su formación y han aprendido de los majores maestros de la época. En los repertorios biográficos son siempre calificados como personas de grandes cualidades morales, piadosos y depositarios de la confianza de la gente. El respeto que gozaban las fetuas, los dictámenes de un muftí, se debía al prestigio que alcanzaba por la fama de sus cualidades morales y de su sabiduría, llegando a viajar la gente de una ciudad a otra para ver a un determinado muftí.

Como afirma L. Gardet[14], el muftí llegó a constituir con el tiempo, un grupo muy característico de lo que podríamos denominar la clase media urbana, paralela a los grupos formados por la clase artesanal, la burocracia desarrollada por el poder estatal o los comerciantes enriquecidos por la expansión califal y los contactos con el Norte de Africa o el Oriente. Los muftíes constituyen un grupo social representativo de la intelectualidad árabe en general, e hispano-musulmana en particular, que se forma en la ciudad musulmana y que progresivamente es adherida y absorbida por el aparato estatal.

Así pues, veremos, en primer lugar, las ciudades elegidas por los muftíes de cada época y el índice de éstos que acoge de los casi 350 extraídos de los repertorios biográficos. En segundo lugar, analizaremos los resultados obtenidos con el desarrollo cronológico del número de muftíes que visita cada lugar. Y en tercer lugar, creemos interesante estudiar las ciudades de origen de estos intelectuales hispano-musulmanes, a fin de facilitar la posibilidad de establecer relaciones de afinidad de algunas regiones andalusíes con respecto a determinados focos de atracción cultural extranjeros.

Aplicaremos el cupo de veinte años como medida cronológica generacional y dividiremos el conjunto de muftíes en tres grandes grupos que guardan entre

12. Acerca de la figura del muftí, *cf.* J. Martos, " Características del muftí en al-Andalus : contribución al estudio de una institución jurídica hispanomusulmana ", *Anaquel de Estudios Arabes*, VII (1996), 127-144.

13. *Cf.* Ibn Al-Faraḍī, *Ta'rīj*, Cairo, 1966, n° 330, 105.

14. L. Gardet, *La cité musulmane. Vie sociale et politique*, Paris, 1954, 136.

sí cierta coherencia : la época omeya (siglos VIII-X), la época de taifas y de hegemonía de los imperios norteafricanos almorávide y almohade (siglos XI-XII) y la época nazarí (siglos XIII-XV) ; todos ellos mantienen actitudes comunes en lo que respecta a los viajes, lo que facilitará una visión macho más global y de conjunto.

En líneas generales, podemos delimitar siete grandes zonas, dentro del mundo medieval musulmán, hacia las que se dirige la emigración cultural andalusí a lo largo de los ocho siglos :

1.- Zona del Magreb (Marraquech, Tánger, Fez, Tlemcén, Ceuta)

2.- Zona de Ifrīqiya (Qayrawán, Túnez, Bugía, Sicilia)

3. - Zona del Alto Egipto (Egipto, El Cairo, Alejandría)

4.- Zona del Ḥiŷ āz (Meca, Medina, peregrinación)

5.- Zona del Sur de Arabia (Yemen, Sanaa)

6.- Zona del Irak (Cufa, Basora, Bagdad)

7. - Zona de Siria (Damasco, Alepo, Ramla, Creta)

De más difícil clasificación son aquellos muftíes en los que el único dato que poseemos es que " viajaron al Oriente ", pues el concepto de este término geográfico varió en las diversas etapas históricas[15] ; no obstante, podemos partir de la base de que, durante la época omeya, este término era sinónimo de Meca-Medina y Egipto — y, en muy pocos casos, lo era de Qayrawán — ; y a lo largo de los siglos XI-XV, " Oriente " es el término empleado para designar a La Meca, como lugar de peregrinación, y sólo en algún ejemplo aislado se refiere a Marraquech.

Época omeya

Durante la época omeya (siglos VIII-X) se registre en la sociedad cultural andalusí uno de los mayores índices de intelectuales que viajan al exterior con el fin de aprender de los diverses maestros, aunque no siempre éste sea el motivo exclusivo del viaje. Por lo que respecta a los muftíes, de un total de 197 biografiados a lo largo de la época (102 para el Emirato y 95 para el Califato, contando también entre ellos a los que vivieron en estos años clasificados como muftíes que emigraron) tenemos noticias fidedignas de que al menos 69 de ellos viajaron (45 muftíes durante el Emirato, es decir, un 44% del total, y 24 durante el Califato, es decir, un 25%). Estos porcentajes no serán superados hasta la época almohade, en donde el índice de muftíes que viajeron se cifra en un 39%.

15. El Oriente (Mašriq) es entendido como la región que comprende los países al este de Tripolitania, en contraposición - a *Magrib*, Occidente, es decir, Egipto, Arabia y el Creciente Fértil, hasta su frontera con Irán ; *cf.* N. Elisseeff, *L'Orient musulman au Moyen Age*, Paris, 1977.

CUADRO I.

Número de muftíes que viajaron en la época omeya.

	n° total de muftíes	viajaron	%
Emirato	102	45	44%
Califato	95	23	24%

A lo largo de las diversas generaciones existentes en estos tres largos siglos, la intensidad de viajes varió con frecuencia, habiendo épocas en las que el porcentaje de muftíes que traspasaban las fronteras andalusíes era del 100% y otras en las que el índice no superaba el 13% del total.

GRÁFICO II.

Evolución de los viajes en la época omeya.

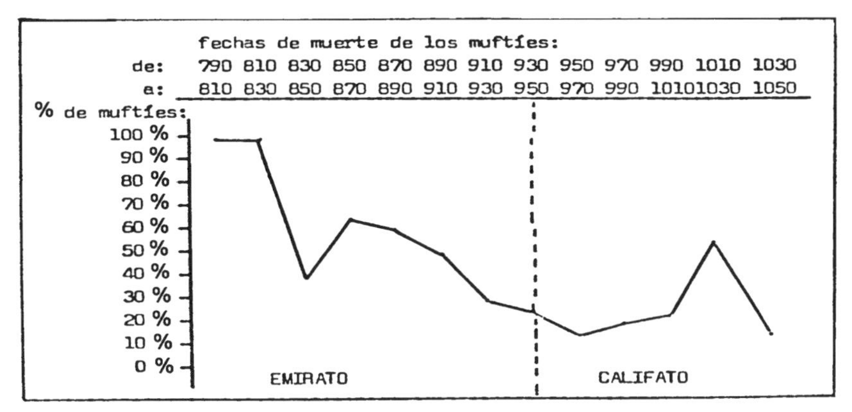

Es en los primeros años del Emirato cuando se registran las mayores cifras de salidas : en las dos primeras generaciones (muftíes fallecidos entre 790 y 830), todos los muftíes viajan, habiendo un descenso entre los muftíes de principios del siglo IX (fallecidos entre los años 830-850) para, posteriormente, mantenerse la línea porcentual durante algunas décadas en cotas que oscilan alrededor del 60%.

Con los muftíes de finales del siglo IX y primera mitad del siglo X (fallecidos entre los años 890-970), estadísticamente, el nivel de los que oyen a los maestros orientales y magrabíes baja progresivamente desde un 60% hasta un 16%, para volver a iniciar una lenta recuperación entre los muftíes de la segunda mitad del siglo X (fallecidos entre los años 990-1030) que alcanza su punto mas alto, con un 57% de ellos, entre los desaparecidos en la veintena de los años 1010-1030.

Por último, la postrer generación de muftíes es la que registra cuantitativamente el número más bajo de viajeros : sólo un 15 % de los 13 muftíes biografiados.

Las regiones geográficas mas visitadas son las de Ifrīqiya, Egipto y el Ḥiŷāz, región esta última que absorbe a todos los que viajan con fines de peregrinación, aparte de los de aprender y asistir a las clases de los maestros más conocidos de la época.

Detalladamente, los destinos de los muftíes durante la época omeya son los siguientes (véase cuadro III).

CUADRO III.

Cronoloyía y direcciones de los viajes en la época omeya.

nº total de muftíes:		(6)	(1)	(5)	(8)	(10)	(27)	(32)	(36)	(25)	(17)	(20)	(7)	(13)	(197)	
fechas de muerte de:		790	810	830	850	870	890	910	930	950	970	990	1010	1030	totales:	
	a:	810	830	850	870	890	910	930	950	970	990	1010	1030	1050	ciud.	reg.
región	ciudad															
-Magreb	-Magreb...	-	-	-	-	-	-	-	-	-	-	-	1	-	1	2
	-Ceuta....	-	-	-	-	-	-	-	-	-	-	-	1	-	1	1%
-Ifrīqi-ya	-Ifrīqiya.	1	-	-	-	1	-	1	1	-	-	-	-	-	4	21
	-Qayrawān.	-	-	-	-	2	8	2	2	-	1	1	-	-	16	10%
	-Túnez....	-	-	-	-	-	-	-	-	-	-	1	-	-	1	
-Egipto	-Egipto...	3	-	1	3	4	8	5	1	2	1	-	-	-	28	28 14%
-Ḥiŷāz	-Ḥiŷāz....	-	-	-	1	-	-	1	-	-	-	-	-	-	2	34
	-La Meca..	-	-	1	-	1	2	2	2	-	1	-	-	-	9	17%
	-Medina...	4	1	2	-	2	-	-	-	-	-	-	-	1	10	
	-Peregrin.	-	-	-	2	-	-	1	4	1	1	2	1	1	13	
-Arabia del Sur	-Yemen.....	-	-	-	-	-	1	-	1	-	-	-	-	-	2	3
	-Sanaa.....	-	-	-	-	-	-	-	1	-	-	-	-	-	1	2%
-Irak	-Cufa......	-	-	-	-	-	-	1	-	-	-	-	-	-	1	6
	-Basora....	-	-	-	-	-	-	1	-	-	-	-	-	-	1	5%
	-Bagdad....	-	-	-	-	-	1	1	1	-	1	-	-	-	4	
-Siria	-Damasco...	-	-	-	-	-	-	1	-	-	1	-	-	-	2	5
	-Alepo.....	-	-	-	-	-	-	-	-	-	1	-	-	-	1	4%
	-Ramla.....	-	-	-	-	-	-	-	-	-	1	-	-	-	1	
	-Creta.....	-	-	-	-	-	-	-	1	-	-	-	-	-	1	
-sin especificar	-Oriente...	-	-	-	1	2	-	2	1	1	-	1	2	-	10	7%
	-viajaron..	-	-	-	-	-	-	-	-	1	1	-	-	1	3	2%

La región más visitada es la del Ḥiŷāz, con un total de 34 muftíes (17% del conjunto de los muftíes omeyas) a los que, en justicia, habría que añadir los 10 clasificados como viajeros al Oriente sin especificar.

De estos 34 muftíes, más de 20 (la suma de los encuadrados en los conceptos de " La Meca ", " Peregrinación " y " Ḥiŷāz "), bien con fines pedagógicos, bien por motivos de cumplimiento del precepto de peregrinaje, aunque sólo en 9 de ellos se especifique claramente que estuvieron en La Meca para aprender de sus maestros.

10 muftíes más tienen como meta de sus Viajes Medina, en especial entre las primeras generaciones de ellos (entre los fallecidos desde el año 790 al año 890), mientras que la generación de muftíes formada por los que murieron en la cincuentena siguiente (entre los años 890 y 950) engloba a la mayor parte de los que eligieron La Meca para aprender y los fallecidos entre los años 950-1050 — los muftíes del Califato — registra y comprende a los que sólo especifican que viajaron a La Meca para peregrinar.

La segunda región más visitada es la Egipcia, de la que únicamente una ciudad, Miṣr, es la aglutinadora de los 28 muftíes que visitan esta zona, ya que no hay comentarios en las biografías de ningún muftí que viajara a Alejandría o a cualquier otra ciudad de esta región.

Con una frecuencia muy parecida (21 muftíes visitantes, un 10% del total) se encuentra la región de Ifrīqiya, que también agrupa casi exclusivamente a todos los muftíes en la ciudad de Qayrawān, si exceptuamos un caso aislado que pasó un tiempo en Túnez y otros cuatro de los que sólo sabemos que " viajaron a Ifrīqiya ".

El resto de las regiones no llega a alcanzar niveles apreciables estadísticamente : la zona de Irak y Arabia del Sur es visitada sobre todo por los muftíes de la última etapa del Emirato (fallecidos entre los años 890 y 950) ; la zona de Siria (5 muftíes, un 4% del total omeya) como meta cultural es descubierta por los muftíes de finales del siglo X (fallecidos entre el año 790 y el año 990) y el Magreb es más visitado por los muftíes que murieron en la veintena que va del año 1010 al 1030. Ninguna de las ciudades de estas regiones visitadas por los muftíes de al-Andalus (Sanaa, Bagdad, Alepo, Basora, Ramla, Creta, etc.) lo es en algún caso por más de un muftí por generación.

Por último, señalaremos que entre los muftíes de los primeros años del Califato tenemos a 2 de los que sólo se indica que " viajaron ", sin especificar ningún detalle más, a los que hay que añadir otro de las mismas características al final del mismo período omeya, muerto entre los años 1030 y 1050.

Un análisis cronológico de las frecuencias de viajes a las grandes - regiones preferidas por los juristas omeyas de al-Andalus (Ifrīqiya, Egipto y Ḥiŷāz) nos mostrará las épocas de mayor o menor auge (véase gráfico IV).

GRÁFICO IV.

Importancia de los puntos de destino de los viajes en la época omeya.

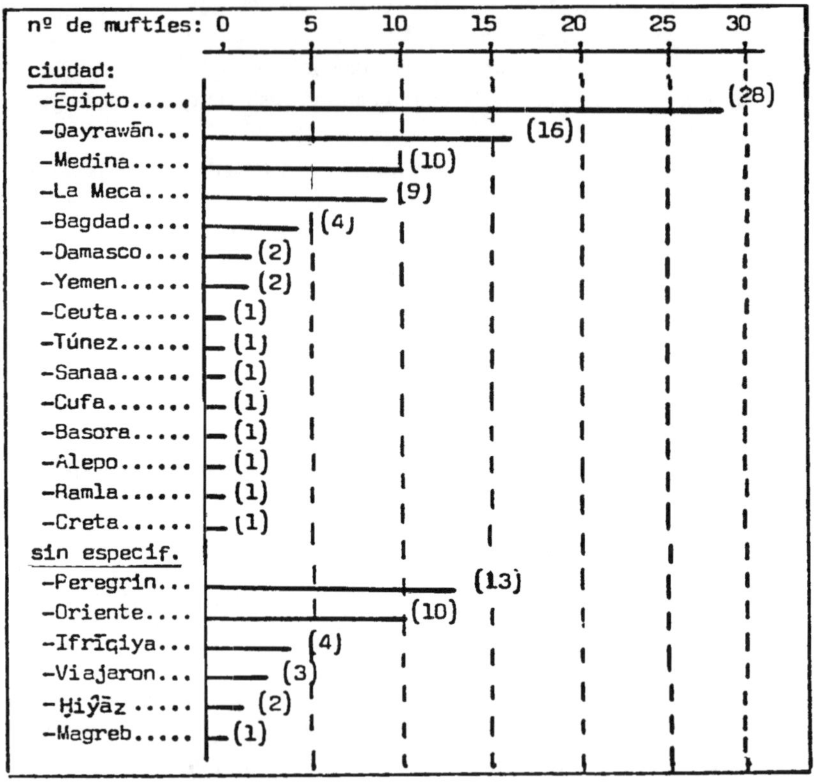

	nº de muftíes: 0 5 10 15 20 25 30
ciudad:	
-Egipto.....	(28)
-Qayrawān...	(16)
-Medina.....	(10)
-La Meca....	(9)
-Bagdad.....	(4)
-Damasco....	(2)
-Yemen......	(2)
-Ceuta......	(1)
-Túnez......	(1)
-Sanaa......	(1)
-Cufa.......	(1)
-Basora.....	(1)
-Alepo......	(1)
-Ramla......	(1)
-Creta......	(1)
sin especif.	
-Peregrin...	(13)
-Oriente....	(10)
-Ifrīqiya...	(4)
-Viajaron...	(3)
-Ḥiŷāz.....	(2)
-Magreb.....	(1)

Las primeras generaciones de muftíes de al-Andalus tienen inevitablemente como meta La Meca-Medina y Egipto, por este orden, alcanzando Qayrawán (Ifrīqiya) sólo un alto índice de muftíes viajeros entre los fallecidos desde el ano 870 al año 910 (30% y 32%), es decir, la época del gran maestro Saḥnūn, sin que tampoco en esta época se llegue a superar el nivel de los muftíes que se dirigen a Egipto.

A partir de esta generación, Ifrīqiya no será en ningún momento una meta relevante de los juristas andalusíes (en ninguna generación se sobrepasa el 10% de los muftíes : 9%, 8,5%, 0%, 6%, 10%, 0% y 0%), como una meta en sí, sino sólo como una etapa del camino hacia Egipto, ciudad que sigue conservando, junto a La Meca, unos niveles respetables de asiduidad, a pesar de receso de muftíes viajeros que se registra en las generaciones de los fallecidos entre los años 910 y 990, particularmente Egipto.

GRÁFICO V.

Evolución de las grandes regiones visitadas por los muftíes en la época omeya.

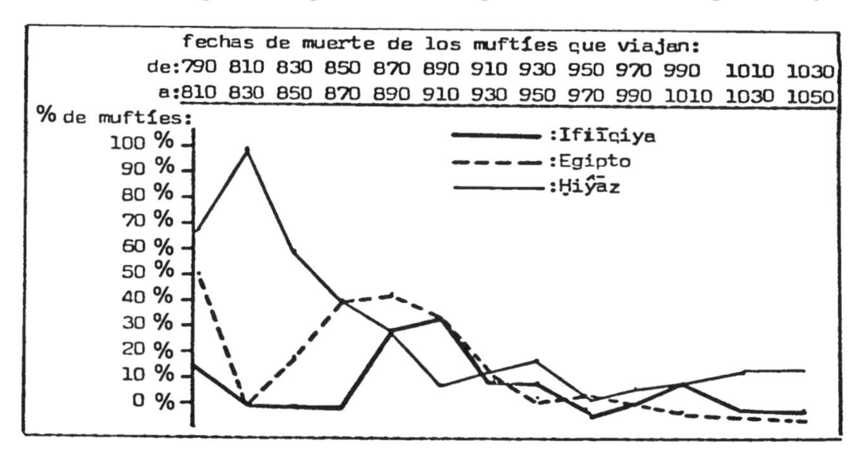

En las dos últimas generaciones de muftíes de la época omeya (final del Califato : muertos entre 1010-1050), los viajes a Ifrīqiya y Egipto descienden de forma absoluta y sólo se tiene constancia de muftíes que visitaron La Meca con el motivo principal de hacer la peregrinación, aunque es de suponer que aprovecharen su estancia para escuchar a los maestros de la época.

De las posibles concomitancias que puedan existir entre los lugares de origen de los muftíes de la España musulmana y los focos orientales o magrebíes de atracción de los juristas, diremos que el conjunto de muftíes que viajan en la época omeya procede en su totalidad de 10 localidades o coras : Córdoba, Toledo, Algeciras, Sevilla, Medina, Sidonia, Elvira, Pechina, Marca Superior, Ecija y Tudmīr.

CUADRO VI.

Ciudades de origen de los muftíes que viajan en la época omeya.

localidad	n° muftíes de la cora	n° muftíes viajeros	%
Córdoba	82	38	46%
Toledo	31	11	35%
Algeciras	9	6	66%
Sevilla	10	3	30%
M. Sidonia	8	2	25%
Elvira	10	2	20%
Pechina	2	2	100%
Marca Sup.	5	2	40%
Ecija	11	1	9%
Tudmīr	2	1	50%

Proporcionalmente, el mayor porcenteje es el obtenido por Pechina, pues los dos muftíes que hemos biografiado de esta cora en la época omeya viajaron, al igual que el 50% de Tudmīr, que en realidad no expresa otra cosa sino que sabemos que uno de los dos muftíes registrados en este época viajó. No obstante, mucho más relevante estadísticamente es el 66% alcanzando por Algeciras (de sus 9 muftíes, viajan los 2/3) o Córdoba, en donde casi la mitad de sus muftíes (un 46%) viajaron fuera de las fronteras andalusíes con el ánimo de escuchar a los maestros del *fiqh* y el *ḥadīṯ*.

Toledo, Sevilla o la Marca Superior se colocan en unos niveles discretos (entre un 40 y 30% : 35%, 30% y 40% respectivamente), y con una media algo más baja les seguirían en proporción de muftíes que viajan, Medina Sidonia y Elvira, la primera con un 25% de sus muftíes (2 de sus ocho) y la segunda con un 20% (2 de sus 10 muftíes) ; Écija se convierte, con un solo muftí que sale al exterior de sus 11 registrados, en la cora o localidad con el menor índice del conjunto de ciudades (un 9%).

Por otra parte, es interesante observar que coras como Badajoz, Beja, Cabra, Carmona, Jaén, Morón o Rayya, puntos geográficos importantes en la época omeya y que cuentan todas ellas con muftíes biografiados, no tienen a ninguno del que nos haya quedado constancia de que viajara.

CUADRO VII.

Relación entre las coras de origen y las ciudades visitadas por los muftíes de la época omeya.

	regiones visitadas:								
coras de origen:	Magreb	Ifrīqiya	Egipto	Ḥiŷāz	Arab.sur	Irak	Siria	Oriente	"viajó"
-Córdoba.	2	11	13	17	1	2	2	4	1
-Algecir.	-	-	2	4	-	-	-	1	1
-Toledo..	-	5	2	3	1	-	-	2	-
-Ecija...	-	-	-	1	-	-	-	-	-
-Sevilla.	-	-	1	1	-	-	-	1	-
-M.Sidon.	-	2	2	-	-	-	-	-	-
-Marca S.	-	-	1	-	-	1	-	-	1
-Elvira..	-	1	2	-	-	-	-	-	-
-Pechina.	-	1	1	-	-	-	-	1	-
-Tudmīr..	-	-	-	-	-	-	-	1	-

Atendiendo a las tres grandes regiones que son focos de atracción : Ifrīqiya, con Qayrawán, Egipto, con Miṣr, y el Ḥiŷāz, con La Meca y Medina, observamos que, en lo que respecta a la primera región, Toledo es la cora andalusí que más muftíes proporcionalmente dirige hacia Qayrawán (5 del conjunto de los 11 biografiados), mientras que Córdoba, Medina Sidonia, Elvira o Pechina lo hacen con unos índices menos significativos ; por el contrario, Sevilla, Écija, Algeciras o la Marca Superior no cuentan con ningun muftí que se sienta atraído por el ambiente culturel qayrawanés.

Egipto, que no cesa de recibir muftíes de al-Andalus durante casi toda la etapa omeya, es escogido mayoritariamente como meta por los muftíes de Córdoba, Elvira, Medina Sidonia y Pechina, mientras que los juristas de Sevilla, Algeciras, Toledo o la Marca Superior sólo se deciden por Egipto como ciudad secundaria en sus itinerarios por Oriente.

La Meca-Medina, la región del Ḥiŷāz, cuenta con un gran número de muftíes procedentes de Córdoba, Algeciras y Écija, y con algo menos de Sevilla y Toledo, mientras que no se detecta presencia alguna de muftíes de Medina Sidonia, Elvira, Pechina, Tudmīr o de la Marca Superior.

Elaborando las tres rutas más frecuentadas (Ifrīqiya-Egipto-Ḥiŷāz, por una parte, Egipto-Ḥiŷāz, por otra, y finalmente, Ifrīqiya-Egipto), podemos afirmar que la primera ruta cultural es seguida especialmente por los muftíes de Córdoba y Toledo (éstos con un peso mayor en Qayrawán y aquéllos en Egipto).

El segundo itinerario, formado por el triángulo Egipto-La Meca-Medina, es el llevado a cabo preferentemente por los muftíes de Sevilla y Algeciras, mientras que el tercer itinerario (Ifrīqiya-Egipto) es frecuentado por los juristas de Medina Sidonia, Elvira y Pechina, que no cuentan con ningún muftí que viajara a La Meca o Medina. En cuanto a los focos exteriores más marginados (Arabia del Sur, Irak, Siria o el Magreb), excepto un par de casos aislados, son los muftíes cordobeses los únicos que se deciden a ampliar el periplo tradicional de la emigración cultural omeya.

En síntesis, podemos concluir que los muftíes omeyas centran sus preferencias en tres puntos : Qayrawán (en especial en la época de Saḥnūn, a mediados del siglo IX, es decir, los muftíes de la 2ª mitad del siglo y muertos entre el año 870 y el año 890, y sobre todo, es elegida esta ciudad por los muftíes de Toledo), Egipto, (que se mantiene a lo largo de toda la época, alcanzando su máxima frecuencia en la primera mitad del siglo IX, es decir, con los muftíes del 2º tercio de este siglo y muertos entre el año 850 y 890, aunque se pierde esta tradición a lo largo de las tres últimas generaciones de muftíes del Califato, y siendo este punto el preferido por los muftíes de Córdoba mayoritariamente) y La Meca-Medina (en particular Medina entre los primeros muftíes y La Meca entre los de las últimas generaciones, con una especial atracción de los residentes en Algeciras).

Las coras con muftíes que viajaran no supera el número de diez, destacando Córdoba, la región Norte con Toledo y la Marca Superior, y el Sur de la Península, con Algeciras, Medina Sidonia y Pechina, siendo las coras que mayor porcentaje de muftíes viajeros tienen Algeciras (66%), Pechina (100%), Tudmīr (50%), Marca Superior (40%), Córdoba (46%) y Toledo (35%).

ÉPOCAS DE TAIFAS, ALMORÁVIDE Y ALMOHADE

En los dos siglos que abarca esta etapa histórica (del año 1031 al año 1232) tenemos constancia de la existencia de un total de 107 muftíes, contando entre ellos a los que ejercieron sus funciones fuera de al-Andalus, agrupados del siguiente modo : período de taifas, 30 muftíes, época almorávide, 46 muftíes, y época almohade, 31 muftíes en conjunto. De ellos, sólo 27 viajaron para oir a los maestros orientales y magrebíes o para hacer la peregrinación.

Desglosando esta cifra, el menor índice de salidas se encuentra en la época de taifas (4 muftíes, un 13%), mientras que la etapa almorávide alcanza un nivel paralelo al del Califato (11 muftíes, un 27%) y la almohade supera en proporción esta fase omeya y se acerca más a los porcentajes obtenidos en el Emirato con 12 muftíes viajeros, lo que representa un 39% del total.

No obstante, aunque en términos estadísticos la frecuencia es mayor que la obtenida en el siglo X, es imposible ocultar un descenso global de los intelectuales y juristas que toman contacto con el exterior, con respecto a los primeros siglos del Islam andalusí.

De acuerdo con estos datos, el esquema resultante de la plasmación de los mismos nos daría el siguiente cuadro estadístico :

CUADRO VIII.

Número de muftíes que viajaron en las épocas de taifas, almorávide y almohade.

	n° total de muftíes	muftíes que viajaron	%
Taifas	30	4	13%
Almorávides	46	11	27%
Almohades	31	12	39%

El gráfico obtenido por la representación de los porcentajes de muftíes que viajan a lo largo de las diversas etapas cronológicas de los siglos XI y XII es el siguiente :

GRAFICO IX.

Evolución de los viajes en las épocas de taifas, almorávide y almohade.

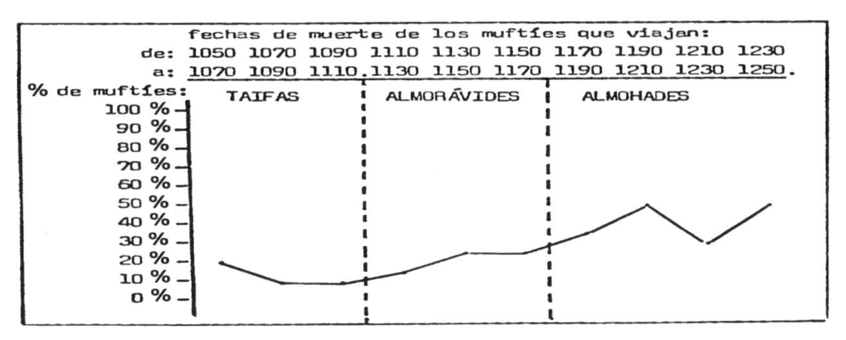

Los menores índices de viajes se registran entre los muftíes de la época de taifas (fallecidos entre los años 1050 y 1110), iniciándose el despegue al acabar esta etapa de transición entre los años de hegemonía omeya y los de predominio norteafricano, con lo que se confirma la tendencia ya observada en el estudio de los maestros y de las disciplinas estudiadas de que los años transicionales son testigos de una paralización cultural.

Con la entrada de los almorávides (muftíes fallecidos entre los años 1110 y 1170) la curva se hace ascendente, con sólo un ligero descenso entre los muftíes que se encuentran a caballo entre las etapas almorávide y almohade.

Los focos de atracción de los juristas andalusíes de estos siglos experimentan también un cambio con respecto a la época omeya :

CUADRO X.

Cronología y direcciones de los viajes en las épocas de taifas, almorávide y almohade.

nº total de muftíes:		(11)	(9)	(10)	(16)	(22)	(8)	(12)	(8)	(7)	(4)	(107)	
fechas de muerte de:		1050	1070	1090	1110	1130	1150	1170	1190	1210	1230	totales	
a:		1070	1090	1110	1130	1150	1170	1190	1210	1230	1250	ciu.	reg
región	ciudad												
-Magreb	-Magreb.....	-	-	-	1	-	-	-	-	-	-	1	12 11%
	-Marraquech	-	-	-	-	1	-	-	2	-	-	3	
	-Tánger....	-	-	1	-	-	-	-	-	-	-	1	
	-Tlemcen...	-	-	-	-	1	-	-	1	-	-	2	
	-Fez.......	-	-	-	-	1	1	1	1	-	-	4	
	-Ceuta.....	-	-	1	-	-	-	-	-	-	-	1	
-Ifrīqiya	-Ifrīqiya..	-	-	1	-	-	-	-	-	-	-	1	3 3%
	-Bugía.....	-	-	-	-	1	-	-	-	-	-	1	
	-Sicilia...	-	-	1	-	-	-	-	-	-	-	1	
-Alto Egipto	-Egipto....	-	1	1	-	1	-	-	-	-	-	3	6 6%
	-Alejandría	-	1	-	-	2	-	-	-	-	-	3	
-Ḥiŷāz	-La Meca...	-	1	-	-	1	1	-	-	1	-	4	16 15%
	-Peregrin..	-	-	-	-	1	-	1	-	-	-	2	
	-Oriente...	1	-	-	1	2	-	2	1	1	2	10	
-Irak	-Bagdad....	-	-	-	-	1	-	-	-	-	1	2	3 3%
	-Basora....	-	-	-	-	1	-	-	-	-	-	1	
-Siria	-Siria.....	-	-	-	-	1	-	-	-	-	1	1	1%
"viajó".............		1	-	-	-	-	-	-	-	-	-	1	1%

Destaca, en primer lugar, una polarización entre el Magreb y el Ḥiŷāz, alcanzando la primera región una importancia fundamental, más relevante aún si tenemos en cuenta que su presencia en los siglos VIII-X es puramente simbólica. El Magreb desplaza y margina a las regiones de Ifrīqiya y Alto Egipto, que siguen siendo zonas de interés para los muftíes (un 3% y un 6% de muftíes viajan a ellas respectivamente), pero con niveles estadísticos mínimos.

La Meca monopoliza exclusivamente a los viajeros de la zona del Ḥiŷāz, viéndose claramente que la práctica de la peregrinación alcanza un gran auge en esta época, siendo precisamente éste el motivo principal aludido por los muftíes para viajar a esta zona (en realidad, la expresión " viajó a Oriente ", que aparece en las biografias de 10 muftíes, tiene en estos siglos un inequívoco significado de " hizo la peregrinación ", razón por la que en el cuadro XIV hemos considerado a esta decena de juristas dentro de la región del Ḥiŷāz).

Por lo que se refiere a la región del Magreb, las ciudades de Fez y Marraquech son las que más atraen las preferencias de los muftíes de al-Andalus, en contraposición a otros centros culturales de la época (Tánger, Tlemcén, Ceuta, etc.). Por último, en lo que respecta a las regiones geográficas con un carácter más secundario desde el punto de vista de la elección de los juristas hispanomusulmanes (Irak, Siria), diremos, brevemente, que mientras la presencia andalusí en la primera región sigue manteniéndose, dentro de la tónica descendente (de un 5% de muftíes en la época omeya, pasa a un 3% en la época que analizamos) y basándose sólo en las ciudades de Bagdad y Basora, la estancia de muftíes hispanomasulmanes en la segunda, Siria, se pierde casi por completo (de un 4% se pasa a un 1%) con la única presencia de un muftí valenciano de los últimos años almohades.

GRÁFICO XI.

Importancia de los puntos de destino de los viajes en las épocas de taifas, almorávide y almohade.

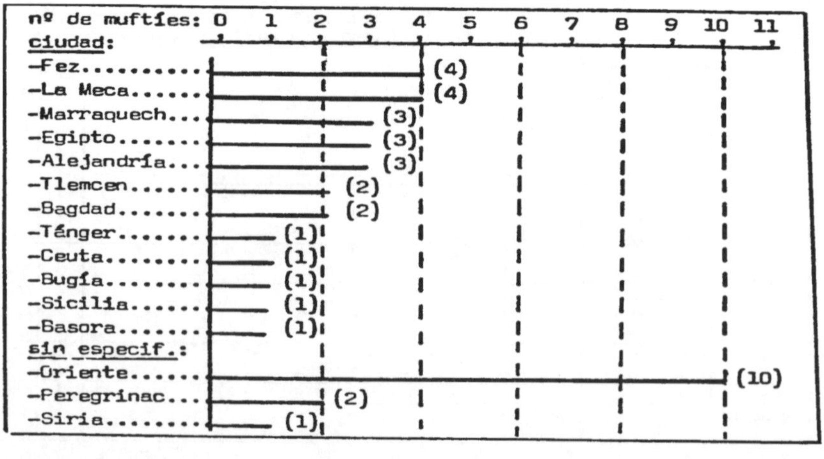

ciudad:	nº de muftíes:
-Fez	(4)
-La Meca	(4)
-Marraquech	(3)
-Egipto	(3)
-Alejandría	(3)
-Tlemcen	(2)
-Bagdad	(2)
-Tánger	(1)
-Ceuta	(1)
-Bugía	(1)
-Sicilia	(1)
-Basora	(1)
sin especif.:	
-Oriente	(10)
-Peregrinac.	(2)
-Siria	(1)

En cuanto al análisis diacrónico de las preferencias geográficas de los andalusíes a lo largo de estos dos siglos (la región del Magreb y La Meca, en primer lugar, y la región de Ifrīqiya y Egipto, en segundo lugar), el diagrama resultante es el siguiente (véase gráfico XII).

A simple vista se puede observer cómo los momentos de auge y decadencia de los viajes a La Meca, por una parte, y los viajes al resto de las regiones, por otra, son en todo momento contrapuestos. Entre los muftíes fallecidos en la veintena del 1090 al 1110, es decir, entre los muftíes de finales del siglo XI, de finales de la época de taifas, es perceptible una subida paralela de los viajes al Magreb, Ifrīqiya y Egipto, plasmándose, por el contrario, un descenso de la orientación migratoria hacia La Meca.

A partir de los muftíes de finales del siglo XI y principios del XII (que murieron entre los años 1130 y 1150), se inicia un predominio de aquellos que viajan preferentemente a La Meca y que durará durante el resto de la época, experimentando sólo un retroceso con la generación de muftíes muertos entre los años 1190 y 1210 (y que vivieron en la segunda mitad del siglo XII).

GRÁFICO XII.

Evolución de las grandes regiones visitadas por los muftíes de las épocas de taifas, almorávide y almohade.

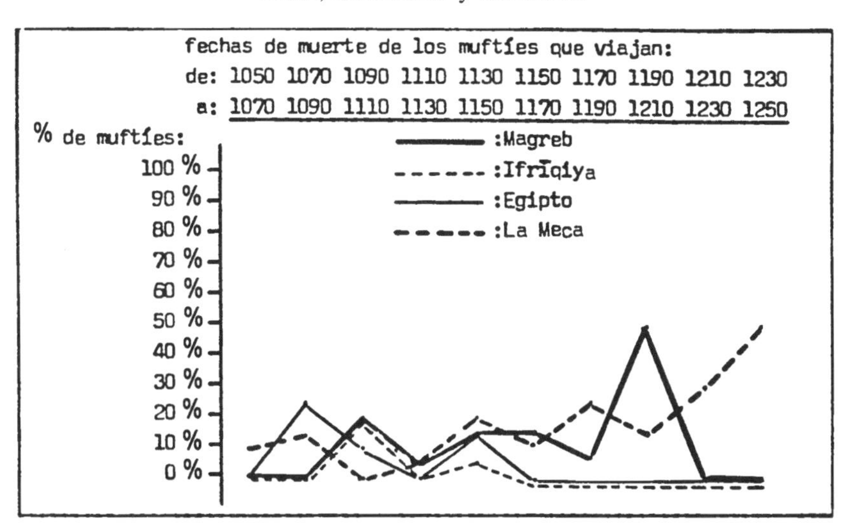

Egipto, como punto de referencia cultural para al-Andalus, desaparece con los juristas de principios del siglo XII (fallecidos a partir del año 1170), aunque es algo más temprana la caída de la región de Ifrīqiya como centro de atracción, caída que es paralela a la experimentada por Egipto.

Por último, las generaciones de muftíes almohades (las cuatro últimas reflejadas en el diagrama) son testigos, en un principio, de una corriente ascendente de los viajes a la región magrebí (de un 8% se salta a un 50%), que acaba por descender rápidamente al nivel 0 en las dos últimas promociones de juristas, dejando ver más claramente la continuidad de La Meca como la ciudad más visitada y el ascenso estadístico que registra en los últimos años (de un 28% de muftíes visitantes se pasa a un 50%).

Analizando los lugares de procedencia de la treintena de muftíes que realiza viajes a lo largo de estos dos siglos (véase cuadro XIII), observamos que éstos se reparten entre 9 ciudades, 4 de la zona levantina (Játiva, Murcia, Valencia y Mallorca), 1 del Sur (Málaga), 3 del Centro (Córdoba, Granada, Sevilla) y 1 del Norte (Toledo).

CUADRO XIII.

Ciudades de origen de los muftíes que viajan en las épocas de taifas, almorávide y almohade.

localidad	n° muftíes de la cora	n° muftíes viajeros	%
Toledo	11	3	27%
Córdoba	21	5	24%
Granada	9	1	11%
Sevilla	13	5	38%
Valencia	12	4	33%
Murcia	14	6	42%
Mallorca	4	1	25%
Játiva	9	1	11%
Málaga	2	1	50%

El mayor índice de viajeros, tanto en números absolutos como porcentuales — si excluimos el poco representativo 50% de Málaga — lo tiene Murcia con un 42% de sus muftíes que deciden aprender de maestros no andalusíes, seguido de Sevilla y Valencia con un 38% y un 33% respectivamente.

Córdoba desciende casi a la mitad (de un 46% a un 24%) si lo comparamos con la etapa omeya, sufriendo también un descenso, aunque no tan acentuado, la cora toledana. Mallorca, Málaga, Elvira y Játiva sólo registran un muftí de esa localidad que traspasará los límites andalusíes, siendo más alarmante este dato en las dos últimas ciudades, al ser residencia de una considerable cifra de muftíes (9 en cada ciudad, con lo que el porcentaje de salidas es de un 11%.

Las provincias o territorios del Sur hispanomusulmán (Algeciras, Medina Sidonia, Almería) o del Oeste (Badajoz, Beja, Cabra, Écija, Carmona), por lo demás, no cuentan con ningún muftí que viajara al extranjero.

CUADRO XIV.

Relación entre las coras de origen y las ciudades visitadas por los muftíes de las épocas de taifas, almorávide y almohade.

coras de origen:	Magreb	Ifrīqiya	Egipto	Ḥiŷāz	Irak	Siria	"viajó"
-Toledo.........	1	-	1	2	-	-	-
-Córdoba........	3	1	1	1	-	-	1
-Granada........	-	-	-	1	-	-	-
-Sevilla........	4	-	-	1	-	-	-
-Valencia.......	-	1	2	3	1	1	-
-Murcia.........	1	-	-	5	-	-	-
-Mallorca.......	-	-	-	1	-	-	-
-Játiva.........	-	-	-	1	-	-	-
-Málaga.........	-	-	-	1	-	-	-

(regiones visitadas)

Relacionando los lugares de origen de los muftíes con los diversos destinos extra-andalusíes, en busca de unas posibles orientaciones geográficas, se puede constatar que las ciudades del Magreb son las elegidas mayoritariamente por los muftíes de Córdoba y Sevilla, mientras que los del Levante y Toledo sólo tienen una presencia mínima en esta región norteafricana al preferir éstos (en particular los muftíes levantinos : Valencia, Murcia, Játiva y Mallorca) a La Meca como meta de sus itinerarios, ciudad que, por otra parte, tiene una escasa presencia de residentes cordobeses y sevillanos.

La Meca se presenta como la única ciudad que acoge muftíes de todas las provincias y tierras de al-Andalus, siendo incluso la única localidad objeto de visita de algunas regiones geográficas (Granada, Mallorca, Játiva, Málaga). En cuanto a la zona del Alto Egipto, el mayor índice de muftíes emigrantes proviene de Valencia, seguido únicamente por Córdoba y Toledo. La ruta que une Egipto con La Meca es la preferida por toledanos y valencianos, mientras que el itinerario entre el Magreb y La Meca es más frecuentado por cordobeses, sevillanos y murcianos. Finalmente, y para completar este panorama de afinidades entre ciudades de al-Andalus y de fuera de sus fronteras, diremos en lo que respecta a los centros culturales de Irak o Siria que éstos sólo son visitados por muftíes de Valencia.

Resumiendo globalmente las tendencias de los viajes de los muftíes que afloran en los siglos XI-XII, es posible apreciar la aparición de un nuevo foco activo de cultura, el Magreb (Fez y Marraquech en particular para los muftíes de al-Andalus), que, junto a La Meca (por razones de peregrinación), monopolizan los itinerarios andalusíes, desplazando a Egipto e Ifrīqiya, elementos dominantes en la anterior época omeya. El mayor movimiento de muftíes que viajan se da con los almohades, siendo prácticamente irrelevante el número de especialistas de la fetua que viajan durante la época de taifas. La Meca se convierte en la ciudad más visitada, especialmente por los levantinos, mientras que Fez y Marrequech lo son por los cordobeses y toledanos.

Epoca nazarí

Los dos siglos y medio de vida que tuvo el reino nazarí granadino contaron al menos con la presencia de unos 25 juristas calificados como muftíes, formando parte también de esta cifra dos alfaquíes que por su labor fuera de las fronteras hispanomusulmanas sus biografías han sido clasificadas en el apartado de los muftíes que emigraron.

De este grupo de 25, sólo tenemos noticias de que viajaron 6 de ellos, lo cual, estadísticamente, nos da un porcentaje semejante al de la época califal, un 24% del total. Estos muftíes que aprendieron fuera de las tierras de al-Andalus mueren todos ellos entre los años 1280 y 1467.

Metodológicamente, el dilatado espacio cronológico y el número relativamente escaso de viajeros, nos ha movido a aumentar en 10 años la duración de las generaciones en que dividíamos las diversas fases históricas estudiadas, pasando de 20 a 30 años por generación, a fin de poder obtener unos índices estadísticos más reales y representativos.

De este modo, el gráfico que resulta tras el despliegue diacrónico de la frecuencia de los viajes es el siguiente :

Gráfico xv.

Evolución de los viajes en la época nazarí.

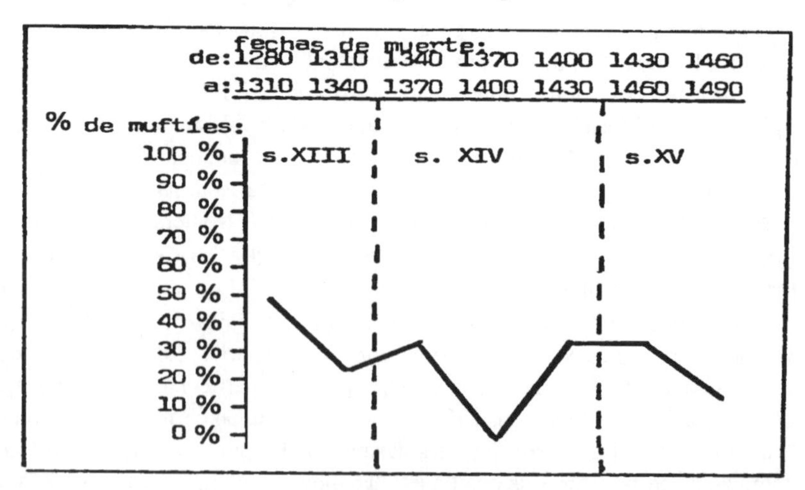

La sucesión de altibajos es la nota predominante, debido en parte a lo poco operativo que resulta desarrollar una cifra tan escasa de muftíes que viajan. No obstante, podemos apreciar que los mayores índices de salida al exterior se producen al principio de la época nazarí (un 50%), a principios del siglo XIV (muertos entre los años 1340-1370 : un 33% de muftíes) y en las dos generaciones de finales del siglo XIV y principios del XV (fallecidos entre 1400 y 1460 : 33% del conjunto de muftíes), siendo estos sesenta años los únicos que presentan una cierta estabilidad en el hábito de viajar.

Por el contrario, las épocas siguientes a estos tres momentos citados registran todas un descenso en el número de muftíes emigrantes, que tiene su más alto exponente a mediados del siglo XIV (muertos entre los años 1370-1400), en donde no ha sido posible localizar nigún muftí que aprendiera directamente de los maestros del Magreb.

La única región que acoge a los juristas de al-Andalus es el Magreb, sin que tengamos noticias de que algún muftí sobrepasara esta zona norteafricana.

Las ciudades elegidas por los granadinos, son tres : Fez, Tlemcén y Ceuta.

CUADRO XVI.

Cronología y direcciones de los viajes en la época nazarí.

nº total de muftíes:	(2)	(4)	(3)	(3)	(3)	(3)	(7)	(25)
fechas de muerte de:	1280	1310	1340	1370	1400	1430	1460	total
a:	1310	1340	1370	1400	1430	1460	1490	
ciudad:								
-Fez..............	-	1	1	-	-	-	1	3 (12%)
-Tlemcén..........	-	-	-	-	-	1	-	1 (4 %)
-Ceuta............	1	-	-	-	-	-	-	1 (4%)
-Magreb...........	-	-	-	-	1	-	-	1 (4%)

Fez en, sin duda alguna, el foco de atracción cultural máximo de los juristas nazaríes (la mitad de los que viajan lo hacen a esta ciudad), seguido de Ceuta y Tlemcén, que sólo registran el paso de un muftí granadino por sus tierras. Es curioso también que ninguna biografía se haga eco de la realización de la peregrinación por algún jurista.

Analizando la época de mayor frecuentación de Fez (véase gráfico XVI), observamos que esta ciudad es un centro intelectual y jurídico para Granada a lo largo de los dos siglos y medio nazaríes, mientras que Ceuta lo es sólo al principio y Tlemcén al final. Únicamente entre los muftíes que vivieron entre la segunda mitad del siglo XIV y primeros años del siglo XV (muertos entre los años 1370 y 1460), y que por tanto aprendieron y tuvieron la oportunidad de viajar en su juventud, es decir, alrededor del 2o y 3o tercio del siglo XIV, es palpable una marginación de esta ciudad.

GRÁFICO XVII.

Evolución de los viajes en la época nazarí.

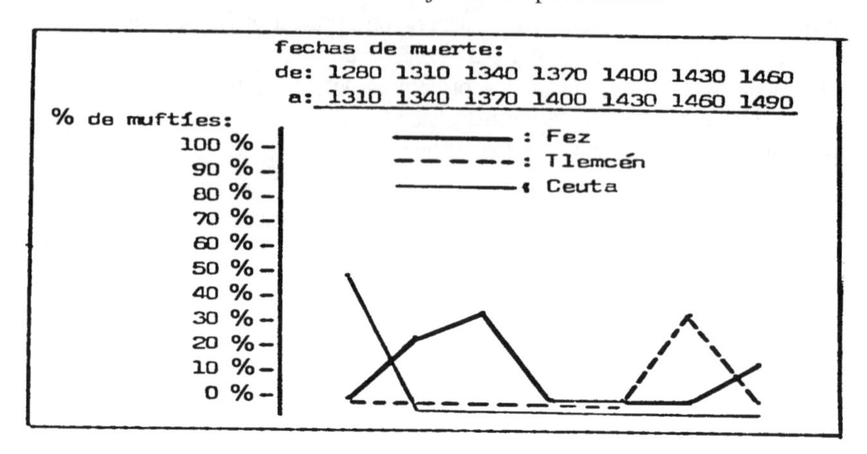

Por último, añadiremos que el lugar de residencia de todos los muftíes granadinos que viajan durante la época nazarí es precisamente la ciudad de Granada, no habiendo ningún muftí de otra localidad (Guadix, Málaya) que visitara el Norte de África.

En síntesis, podemos decir que durante la época nazarí, el índice de muftíes que viajan es discretamente bajo, siendo el Magreb la región que les sirve de meta, especialmente la ciudad de Fez. Las épocas de mayor intensidad en la realización de viajes con fines culturales se localiza en los últimos años del siglo XIII y primeros del siglo XIV y en la segunda mitad de este siglo, siendo todos los muftíes que viajan residentes de la ciudad de Granada.

Resumiendo globalmente todos los datos existantes acerca de los viajes y las relaciones culturales de los muftíes de al-Andalus, podemos concluir :

1.- El índice media de muftíes que viajan gira alrededor de un 25%, alcanzándose los mayores porcentajes durante el Emirato y los almohades (44% y 39% respectivamente) y el menor durante la época de taifas (un 13%).

2.- Los puntos de destino de estos muftíes son : durante la época omeya (siglos VIII-X) Egipto, Meca-Medina y, en menor medida, Qayrawán ; durante los siglos XI-XII (taifas, almorávides y almohades), Fez y Marraquech desplazan a Egipto e Ifrīqiya y en el Oriente, La Meca monopoliza los viajes a esta zona ; y durante el período nazarí, Fez será la ciudad preferida de los muftíes granadinos por excelencia ; los centros culturales de las regiones del Yemen, Siria o Irak, sólo serán frecuentados en los siglos IX-XII y muy minoritariamente.

3.- El mayor aporte de muftíes viajeros lo hacen, en la época omeya, lógicamente, Córdoba y, en menor medida, Toledo, mientras que durante la etapa de taifas y de hegemonía norteafricana, la zona levantina supera al resto de al-Andalus.

4.- Proporcionalmente al número de sus muftíes, las ciudades que más salidas al exterior registran son, en los siglos omeyas, Algeciras y Córdoba principalmente, a quienes acompañan Murcia y Almería ; durante los siglos XI-XII, serán Murcia y Sevilla las que alcancen los índices más altos junto a Málaga ; por último, en la etapa nazarí, todos los muftíes que viajan tienen su residencia en la ciudad de Granada.

5.- Las principales rutas seguidas por los muftíes de al-Andalus son : siglos VIII-X (época omeya) :

a) Ifrīqiya - Egipto - Ḥiŷāz : frecuentada por los de Córdoba y Toledo.

b) Egipto - Ḥiŷāz : preferida por los muftíes de Sevilla y Algeciras.

c) Ifrīqiya - Egipto : seguida por los de Medina Sidonia, Elvira y Pechina.

Siglos XI-XII (taifas, almorávides, almohades) :

a) Magreb - Ifrīqiya - Egipto - La Meca : seguida por Toledo, Córdoba, Valencia.

b) Magreb - La Meca : elegida por los muftíes de Sevilla y Murcia.

Siglos XIII-XV (época nazarí) :

a) Granada - Tlemcén - Fez : frecuentada en el siglo XIII.

b) Granada - Fez : seguida por los muftíes del siglo XIV.

c) Granada - Ceuta - Fez : preferida por los muftíes del siglo XV.

Computer applications to the history of the medieval exact sciences : suggestions for future research

E.S. Kennedy

The Computer, its Advantages, and a Disadvantage

The leading characteristics of the digital computer are :

1. The capacity to store masses of information of all varieties : verbal, numerical, graphical, and so on, and of an arbitrary degree of complexity.

2. It can process these data at high speed, subjecting them to linguistic, mathematical, statistical, or other analysis limited only by the ingenuity of the programmer.

3. Finally, it can present the results of these investigations, in print or pictures, in a manner suited to the needs of the recipient. Again the effectiveness of the output depends less upon the power of the machine than on the sorting ability of the programmer.

For carrying out all three functions, one or more computer languages are employed.

The upshot of this has been a boon to the historian of science by the solution of problems which without the computer were either very difficult or unsolvable. An example is the development of sexagesimal calculators by Benno van Dalen, Honorino Mielgo, and Glen Van Brummelen. These, to my knowledge, have not been published, but they have resulted, for instance, in the final abandonment of the sexagesimal multiplication table invented two millennia ago by Mesopotamian priests and resurrected by O. Neugebauer and his disciples. For the latter, an immense amount of time previously spent in laborious computation has been saved by the provision of these electronic calculators. Some of them are also capable of invoking trigonometric functions in sexagesimals. These can be introduced into calculations for, say, checking numerical results given in medieval texts.

Of great utility for identifying the functions embedded in numerical tables, together with their parameters, is the publication[1] by van Dalen.

Another example is the program written by Mielgo for portraying configurations on the sphere in orthographic projection. It is very useful for drawing figures, e.g. modern versions of the figures found in manuscripts on spherical astronomy.

Thus far, everything seems to favor the use of the computer. It is necessary to state, however, that very serious difficulties may arise, due to the headlong speed of advance in computer technology. For instance, any particular computer language, regardless of how powerful it may have been at the time of writing, is rapidly made obsolete by new technology. In many cases the manufacturer issues successive updated versions of the language, for each of which the preceding version is compatible. This may be in the sense that the updated version has a special program which will translate obsolete programs so that they are acceptable. But the time eventually comes when the basic concept of the language is itself so obsolete that it and all its versions disappear. If a project is wholly dependent upon a single language, it may find itself unable to proceed. But in the history of science there are many important problems for which solutions are dependent upon continued operation fed by increments of new data. Hence, in considering the setting up of such projects, the role of the computer scientist, to keep up with his field and forestall or overcome difficulties as they arise, is at least as important as that of the historian. But the historian is also essential, and it is improbable that a single person would be capable of performing both functions. Depending upon the magnitude of the operation, one or more programmers may be necessary. All in all, it looks as though a satisfactory solution will require a permanent organization within a sponsoring institution, probably educational, to guarantee continuity, and the very considerable funding which will be required.

Described below are four suggested projects. All of them demand, to a greater or lesser degree, the type of organization just sketched.

A Bibliography of the Islamic exact sciences

The need for an up to date bibliography of the field is obvious. The numerous works of Professor Fuat Sezgin are of great value, but a number of years have elapsed since the appearance of the last volume dealing with science[2]. For generations of our predecessors Heinrich Suter's *Die Mathematiker und Astronomen der Araber...* was the ideal bibliography. By 1983 it had been for

1. B. van Dalen, *Ancient and Medieval Astronomical Tables : Mathematical structure and parameter values,* Utrecht University, 1993.

2. F. Sezgin, *Geschichte des arabischen Schrifttums, Band VII Astrologie, Meteorologie und Verwandtes,* Leiden, 1979.

years hopelessly out of date. But in that year it was supplanted by the excellent *Matematiki i Astronomy Musul'manskogo...Trudy* by G.P. Matvievskaya and B.A. Rosenfeld. This was out of date by the time it appeared, as the authors recognized. It is suggested that this " Russki Suter " be in turn replaced by a " computerized Suter ", to be followed by others at suitable intervals. Meanwhile the attendant database would be continuously able to respond to inquiries from historians certain to obtain the latest bibliographical information. Conversely, individual historians will be glad to contribute new data.

A Mathematical, Astronomical, and Astrological Dictionary

For mathematics there already exist technical dictionaries in Arabic and Persian. But the branches of medieval mathematics are extensive and numerous, each with its own vocabulary, including many terms which are peculiar to the branch and not common to the subject as a whole. As a consequence, no single compiler can be expected to control all the branches.

As to published sources, aside from the extant dictionaries mentioned, it is the fortunate custom that an editor issuing a critical edition of a medieval mathematical text append to it a glossary of technical terms. The classical example is the monumental three volume edition of al-Battānī's *zīj*[3], which contains an invaluable glossary. For some of us, however, the fact that the meanings are in Latin is a considerable drawback.

Additionally, there must be a large number of unpublished secondary sources, for it can be assumed that scholars working from primary sources, the manuscript texts, will over the years accumulate files which amount to personal dictionaries. Clearly these should be collected, and their contents added to the master file so that they will become available to all.

New sources will continue to appear as more texts are worked over. Sometimes this happens in wholly unexpected ways. For instance, the author recently was surprised to discover that the mathematical vocabulary of Arab Spain and the Maghrib is frequently radically different from that of the Near and Middle East.

Practically everything just said about mathematics applies also to astronomy. Moreover, mathematics is such a basic and pervasive tool for the astronomer that it is sometimes difficult to say where the one discipline begins and the other ends. This is particularly the case with trigonometry and computational mathematics. Therefore material from astronomical sources should be consolidated in a single database with mathematics.

Astrology is a pseudoscience, but this is not a valid reason for ignoring it. What is more to the point here is that any serious astrologer must command a

3. C.A. Nallino, *Al-Battānī sive al-Batenii Opus Astronomicum*, Milan, 1899-1907 (3 vols).

great deal of astronomy, hence also mathematics. Conversely, certain mathematical procedures of great interest are to be found uniquely in astrological treatises. An example is the calculation of the allegedly exact time of a nativity by the *namūdar* of Hermes. It follows that astrological technical terms should also be part of the dictionary database.

Something must be said about languages. The vast bulk of the texts in all three components are in Arabic. Therefore this should be the language employed for the words being defined. There are many texts also relevant to all three components, but in Persian. Fortunately this language is almost always written in Arabic characters. The bulk of the Persian technical vocabulary is taken from Arabic, but treatises written in Persian also utilize a very large number of technical terms of Persian origin. Some, particularly those from astrology, have entered Arabic as loan words. All should be included in the list of words to be defined.

Concerning the main text of the dictionary, the meanings, many historians, while competent to read the Arabic characters, nevertheless continue to have difficulty with Arabic running text. This is common among persons who have neither Arabic nor Persian as their mother tongue. For the benefit of such people, a language for the dictionary text more nearly universal is advocated, say English.

HOROSCOPES

Every horoscope is an astrological document, but its contents may be of great interest and utility to historians of astronomy and mathematics. For the information it contains comprises the celestial longitudes of the planets and the ascending and descending lunar nodes, plus the ascendent, the horoscope proper, being the longitude of the ecliptic point crossing the eastern horizon at the instant and locality for which the horoscope is cast. The horoscope may in addition give, for example, the longitudes of the twelve astrological houses or the locations of prominent fixed stars. The precision of the determinations varies widely, from seconds of arc to the zodiacal sign only. However, even if some planets are missing, the information supplied will be redundant, because the celestial motions involved vary in speed between the slow movements of precession and of the planet Saturn, and the speedy daily rotation amounting to fifteen degrees per hour. Hence almost any horoscope may be dated unless, it is faked, and its planetary mean motions inferred. From the latter, the horoscope may be associated with a particular center of astronomical activity, and facts concerning transmission established. If the longitudes are precise and consistent, the geographical position for which the horoscope was cast can be determined.

Horoscopes in profusion exist, dating from the fourth century B.C. Collections of them also exist, notably that amassed by O. Neugebauer and H.B. van

Hoesen[4]. They need combining into a single database, capable of accepting the multitude of additional horoscopes which is sure to appear as time passes. From a sufficient number of horoscopes cast by a single person it is possible to infer precise descriptions of the planetary models which characterize the system used by the caster.

Arrangements should be made so that the collection can be sorted in many ways, including chronological order, planetary positions, persons or schools computing the horoscopes, particular techniques for, say, calculating the astrological houses, and so on.

Geographical Tables

Scattered through the medieval literature, mostly in *zījes*, but also in geographical works and inscribed on astrolabes, are lists of geographical localities. These are for the most part cities, and to each its longitude and latitude have been appended. Something like eighty such sources are thus far known, and have been entered into a database[5]. Their size varies greatly, the largest containing about 950 names, but about three quarters of the rest have fewer than three hundred entries. The total of different localities named is about 2500.

The collection is of utility to historians of mathematical geography, since it supplies information concerning the precision of observations, the location of base meridians, the times of emergence and locations of centers of geographical research, and the transmission of geographical knowledge.

The project is in good hands, for it has been taken over by Dr. Merce Comes of the Department of Arabic, University of Barcelona. It is the only one of the five which is actually underway.

Astronomical Handbooks (*Zījes*)

There exist thousands of medieval manuscripts dealing with the exact sciences, most of them unread in modern times. Among them, the most rewarding category of historical sources may well be the writings known as *zījes*. A *zīj* is a collection of the numerical tables needed for solving the standard problems of astronomy and astrology, preceded by explanations of how the tables are to be used. Worked examples are frequent, and descriptions of operations are occasionally followed by proofs. The names of well over two hundred *zījes* are known[6], of which more than half are extant. Of these, more or less complete

4. O. Neugebauer, H.B. van Hoesen, *Greek Horoscopes,* Philadelphia, 1959.

5. E.S. and M.H. Kennedy, *Geographical Coordinates of Localities from Islamic Sources,* Frankfurt a. M., 1987.

6. E.S. Kennedy, " A Survey of Islamic Astronomical Tables ", *Transactions of the American Philosophical Society*, vol. 46, Part 2 (1956). The work was reprinted in 1989, with the addition of a preface.

descriptions of twenty-one have been published. Nine *zījes* have been edited, translated, and published. The number found to be extant has been steadily increasing over the years, as has information about the rest.

It is suggested that the information outlined above be converted into a continuing database. It would then be possible to issue, at intervals of years, an up to date survey of *zījes*. As noted above in the case of the general bibliography, the interchange of information between individual scholars and the project organization would automatically ensue.

A start is already being made in assembling the database, as follows. For each *zīj* it is possible to calculate and assemble a set of base parameters : planetary mean motions, apogees, initial positions, maximum equations, and so on. By comparing parameter sets between many pairs of *zījes*, it is possible to establish relations between them, thence to infer the dates and locations of centers of astronomical activity, the transmission of astronomical theory from one place to another, and so on. Dr. Benno van Dalen has undertaken to compile a parameter file, not only from *zījes*, but including already existing parameter collections, the earliest being that of O. Neugebauer. This is a first step in the right direction.

The number of *zījes* is large, but the number published is small. Their contents are predominantly numerical tables, rather than running text, and it is from tables, not text, that the parameter sets which characterize individual *zījes* can be inferred. The reading of manuscript numeral forms, in contrast to text, is easy. It can be mastered quickly, even by readers who have no Arabic. Moreover the printing of facsimiles is now cheap and easy. It follows that publishers should be encouraged to follow the example of Professor Fuat Sezgin and turn out facsimiles of important *zījes*.

UN COMPLÉMENT ARABE AUX *DONNÉES* D'EUCLIDE : LE *KITĀB AL-MAFRŪḌĀT* DE ṮĀBIT IBN QURRA

Hélène BELLOSTA

L'intérêt pour le problème de l'analyse qui se manifeste dans le monde arabe, à partir du IXe, mais surtout aux Xe et XIe siècles, a-t-il été à l'origine d'une lecture renouvelée des *Données* d'Euclide (propédeutique à l'analyse) par les mathématiciens arabes ? Les modifications que font subir ceux-ci à la terminologie euclidienne sont-elle le signe de mutations plus profondes, et les besoins des astronomes, ainsi que la naissance et le développement de l'algèbre, ont-ils joué un rôle dans cette évolution ? Quelques exemples tirés d'une oeuvre de Ṯābit Ibn Qurra (mort en 901) le *Kitāb al-mafrūḍāt* permettront de mettre en évidence le fait que, plutôt que de positivités culturelles séparées mieux vaudrait ici parler, comme Cavaillès, de " révision perpétuelle des contenus par approfondissement et rature ".

LES *DONNÉES* D'EUCLIDE

Les *Données* d'Euclide constituent une propédeutique à l'analyse ; c'est pour cette raison que Pappus classe cet ouvrage, dans l'introduction au Livre VII de *La Collection mathématique*, parmi les oeuvres constituant la *Collection analytique*. En effet, faire l'analyse d'un problème (et exclusivement d'un problème, *i.e.* d'un énoncé dans lequel on demande de trouver, ou de montrer que l'on peut trouver, un objet possédant certaines propriétés, c'est-à-dire de démontrer une proposition existentielle) c'est montrer que lorsque certains objets sont " donnés " (par hypothèse) ou " connus " d'autres objets sont alors " donnés " (par démonstration) ou " connus ". Une synthèse classique (de problème) se compose alors de deux parties : dans la première on construit l'objet recherché, dont l'analyse a permis d'affirmer qu'il est donné (ou connu), dans la seconde on montre que l'objet ainsi construit est bien solution du problème. C'est en ce sens que *Les Données* d'Euclide (comme plus tard les ouvrages de ses successeurs arabes) sont une propédeutique à l'analyse : c'est un ouvrage dans lequel sont établis des résultats qui permettent d'abréger l'analyse.

Cette formulation de l'analyse en termes de *données* soulève immédiatement la question du ou des différents sens du terme " donné " : que signifie en effet, d'abord pour Euclide, puis pour ses successeurs, le fait d'affirmer ou de démontrer que certains objets sont *donnés* ou *connus* ?

Pour Euclide un objet mathématique peut : être donné *de grandeur* (aires, lignes, angles), être donné *de position* (points, lignes, angles), être donné *d'espèce* (figures rectilignes). Euclide définit en outre la notion de rapport donné. Comme souvent dans les *Éléments*, les définitions que donne Euclide ne sont pas opératoires, c'est-à-dire ne peuvent être utilisées, et ne le sont d'ailleurs pas, dans les démonstrations qui suivent (il en va de même pour les définitions du point, de la droite, du rapport[1]).

Les *Données* ont suscité beaucoup moins d'intérêt, et été beaucoup moins étudiées et commentées que les *Éléments* ; ceci est vrai tant dans le monde gréco-hellénistique (pour lequel on ne possède que le commentaire du philosophe Marinus de Néapolis, disciple de Proclus, au Ve siècle[2]), que dans le monde arabe, ou plus tard en Europe.

En ce qui concerne l'Antiquité, la rareté des analyses dans les oeuvres des mathématiciens gréco-hellénistiques (Descartes ira même jusqu'à leur reprocher d'avoir délibérément caché celles-ci), pourrait rendre partiellement compte de ce fait. L'intérêt pour le problème de l'analyse qui se manifeste aux Xe et XIe siècles dans le monde arabe[3], s'il n'a guère suscité de commentaires de cet ouvrage (on ne possède à ce jour aucun commentaire des *Données* en langue arabe), a cependant incité un certain nombre de géomètres à tenter

1. Définition 1 : *Des espaces, des lignes, et des angles, auxquels nous pouvons trouver des grandeurs égales, sont dits donnés de grandeur* ; Définition 2 : *Une raison est dite donnée, quand nous pouvons lui en trouver une qui soit la même* ; Définition 3 : *Des figures rectilignes, dont chacun des angles est donné, et dont les raisons de leurs côtés entre eux sont données, sont dites données d'espèce* ; Définition 4 : *Des points, des lignes, et des angles qui conservent toujours la même situation, sont dits donnés de position* ; Peyrard, *Les Oeuvres d'Euclide*, Paris, 1993 ; *Les Données*, 517-603.

2. M. Michaux, *Le commentaire de Marinus aux* Data *d'Euclide*, Louvain, 1947. Le texte de Marinus est intégralement cité par J. Itard dans son introduction à la traduction des oeuvres d'Euclide par Peyrard ; Peyrard, *Les Oeuvres d'Euclide, op. cit.*, XVI-XXIII.

3. Aux Xe et XIe siècles, les plus grands géomètres de l'époque, al-Sijzī, mais surtout Ibn Sinān et Ibn al-Haytham (l'Alhazen du Moyen Age latin) ont consacré au problème de l'analyse et de la synthèse d'importants traités théoriques. Leurs oeuvres s'inscrivent dans le mouvement très actif de recherche mathématique engagé à cette époque, qui a abouti, non seulement à d'importantes découvertes, mais aussi à de profondes modifications des disciplines mathématiques et de leurs rapports mutuels, rendant ainsi nécessaire la reprise de ce thème classique de la philosophie des mathématiques qu'est le problème de l'analyse et de la synthèse ; H. Bellosta, " Ibrāhīm Ibn Sinān : on analysis and synthesis ", *Arabic sciences and philosophy*, vol. I, n° 2 (1991), 211-232 ; R. Rashed, " L'analyse et la synthèse selon Ibn al-Haytham ", *Mathématiques et Philosophie de l'Antiquité à l'Âge classique*, Paris, 1991, 131-162 ; R. Rashed, " La philosophie des mathématiques d'Ibn al-Haytham ", *Mélanges de l'Institut Dominicain d'Études Orientales du Caire*, Louvain, 1991, 31-231.

d'étendre les propriétés des *Données*[4]. Ce peu d'intérêt pour *Les Données* d'Euclide s'explique peut-être mieux par le fait que, à l'exception de quelques définitions et propositions, dont l'usage est en outre rare[5], la majeure partie des propositions qui y figurent se présente essentiellement comme une reformulation, en termes de " données " de propositions figurant dans les *Éléments*, et leur démonstration en est souvent une conséquence immédiate.

LES TRADUCTIONS ARABES DES *DONNÉES*

Les Données ont été traduites en arabe au IXe siècle par Isḥāq Ibn Ḥunain, et cette traduction a été revue par Ṯābit Ibn Qurra[6]. On a en outre une rédaction des *Données* par Naṣīr al-Dīn al-Ṭūsī au XIIIe siècle, faite à partir de la version Isḥāq/Ṯābit.

À côté du nombre impressionnant de versions diverses et de manuscrits des *Éléments*, on ne peut qu'être frappé par le très petit nombre de manuscrits des *Données* : nous possédons à l'heure actuelle un unique manuscrit de la version Isḥāq/Ṯābit, et environ 25 manuscrits de la rédaction de d'al-Ṭūsī, ce qui témoigne malgré tout d'une relative diffusion de l'oeuvre. La rédaction d'al-Ṭūsī[7], élégante et concise, tout en respectant l'esprit et les démonstrations du texte initial, fait partie de la célèbre collection intitulée " les intermédiaires " *(al-mutawassiṭāt)* ou " les petites astronomies ", collection qui contient également quelques ouvrages mathématiques destinés à l'enseignement ; comme c'est également le cas pour d'autres ouvrages, la rédaction de Naṣīr al-Dīn al-Ṭūsī a quelque peu éclipsé la version initiale d'Isḥāq/Ṯābit.

Que l'on lise la traduction de Ṯābit ou celle de Naṣīr al-Dīn al-Ṭūsī, on est frappé d'emblée par le fait que, excepté dans le titre (*Kitāb al-Mu'tiyāt*) on passe de la terminologie *donné* (*mu'ṭan*) à la terminologie *connu* (*ma'lūm*)[8]. Une possible signification de ce glissement sémantique est peut-être tout d'abord une volonté de distinguer *mafrūḍ* (donné par hypothèse) et *ma'lūm* (connu par démonstration) ; cette terminologie a également l'avantage de permettre (vraisemblablement sous l'influence de l'algèbre) d'opposer ce qui est *connu* et ce qui est *inconnu* c'est-à-dire pas encore connu (*ġaïr ma'lūm* ou

4. L'ouvrage le plus remarquable sur le sujet étant le traité sur *Les Connus* d'Ibn al-Hayṯam ; R. Rashed, " La philosophie des mathématiques d'Ibn al-Haytham, II. *Les Connus* ", *Mélanges de l'Institut Dominicain d'Études Orientales du Caire*, Louvain, 1993, 87-275.

5. Il s'agit des définitions 11 et 12 et des propositions 10-11, 13 à 21 qui s'y rapportent ; voir H. Bellosta, " Ibrāhīm Ibn Sinān : on some neglected propositions of Euclid's *Data* ", à paraître dans *Arabic sciences and philosophy*.

6. *Tarǧamahu Isḥāq, aṣlaḥhu Ṯābit*, selon l'expression consacrée.

7. Sur les " rédactions " d'al-Ṭūsī, voir R. Rashed, *Les mathématiques infinitésimales du I^e^ au Xe siècle*, vol. I, Londres, 1993, 1996, 7-11 (2 vols).

8. *'a'ṭā* donner quelque chose à quelqu'un, accorder, agréer (une demande, une prière), *faraḍa* établir quelque chose comme précepte, assigner (un traitement), supposer que quelque chose est donné, *mafrūḍ* donné, déterminé, prescrit, *mafrūḍāt* données (d'un problème).

mağhūl) (c'est-à-dire d'opposer ce qui est déterminé à ce qui est indéterminé, la *chose* des algébristes). Ceci a en tout cas l'avantage immédiat de rendre la formulation des démonstrations moins déconcertante de prime abord[9].

Un complément arabe aux *Données* : Le *Kitāb al-mafrūḍāt* de Ṯābit Ibn Qurra

Ṯābit, qui connaissait bien *Les Données* d'Euclide puisque, on l'a vu, il en a amendé la première traduction arabe, est l'auteur, entre autres, d'un petit traité intitulé *Kitāb al-mafrūḍāt* (*Le Livre des données*, ou *Le Livre des hypothèses*) qui pourrait se présenter comme un complément aux *Données* d'Euclide ; nous en avons également une rédaction par Naṣīr al-Dīn al-Ṭūsī (*taḥrir kitāb al-mafrūḍāt*), témoignant d'un certain intérêt de celui-ci pour cette oeuvre. L'étude du traité de Ṯābit va nous permettre de préciser un des points les plus faibles des *Données* qu'est la notion de donné de grandeur : " des espaces, des lignes, et des angles, auxquels nous pouvons trouver des grandeurs égales, sont dits donnés de grandeur ". Le problème que pose cette définition est en effet lié au problème du statut de la grandeur chez le géomètre.

Nous ne possédons à l'heure actuelle qu'un unique manuscrit de la version Isḥāq/Ṯābit, et une quinzaine de manuscrits de la rédaction de Naṣīr al-Dīn al-Ṭūsī[10]. Ce traité comporte 36 propositions. Si les 12 premières ont trait à des problèmes de construction (par exemple trisecter un angle droit) et si les propositions 13 et 14 sont des théorèmes, les 22 propositions restantes sont exprimées dans la terminologie des " connus " : si tels et tels éléments d'une figure sont connus, alors tels et tels autres éléments le sont également. Ces propositions se répartissent de la façon suivante : les propositions 20 et 26 se ramènent en fait à des résolutions de systèmes (si la surface (*i.e.* le produit) de deux droites est connue, ainsi que leur différence alors ces deux droites sont connues, et si le produit de deux droites est connu, ainsi que leur rapport, alors ces deux droites sont connues) ; dans les autres propositions, que l'on pourrait qualifier, si l'on ne craignait pas que ce soit anachronique, de problèmes de relations métriques dans le triangle (propositions 16, 17, 22, 23, 24, 25, 27, 31, et 35), le quadrilatère (propositions 15, 18, 19, 21, et 28) ou le cercle (propositions 29, 30, 32, 33, 34, et 36), il s'agit de montrer que si certains éléments (grandeurs de certains segments, angles) d'une figure sont connus alors les grandeurs d'autres segments sont également connues. Prenons comme exemples les propositions 16, 17, 27, et 35.

9. Notons que cette terminologie n'est pas passée en Europe et que les géomètres des XVe et XVIe siècles, Viète, E. Pascal, ou Fermat, dans leurs analyses, font quant à eux, usage de la terminologie *donné*.

10. Mon étude repose sur la rédaction d'al-Ṭūsī, et plus précisément sur les manuscrits : Mss. B.N. 2467, ff. 68v-72v, Berlin Qu 1867, ff. 151v-155v, et Leyde, Or 14, 470-481.

PROPOSITION 16 : *soit* ABC *un triangle rectangle isocèle de sommet* A, *si un côté est connu, les autres le sont* (proposition non démontrée et qualifiée d'évidente).

PROPOSITION 17 : *soit* ABC *un triangle rectangle en* A, *tel que* $\angle C = \frac{\pi}{6}$, *si un côté est connu, les autres le sont.*

Ces deux propositions suscitent immédiatement la question suivante : quel est l'intérêt pour Ṯābit de les établir puisqu'elles se déduisent immédiatement des propositions 40, 50, et 55 des *Données* d'Euclide ?[11].

Preuve :

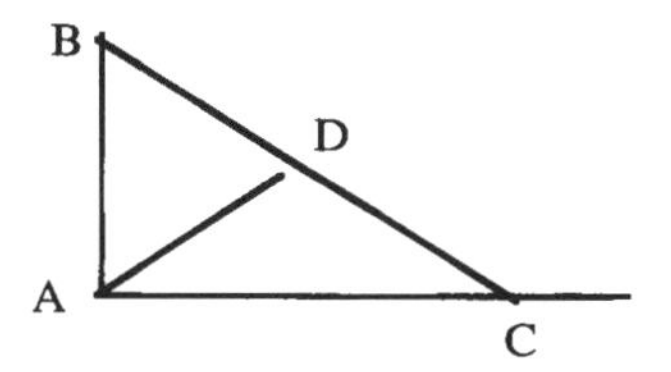

construction auxiliaire $D \in [BC]$ tel que $\angle BAD = \frac{\pi}{3}$ (construction utilisée dans la proposition 1 pour trisecter un angle droit),

$[ABD$ équilatéral, $\angle DAC = \frac{\pi}{6}$, et DAC isocèle de sommet $D]$

$\Rightarrow [DC = DA = DB = AB]$;

1. $[BC$ connue$] \Rightarrow [AB = \frac{1}{2} BC$ connue$] \Rightarrow [AC$ connue[12]] ;
2. $[AB$ connue$] \Rightarrow [BC = 2\,AB$ connue$] \Rightarrow [AC$ connue$]$;
3. $[AC$ connue, $BC^2 = 4\,AB^2$, $BC^2 = AB^2 + AC^2] \Rightarrow [AC^2 = 3\,AB^2$ connue$]$
$\Rightarrow [AB$ connue$] \Rightarrow [BC$ connue$]$.

Si Ṯābit éprouve le besoin de démontrer un cas particulier (ABC rectangle et $\angle C = \frac{\pi}{6}$) d'une proposition qui se déduit immédiatement des *Données* d'Euclide et qui serait la suivante : *si un triangle de forme connue a un côté connu, les autres côtés sont connus,* c'est vraisemblablement parce que le résultat obtenu par ce dernier ne le satisfait pas. En effet si les propositions 40, 52, et 55 des *Données* permettent d'affirmer que les côtés du triangle ABC sont connus, elles ne permettent en aucune façon de les déterminer (de les calculer) simplement. La méthode de Ṯābit permettrait par contre à un astronome (ou un

11. Proposition 40 : *si chacun des angles d'un triangle est donné de grandeur, le triangle est donné d'espèce* ; Proposition 52 : *si sur une droite donnée de grandeur on décrit une figure donnée d'espèce, la figure décrite est donnée de grandeur* ; Proposition 55 : *si un espace est donné d'espèce et de grandeur, ses côtés seront donnés de grandeur*. La définition que donne Euclide d'un espace ou d'une figure, donné de grandeur (Définition 1), n'est pas très claire ; cependant l'usage qui en est fait dans les *Données* permet de la préciser : des espaces de grandeur donnée sont en fait des espaces dont la surface est donnée.

12. Note marginale dans le manuscrit de Berlin : car $AB^2 = 1/4\,BC^2$ et $AC^2 = 3/4\,BC^2$.

algébriste), qui connaîtrait une mesure de l'un des côtés de ce triangle, de calculer les deux autres[13].

Qu'en est-il cependant pour un géomètre, et les choses sont-elles aussi simples pour lui ? Si la détermination de BC à partir de AB (ou de AB à partir de BC) ne pose pas de problème, que peut faire le géomètre d'une égalité comme $AB^2 = 3\ AC^2$, et n'est-il pas ramené à la situation du malheureux esclave de Ménon, à qui après avoir demandé de déterminer (*calculer*) le côté d'un carré double du carré initial, Socrate demande : " alors, avec quelle ligne ? Tâche de me le dire exactement, et si tu ne veux pas faire le calcul, montre-la-nous "[14], la construction géométrique (*montre*) venant à point nommé remplacer le calcul impossible (*calcule*).

PROPOSITION 27 : *si les côtés du triangle* ABC *sont connus, le diamètre du cercle circonscrit est connu.*

Cette proposition, d'un usage courant chez les géomètres, ne figure pas dans *Les Données* d'Euclide ; elle pourrait cependant être déduite de la proposition 42 et de la définition 7 des *Données*[15] (à condition toutefois d'admettre, ce qui semble aller de soi, mais que ne dit cependant jamais Euclide, que si un segment de cercle est de grandeur donnée, le rayon du cercle dont il est un segment est également de grandeur donnée)

Preuve :

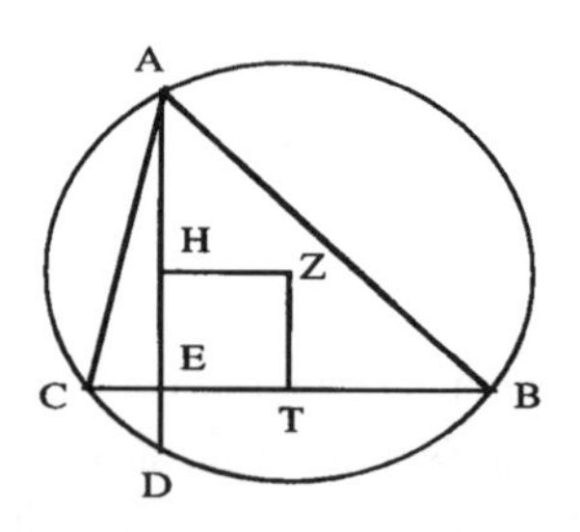

[ABC connu] => [AE, BE et EC connues (lemme II)],

[$AE.ED = EC.EB$, AE, BE et EC connues] => [ED connue (*Data* P. 57)]

=> [AD connue (*Data* P. 3)],

[$AD = 2\ AH$ et $BC = 2\ BT$ connues] => [AH et BT connues],

[BE et BT connues] => [$ET = HZ$ connue (*Data* P. 3 ou P. 4)],

[HZ et AH connues] => [AZ connue (lemme I)].

13. Voir les méthodes de Ptolémée et de Héron pour le calcul des racines. T.L. Heath, *A history of Greek mathematics*, vol. I, New York, 1981, 61-63, vol. II, 323-326 (2 vols).

14. *Ménon*, 83d-84a.

15. Proposition 42 : *si les côtés d'un triangle ont entre eux une raison donnée, ce triangle sera donné d'espèce* ; Définition 7 : *des segments de cercles sont dits donnés de grandeur quand les angles qu'ils comprennent et les bases de ces segments sont donnés de grandeur.*

Dans cette démonstration, Ṯābit utilise ici, et sans les démontrer, deux lemmes ; tout d'abord le *lemme I : si deux côtés d'un triangle rectangle sont connus, le troisième côté est connu.* Ce lemme, qui est l'un des outils fondamentaux de l'*Almageste* de Ptolémée (et de façon plus générale des astronomes) pour tout ce qui concerne le calcul des cordes, ne figure pas sous cette forme dans *Les Données* ; il pourrait cependant être déduit de *Éléments* I 47 (théorème de Pythagore), puis des propositions 3 ou 4, et 55 des *Données*[16].

L'usage d'*Éléments* I 47 permet en effet à l'astronome ou à l'algébriste comme al-Ḫwārizmī, grâce au théorème de Pythagore, de calculer le troisième côté : al-Ḫwārizmī par exemple, pour calculer l'aire a d'un triangle équilatéral de côté 10 procède de la façon suivante[17] :

il calcule d'abord la hauteur AH issue de A :

$$AH^2 = AB^2 - BH^2 = 100 - 25 = 75,$$

d'où $AH = \sqrt{75}$ (*ǧiḏr*), d'où $a = \sqrt{75}.5$ soit à peu près 43 (*ṯalaṯa wa arba'ūn wa šay qalīl*).

Ce lemme pose en revanche au géomètre, du fait de l'existence des irrationnelles, un problème qu'évite le premier. Le géomètre ne peut que constater, ce qui découle de la proposition 55 des *Données*, que si la surface d'un carré est connu (carré de grandeur connue) son côté est connu.

Ṯābit utilise également à deux reprises dans cette proposition le *lemme II : si les côtés d'un triangle sont connus, la hauteur et le segment joignant le pied de cette hauteur à l'une des extrémités de la base, sont connus.*

Il ne démontre pas ce lemme qu'il utilise pourtant à 5 reprises dans son ouvrage. Ce lemme pourrait être déduit des *Éléments* et des *Données* de la façon suivante :

soit ABC le triangle AH la hauteur issue de A.

$$AC^2 = AB^2 + BC^2 \pm 2\ BH.BC \text{ (\textit{Éléments} II 12 ou II 13),}$$

$$[AB,\ AC,\ BC \text{ connues et } AC^2 = AB^2 + BC^2 \pm 2\ BH.BC]$$

$$\Rightarrow [BH.BC \text{ connue (\textit{Data} P. 3 et P. 4],}$$

[$BH.BC$ connue et BC connue] => [BH connue (*Data* P. 57[18])].

On ne peut manquer de trouver étrange le fait que Ṯābit ne démontre pas une proposition qu'il utilise à plusieurs reprises et qui se déduit beaucoup

16. Proposition 3 : *si tant de grandeurs données qu'on voudra sont réunies, la grandeur composée de ces grandeurs sera donnée* ; Proposition 4 : *si d'une grandeur donnée on retranche une grandeur donnée, la grandeur restante sera donnée.*

17. On trouve le même exemple numérique dans *Les Métriques* de Héron d'Alexandrie ; F. Rosen, *The algebra of Mohammed ben Musa*, Londres, 1830, 58-59 (texte arabe) ; A.P. Youschkevitch, *Les mathématiques arabes (VIIe-Xe siècles)*, Paris, 1976, 49.

18. Proposition 57 : *si un espace donné est appliqué à une droite donnée, dans un angle donné, la largeur de l'application est aussi donnée.*

moins aisément des *Données* que la proposition 17 dont il donne, on l'a vu, une démonstration. Il nous semble nécessaire d'évoquer ici la démonstration qui figure dans le *Kitāb al-ğabr* d'al-Ḫwārizmī (*Bāb al-misāha*), et que Ṯābit ne pouvait manquer de connaître :

al-Ḫwārizmī calcule AH et BH dans le cas où $BC = 14$, $AB = 13$, et $AC = 15$[19] ; les valeurs numériques données aux grandeurs connues jouent ici le rôle de paradigmes permettant, en l'absence de symbolisme, une résolution algébrique ; al-Ḫwārizmī ramène le problème à la résolution d'une équation du premier degré de la façon suivante :

on pose $x = BH$ (*fa ğa'alnā al-šay...*), que tant al-Ḫwārizmī que Ṯābit qualifient de *masqaṭ ḥiğr*[20] (surplomb) ;

$$AH^2 = AB^2 - BH^2 = 169 - x^2 \ (x^2 = \textit{māl}),$$

$$CH = BC - BH = 14 - x, \ CH^2 = 196 + x^2 - 28x,$$

$$AH^2 = BC^2 - CH^2 = 29 + 28x - x^2,$$

$$\text{d'où } 169 - x^2 = 29 + 28x - x^2, \text{ et } 28x = 140, \text{ soit } x = 5,$$

$$AH^2 = 169 - 25 = 144, \text{ et } AH = 12,$$

$$\text{soit } a = 12 \text{ x } 7 = 84.$$

Observons que cette démonstration *algébrique* (qui revient à calculer AH en fonction de l'inconnue BH, de deux manières) dont l'objet est bien, connaissant les trois côtés d'un triangle de déterminer (calculer) la hauteur et le surplomb, c'est-à-dire, pour user d'une terminologie *géométrique*, de montrer que si les trois côtés d'un triangle sont connus la hauteur et le surplomb le sont également, n'est nullement exprimée en termes de données (ou de connues) : le calcul numérique effectif de la *chose* (l'inconnue) remplace ici la démonstration du fait que le segment BH est connu.

On pourrait ainsi être tenté de penser que Ṯābit, qui connaissait vraisemblablement cette démonstration, s'est alors dispensé d'en donner une version géométrique en termes de connues.

PROPOSITION 35 : ABC *triangle tel que* $\angle A = \frac{\pi}{3}$, AB *connue ainsi que le diamètre du cercle circonscrit, alors* BC *et* AC *sont connues.*

19. On trouve encore le même exemple numérique dans *Les Métriques* de Héron, mais la méthode de Héron, qui équivaut à la méthode géométrique que nous avons suggérée, est différente de celle d'al-Ḫwārizmī ; F. Rosen, *The algebra of Mohammed ben Musa*, 59-61 (texte arabe) ; T.L. Heath, *A history of Greek mathematics*, vol. II, 321 ; A.P.Youschkevitch, *Les mathématiques arabes*, 49.

20. *masqaṭ ḥiğr* : lieu où tombe le mur ? surplomb du mur, entre l'aplomb du mur et le pied du mur ; je n'ai pour ma part rencontré cette terminologie que dans ces deux textes, ainsi que dans le *Traité sur Les Cadrans Solaires* du même Ṯābit, où ce terme désigne alors le pied de la hauteur ; c'est d'ailleurs peut être dans la terminologie utilisée pour les cadrans solaires qu'il faut voir l'origine de ce terme, qui semble en tout état de cause relever d'un contexte de mathématiques pratiques ; Thâbit Ibn Qurra, *Oeuvres d'astronomie*, in R. Morelon (éd.), Paris, 1987, 160.

Preuve :

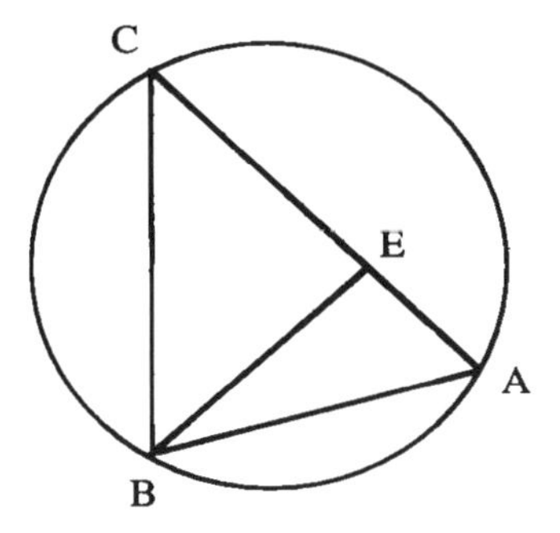

construction auxiliaire : *BE* hauteur issue de *B*.

[$\angle A = \frac{\pi}{3}$, et le diamètre du cercle est connu] => [*BC* connue (*Data* P. 88)],

[$\angle A = \frac{\pi}{3}$, *BAE* rectangle, et *AB* connue] => [*BE* connue (P. 17)],

[*BC* et *BE* connues] => [*CE* connue (lemme I)] => [$AC = CE + EA$ connue].

Notons ici encore (comme dans le cas des propositions 16 et 17) que nous sommes en présence d'un cas particulier d'une proposition qui se déduirait aisément des propositions 89, 40, 52 et 55 des *Données*[21], et qui serait la suivante : *si un triangle inscrit dans un cercle de grandeur connue a un angle et un côté adjacent connus alors les deux autres côtés sont connus.* Si l'usage de la proposition 88 des *Données* ne permet pas la détermination (le calcul) de *BC*, une table de cordes permettrait cependant à un astronome d'en déterminer une valeur approchée ; notons également que l'utilisation de la proposition 17 permettrait, en lieux et places de cette proposition, la détermination de *BC* : en effet si *O* est le centre du cercle circonscrit au triangle *ABC*, et *H* le milieu de *BC* le triangle *OHB* satisfait aux hypothèses de la proposition 17 ($\angle BOH = \frac{\pi}{3}$, et *OB* est connue) donc $BH = \frac{1}{2}BC$ est connue et *BC* est connue.

CONCLUSION

Les propositions que nous venons d'étudier (et les autres propositions du traité de Ṯābit), qui portent sur des calculs sur les triangles (rectangles ou autres), des relations entre le diamètre du cercle circonscrit et les côtés de ces triangles, des relations entre diverses cordes d'un cercle, ou entre côtés et diagonales de quadrilatères (inscriptibles ou non), sont de celles qui interviennent le plus souvent dans les calculs des astronomes. Nous avons vu que Ṯābit

21. Proposition 88 : *si dans un cercle donné de grandeur on mène une ligne droite qui retranche un segment comprenant un angle donné, la droite menée sera donnée de grandeur* ; Proposition 89 : *si dans un cercle donné de grandeur on mène une ligne droite donnée de grandeur cette droite retranchera un segment qui comprendra un angle donné.*

donne de ces propositions, dont certaines sont des conséquences immédiates de théorèmes des *Données* d'Euclide, des démonstrations qui permettraient à un astronome de déterminer effectivement les diverses grandeurs cherchées, sans se contenter d'affirmer, ce qui découle souvent immédiatement des *Données*, que celles-ci sont connues (ou données). Que ce soit pour les nouveaux besoins de l'astronomie et de la trigonométrie balbutiante, ou sous l'influence de l'algèbre, il semble bien que s'opère là, un glissement sémantique entre *connu = constructible* (pour Euclide[22]), et *connu = calculable*, et qu'une exigence de calculabilité se substitue chez Ṯābit à une exigence de constructibilité (qui était celle d'Euclide).

Le traité de Ṯābit nous paraît ainsi se situer au confluent de deux traditions : une tradition purement géométrique, qui est celle des *Données* d'Euclide, et une tradition relevant de mathématiques que l'on pourrait qualifier de mathématiques pratiques, et qui est celle des *Métriques* de Héron d'Alexandrie, ou du chapitre consacré à des calculs d'aires (*Bāb al-misāḥa*) du *Kitāb al gabr* d'al-Khwārizmī (dont l'influence sur le traité de Ṯābit semble manifeste dans certains usages terminologiques[23], et peut être, même si cela peut sembler paradoxal, dans certaines absences de démonstration). Ce qui oppose et différencie ces deux traditions semble bien être le statut de la grandeur (essentiellement de la longueur) : alors que pour le géomètre (Euclide et ses successeurs) la grandeur (du fait de l'existence des irrationnelles) ne saurait être un nombre (nombre entier ou rationnel), il en va différemment dans la tradition des mathématiques pratiques (qui est celle des astronomes, du *Bāb al-misāḥa* d'al-Khwārizmī, et en un sens des algébristes) ; pour l'astronome seules interviennent des mesures approchées des grandeurs, qui sont alors des nombres *(i.e.* des nombres rationnels), et l'algébriste quant à lui " opère sur les *inconnus* (*al-mağhūlāt*)... comme l'arithméticien opère sur les *connus* (*al-ma'lūmāt*) "[24] ; si le statut de la grandeur pose chez Euclide et ses successeurs un grand nombre de problèmes tant philosophiques que mathématiques, la position sur le sujet du tenant des mathématiques pratiques n'en soulève guère : une fois admise la possibilité de mesurer ou d'approximer une grandeur par un

22. Marinus, pour justifier sa propre définition de " donné " comme ce qui est à la fois *connu et constructible*, relève à ce propos " l'usage constant qu'il (Euclide) fait du mot construisons, et bien qu'il néglige de mentionner le connu, le considérant comme une conséquence du constructible " ; Peyrard, *Les Oeuvres d'Euclide*, XII.

23. Nous avons déjà relevé l'usage du terme *masqaṭ ḥiğr* (voir note 20), on pourrait également citer l'usage de la terminologie *taksīr^{u}-hu ma'lūm* (du verbe *kassara* mesurer une superficie) " dont la surface est connue " venant en lieux et places de la terminologie euclidienne *ma'lūm al-qadr* " de grandeur connue ", et qui, me semble-t-il, confirme la présence de l'idée de mesure dans ce traité. Notons que cette terminologie utilisée par al-Ḫwārizmī dans la partie géométrique consacrée à certains calculs d'aires (*Bāb al-misāha*) de son *Kitāb al-ğabr*, est également utilisée par les Banū Mūsā, dans le *Traité sur la mesure du cercle*, la terminologie de Ṯābit dans son traité sur la mesure de la parabole étant plutôt *misāḥa*. F. Rosen, *The algebra of Mohammed ben Musa*, 57, 59 (texte arabe) ; R. Rashed, *Les méthématiques infinitésimales,* vol. I, 83, 89, 135, 137.

24. Al-Samaw'al, *Al-Bahir en algèbre d'As-Samaw'al*, in S. Ahmad, R. Rashed (éds), Damas, 1972, 9.

nombre (rationnel), ou ce qui revient presque au même de manipuler toute grandeur comme un nombre, celui-ci ne considère plus que des nombres ; les grandeurs ne l'intéressent qu'en tant qu'elles sont mesurables, c'est-à-dire approximables par un rationnel[25]. C'est cette différence de statut de la grandeur (qui recouvre en outre l'opposition valeur exacte/ valeur approchée) qui caractérise l'opposition entre ces deux traditions.

Il semble que pour Ṯābit se dessine ainsi une troisième voie visant à déterminer *(i.e.* calculer) *more geometrico*, c'est-à-dire avec les contraintes de la géométrie euclidienne, les grandeurs des éléments inconnus de la figure (et non de les construire). Cependant, lors de ce passage de constructibilité à calculabilité, le géomètre se heurte vite aux contraintes inhérentes à la géométrie même, que nous avons évoquées et que nous allons essayer de préciser. Les géomètres refusent en effet les facilités de l'algèbre, et restent volontairement, par souci de rigueur théorique, dans un cadre strictement euclidien ; si l'algébriste (ou l'astronome) détermine effectivement les inconnues au cours de l'analyse, c'est qu'il s'en donne les moyens en calculant, sans aucune justification théorique, sur celles-ci comme l'arithméticien calcule sur les nombres ; les géomètres eux sont soumis à des contraintes plus grandes : grandeurs et rapports n'étant pas des nombres, il leur est difficile de déterminer effectivement le côté d'un carré dont la surface est connue, il leur est impossible de considérer le rapport d'une surface et d'une longueur (c'est-à-dire de substituer une détermination effective des grandeurs à l'utilisation des propositions 55 ou 57 des *Données*, toutes choses que se refuse à faire Ṯābit géomètre mais que fait couramment Ṯābit astronome[26]), il leur est impossible de multiplier une longueur ou une surface par un rapport. La persistance de la terminologie de " connu " dans les analyses de ce type de problèmes de géométrie (absente des problèmes d'algèbre, dans lesquels on détermine effectivement les inconnues, voir la démonstration d'al-Ḫwārizmī, ou des problèmes d'astronomie), ne serait alors qu'un pis aller de géomètre, éludant la constatation du fait que ces grandeurs dont on affirme qu'elles sont connues ne peuvent en fait être déterminées (calculées) mais seulement construites. La terminologie de connu sera d'ailleurs définitivement abandonnée dans ce type de problèmes lorsque grandeurs et rapports seront définitivement assimilés à des nombres (rationnels ou irrationnels), et que, après avoir supprimé les contraintes d'homogénéité dans l'écriture des équations ou des égalités, et fait abstraction de l'idée de dimen-

25. C'est d'ailleurs ce que relève Marinus dans son *Commentaire aux* Data *d'Euclide* : il évoque, pour les réfuter ensuite, un certain nombre d'interprétations du concept de " donné ", et attribue en particulier à Ptolémée l'opinion que " donné " pourrait signifier rationnel : " certains ont affirmé que c'était la même chose que le rationnel, ainsi que semble le faire Ptolémée, en appelant donné ce dont la mesure est connue exactement ou à très peu près (on retrouve le *wa šay qalīl* d'al-Ḫwārizmī) ". F. Peyrard, *Les Oeuvres d'Euclide*, XVII.

26. Thâbit Ibn Qurra, *Oeuvres d'astronomie*, in R. Morelon (éd.), Paris, 1987, 283.

sion, on pourra effectivement déterminer (calculer) les grandeurs inconnues, c'est-à-dire après Descartes.

ḲĀḌĪZĀDE AL-RŪMĪ ON SAMARḲANDĪ'S AS̱HKĀL AL-TA'SĪS : A MATHEMATICAL COMMENTARY

Gregg DE YOUNG

The commentary is one of the most common forms of composition employed within the medieval mathematical tradition. What is the role of the commentator in the commentary process ? Does he merely explicate terms and concepts ? Does he provide intellectual and/or historical context to make the original text more comprehensible to his readers ? How closely is he bound, either verbally or conceptually, to the text on which he is commenting ? Given the large number of surviving commentaries produced within the Arabic/ Islamic intellectual tradition (especially during the medieval or " scholastic " period), to suggest generalized answers is hazardous. In this paper we examine one commentary within the Arabic/Islamic Euclidean tradition in order to gain more insight into the commentary process. For this purpose, we have selected the commentary by Ḳāḍīzāde al-Rūmī on the *As̱h kāl al-Ta'sīs* of Samarḳandī. Both the original treatise and its commentary have proved of enduring interest within this branch of Euclidean studies, continuing to be copied and studied for centuries. Surviving manuscripts can be found in numerous repositories from Morocco to India.

S̱HAMS AL-DĪN MOḤAMMED B. AS̱HRAF AL-ḤUSAYNĪ AL-SAMARḲANDĪ[1]

Samarḳandī was a contemporary of Naṣīr al-Dīn al-Ṭūsī and Ḳuṭb al-Dīn al-S̱hīrāzī, active about the last half of the 7th/ 13th century. He was considered a skilled logician and mathematician but was not associated with the scholars gathered at the Maragha Observatory during the lifetime of Ṭūsī.

The *As̱h kāl al-Ta'sīs* is his best known mathematical work. It consists of 35 propositions drawn from Euclid's *Elements* : 29 from Book I, five from the

1. H. Dilgan, " Al-Samarkandi ", in C. Gillispie (ed.), *Dictionary of Scientific Biography*, vol. 12, New York, 1981, 91 ; L.B. Miller, " Al-Samarḳandī ", *Encyclopaedia of Islam*, vol. 8, 2nd ed., Leiden, 1995, 1038-1039.

beginning of Book II, and the first proposition of Book VI. These 35 propositions are in some sense — at least according to the titre assigned to this treatise — fundamental propositions. Before analysis of the commentary by Ḳāḍīzāde al-Rūmī, we overview the text on which he comments.

Ashkāl al-Ta'sīs

Samarḳandī's introduction explains why he extracted these propositions from among the 465 in Euclid's treatise[2]. They are, he says, the foundation of " geometry " — the relation among plane figures and their constituent parts. These extracted propositions address two basic topics : (1) triangles and the equality or inequality of their parts (sides and angles) and (2) parallel lines and the study of quadrilaterals whose sides are parallel (parallelograms, rectangles, squares). And what, for Samarḳandī, constitutes a " fundamental " proposition ? In general, it seems to be one that is used as the basis for further propositions. For example, if Euclid demonstrates the converse of a theorem merely for the sake of completeness but never uses that result in demonstrating a later proposition, Samarḳandī does not include it within his " fundamental " propositions.

Samarḳandī's text begins with definitions. These are, for the most part, those of Euclid, Book I. From these, however, he omits definition 7, apparently subsuming plane surfaces under the more general rubric of surfaces. In its place, he inserts the definition of a solid and its terminus from Book XI. He then omits definition 9 (a rectilinear angle). Definition 13 (the circle) is also omitted — it does not figure in any propositions that he considers " fundamental ". Euclid's definitions 15-21 (classification of rectilinear figures and triangles based on the characteristics of their sides and angles) are also gone from Samarḳandī's text. Apart from the square and rectangle, Euclid does not again refer to these definitions and so Samarḳandī does not include them as " fundamental ". If, however, Samarḳandī is indeed omitting any definition not used by Euclid or himself, it is difficult to understand why he retains those definitions that identify various classes of quadrilaterals.

Following these definitions, Samarḳandī turns to axioms and postulates. He omits the fifth (or parallel line) postulate of Euclid from his list since he believes it can be demonstrated geometrically. (It has become Proposition III of the *Ashkāl al-Ta'sīs.*) He also adds two postulates not found in Euclid. The first, that two straight lines do not encompass an area, was discussed (and rejected) by Proclus[3]. The other, that a straight line can only be continued in a straight line by one line, seems to have been inserted by Samarḳandī himself.

2. T.L. Heath, *The Thirteen Books of Euclid's Elements*, New York, 1956 (3 vols). All references to Euclid's text are to this edition.

3. *Idem*, 232.

Samarḳandī has re-ordered the extracted propositions from their original position in Euclid's text. (See the Table of Correspondence in the Appendix.) Moreover, many 'construction propositions' have been removed. For example, Euclid's first three propositions (to construct an equilateral triangle ; to construct a line equal to a given straight line ; to cut off from a given segment the equal of a smaller segment) have been excised. Similarly, he has omitted Euclid I, 9 (to bisect a given line segment), and Euclid I, 10 (to bisect an angle). From the end of Book I he also drops several constructions : Euclid I, 42 (to construct within a given rectilinear angle a parallelogram equal to a given triangle) ; Euclid I, 44 (to apply a given parallelogram to a specified line segment) ; Euclid I, 45 (to construct within a given rectilinear angle a parallelogram equal to a given rectilinear figure) ; Euclid I, 46 (To construct a square on a given line segment).

Samarḳandī has not removed all constructions, though. He retains Euclid I, 11 and 12 (to erect a perpendicular from a point to a given line segment and to erect a perpendicular from a point on a given line segment), Euclid I, 22 (to construct a triangle from three specified line segments) and Euclid I, 23 (to construct an angle equal to a given rectilinear angle). From these retained constructions, the other constructions can be produced. Samarḳandī seems to assume that some construction techniques need no formal demonstration.

Ian Mueller has recently argued that Euclid I, 45 is the *logical* culmination of Book I[4]. Heath had earlier claimed that this proposition was essential to the development of " geometrical algebra "[5]. It is then noteworthy that Samarḳandī has chosen to omit it from his " fundamental " propositions. Samarḳandī clearly chose his propositions using different criteria.

In general, construction operations are not prominent in the demonstrations of Samarḳandī. His Proposition I (corresponding to Euclid I, 13) demonstrates that when a line falls upon a given line, the angles formed at the sides of the incident line are either two right angles or their sum equals two right angles. When the two are *not* right angles, Euclid says we must erect a perpendicular at the point of meeting. Samarḳandī does not call for this construction. Instead, he says in essence that, since there exists a perpendicular to that line at the point, let us assume it and proceed with the demonstration. (Because he does not need to construct the perpendicular he can re-order this proposition to the first position).

Samarḳandī also frequently uses transferral of all or part of the geometric figure. It is true that Euclid had resorted to this technique occasionally, as in Euclid 1, 4 (if two sides and the included angle of one triangle are equal to the corresponding two sides and included angle of another triangle, the two trian-

4. I. Mueller, *Philosophy of Mathematics and Deductive Structure in Euclid's Elements*, Cambridge, 1981, 16ff.

5. *Cf.* T.L. Heath, *The Thirteen Books of Euclid's Elements*, *op. cit.*, vol. 1, 346-347.

gles are equal to one another and their remaining parts are also equal to one another). Euclid says we should superimpose the given parts of one triangle on the corresponding elements of the other triangle and claims that when this is done, the remaining parts will also coincide. Heath has argued that Euclid seems reluctant to use this implicit transport of geometric entities[6]. In the *Aṣẖ kāl al-Ta'sīs,* Samarḳandī certainly employs the superposition technique more often than did Euclid.

The main re-ordering between Samarḳandī and Euclid occurs in the first nine propositions of the *Aṣẖ kāl al-Ta'sīs.* There are other interesting variations as well. In Proposition V (Euclid 1, 24 and 25), for example, we not only find Samarḳandī combining two propositions (in Proposition XVIII he combines Euclid 1, 27 and 28), but he has also reversed Euclid's phrasing. Whereas Euclid argued that the larger angle will subtend the longer side, Samarḳandī enunciates the problem by saying that the smaller angle will subtend the shorter side. The rationale for this change in verbal argument is not clear. Obviously, it entails no essential change in the logic of the proposition. (Neither Ṭūsī nor Abharī, the closest lines of influence on Samarḳandī, have a similar re-phrasing.)

Only in Proposition III does Samarḳandī give indication of his sources/predecessors : Ibn al-Hayṯẖam, Omar al-Ḵẖayyāmī, al-Jawharī, Naṣīr al-Dīn al-Ṭūsī, Aṯẖīr al-Dīn al-Abharī, Ḳāḍī Hameh. The order of names is not completely chronological. What prompted this choice of name ordering is not clear — but clearly Samarḳandī knew the major geometrical works of his time.

ḲĀḌĪZĀDE AL-RŪMĪ[7]

Ḳāḍīzāde al-Rūmī (died *ca.* 1436) was born in Anatolia (al-Rūm) near Bursa where autographs of some of his most important treatises are reputed to exist. He soon migrated to the court of Ulugh Beg in Transoxiana. He spent the most productive years of his life in Samarḳand in various administrative positions. Much of his extant writing is commentaries on earlier treatises, notably Ptolemy's *Almagest* and Samarḳandī's *Aṣẖ kāl al-Ta'sīs.* If numbers of surviving copies are an indication, this latter commentary was even more popular than its original.

ḲĀḌĪZĀDE'S COMMENTARY[8]

The commentary on the *Aṣẖ kāl al-Ta'sīs* covers the entire text and is more extensive than the original treatise by Samarḳandī. It is a full-fledged commen-

6. *Cf.* T.L. Heath, *The Thirteen Books of Euclid's Elements*, *op. cit.*, vol. 1, 249.

7. H. Dilgan, " Qāḍī Zāde al-Rūmī ", in C. Gillispie (ed.), *Dictionary of Scientific Biography*, New York, 1981, 227-229.

8. This section of the paper is based on M. Souissi (ed.) *Aṣẖkāl al-Ta'sīs lil-Samarḳandī Ṣẖarḥ Qāḍī Zāde al-Rūmī*, Tunis, 1984.

tary, not a collection of marginal annotations (*ḥas̲h̲ īya*). Some commentators use the method of direct quotation (" He said ... ") followed by the commentator's thoughts (" I say ... "). Ķāḍīzāde, however, skilfully interweaves the entire text of Samarķandī and his own comments so that they form an almost completely seamless whole.

Ķāḍīzāde considerable expands the introductory discussion of Samarķandī. He does not re-introduce the definitions omitted by Samarķandī. Instead, he adds non-rectilinear figures to the existing definitions, although they are not used in either the *Elements* or his commentary. This material resembles the discussion of geometry in *Rasā'il* of the Ik̲h̲ wān al-Ṣafā'[9].

Samarķandī, as we know, had added two postulates. Neither appears in the Arabic versions of the *Elements*. Ķāḍīzāde asserts that both are demonstrable (he gives the demonstrations) and therefore are not true postulates. The *Iṣlāḥ* of the Elements composed by At̲h̲īr al-Dīn al-Abharī has similar demonstrations (indication that Ķāḍīzāde's work has close ties with that of Abharī). Ķāḍīzāde, however, does not attribute these demonstrations to Abharī.

Ķāḍīzāde seems more willing than his predecessor to allow transporting geometric entities without distortion. For example, in explicating Samarķandī's Third Postulate, he describes construction of a circle (which Samarķandī ignored) from the rotation of the radius about the fixed center. Similarly, commenting on Proposition I, he explains that the incident line can be considered to rotate about the point of contact until the two angles become equal (there is an implicit assumption of continuity here). Whether or no Euclid was diffident about using this argument, Samarķandī, and certainly Ķāḍīzāde, felt not such compunction. They use the technique again and again without hesitation.

As already noted, Samarķandī omits Euclid's fifth (parallel lines) postulate because he considers it demonstrable. It has become his Proposition III. The demonstration is superficially simple : on the side on which the interior angles are *less* than two right angles, the lines are moving closer and closer together, the one approaching the other. In other words, the distance between the two lines is continually decreasing. Thus, they must at some point, when the distance between them decreases to zero, meet. Ķāḍīzāde, in his gloss, explains that this " demonstration " is intended to answer those who object based on an infinity argument : because any line segment is infinitely divisible, the line segment measuring the distance between the two lines is also. In this case, the two lines may be said to continually approach one another but never meet since a distance will always exist between them. If one were to employ this interpretation, however, other postulates also become impossible. We will not, then, be able to extend a given line segment to reach a specified point because the distance between the point and the end of the line segment is infinitely divisible,

9. Y. Marquet, " Ik̲h̲wān al-Ṣafā' ", *Encyclopaedia of Islam*, vol. 3, 2nd ed., Leiden, 1986, 1071-1076.

so that the extension cannot be actually completed. If his argument were allowed, the entire structure of the *Elements* would collapse.

Ḳāḍīzāde's commentary includes much direct explication of Samarḳandī's terse treatise. Ḳāḍīzāde makes certain that the reader not only is referred tot he appropriate proposition to justify each step of Samarḳandī's demonstration, he typically repeats the enunciation of the proposition so that the reader does not have to look back continually in the treatise. He also takes greater care than Samarḳandī to make clear the antecedents of pronouns referring to geometric elements in the proposition. Such features make his commentary a valuable educational tool.

Besides explication, Ḳāḍīzāde discusses additional cases of propositions. Examples occur in Propositions XIV, XV, XXIII, and XXVII. Such discussions are also prominent in the *Taḥrīr* of Ṭūsī. The cases discussed by Ḳāḍīzāde are not found in the work of Ṭūsī, however. Discussing Proposition XV, Ḳāḍīzāde includes a case earlier noted by Proclus[10], although it is not here attributed to him, so it is not clear whether this is a case of independent discovery or of plagiarism. The purpose for discussing such additional cases is not clear. It sometimes seems that Euclid, while aware of the existence of multiple cases, chose to demonstrate only the more difficult, leaving the others to challenge intelligent readers. Perhaps the continued interest in multiple cases merely provided a way to demonstrate versatility as a mathematician.

Ḳāḍīzāde seems indebted especially to the *Taḥrīr* of Naṣīr al-Dīn al-Ṭūsī and to the *Iṣlāḥ* of Athīr al-Dīn al-Abharī. These two are among Samarḳandī's sources, but Ḳāḍīzāde cites them more frequently. In discussing Samarḳandī's Proposition VII (Euclid 1, 6), he quotes the demonstration of Abharī (" the Master of the *Iṣlāḥ* "), then suggests a shorter version. This later seems to be his own contribution. In his consideration of Samarḳandī's Proposition XVIII (Euclid 1, 27 and 28), he quotes Abharī's demonstration of Euclid's parallel lines postulate. There are several references to Ṭūsī (" the Master of the *Taḥrīr* ") but no direct quotations from his *Taḥrīr*. Perhaps Ḳāḍīzāde considered it so well known that a reference would be sufficient to recall to the reader what Ṭūsī had said, while Abharī's work was less well-known and so needed to be cited fully.

At the end of Samarḳandī's Proposition VI, Ḳāḍīzāde re-introduces the construction propositions corresponding to Euclid I, 1-3 which had been omitted. They are, he says, necessary to demonstrate what follows. He also includes explicit directions for bisecting a line segment, something Samarḳandī had omitted from his text. Since Ḳāḍīzāde is correct in stating that these are required for full demonstration of these propositions, we must ponder why Samarḳandī, himself an astute mathematician, omitted them in his treatise. A

10. *Cf.* T.L. Heath, *The Thirteen Books of Euclid's Elements*, *op. cit.*, vol. 1, 293.

possible answer is that he did not intend to construct a theoretical work but rather to draw together propositions with practical applications in daily life.

Thus it seems that a commentator, such as Ķāḍīzāde, might ignore the original intent of the author whose work he claims to explicate. He seems free to add materials of his own, not merely for the sake of explication but to exhibit his own prowess.

Appendix

Table of Correspondence

Samarķandī's Proposition	Corresponding Proposition in Euclid
1	I, 13
2	I, 14
3	Postulate 5
4	I, 4
5	I, 24 and 25
6	I, 5
7	I, 6
8	I, 8
9	I, 11
10	I, 12
11	I, 15
12	I, 16
13	I, 18
14	I, 19
15	I, 22
16	I, 23
17	I, 26
18	I, 27 and 28
19	I, 29
20	I, 32
21	I, 33
22	I, 34

Samarḳandī's Proposition	Corresponding Proposition in Euclid
23	I, 35
24	I, 36
25	I, 37
26	I, 38
27	I, 41
28	VI, 1
29	I, 43
30	I, 47
31	II, 1
32	II, 2
33	II, 4
34	II, 5
35	II, 6

THE SOLUTION OF APOLLONIUS' PROBLEM IN THE MEDIEVAL ARAB EAST

Irina LUTHER

The problem to construct a circle tangent to three given circles was evidently for the first time formulated and solved by Apollonius of Perga (260-170 B.C.) in his work *On tangency*, which is mentioned only in *Mathematical collection* of Pappus of Alexandria (III A.D).

Almost two thousand years the problem was of great interest for many distinguished mathematicians[1]. Some of them (Pappus of Alexandria, François Viète (1540-1603), René Descartes (1596-1650), Pierre Fermat (1601-1665), Isaac Newton (1643-1727), Guillaume François Antoine L'Hospital (1661-1704), Leonhard Euler (1707-1783), Johann Lambert (1728-1777), Lazare Carnot (1753-1823), Joseph Gergonne (1771-1859) and others) had got their own solutions of the so-called Apollonius' problem. But nobody of them had succeeded in reconstruction of the Apollonius' solution[2].

Nothing was known up to now about the attempts of medieval Arabic scholars to solve this problem. Nevertheless such attempts, as we can assert now, had been undertaken. The only example of this kind has been discovered in the treatise *Al-masā'il al-mukhtāra*[3] of Ibrāhīm ibn Sinān (908-946), a grandson of the prominent medieval scholar Thābit ibn Qurra (836-901).

1. On the history of Apollonius' problem see : M. Chasles, *Aperçu historique sur l'origine et le développement des méthodes en géometrie*, Paris, 1889.

2. Recently, the Russian historian of mathematics Albert Khabelashvili has suggested his own reconstruction of Apollonius' solution. He supposed that Apollonius had come to this problem while considering a cross-section of four straight circular cones when all cones have the equal angles at their vertices and one cone is inverted with respect to the others and tangent to the others. As a result of such section we have four circles in the same plane, one of which is tangent to three others.

3. According to A.S. Saidan the title is " Selected problems ", according to J.P. Hogendijk — " The exquisite problems ". The last is more correct. The Arabic word *al-masā'il* means " problems, examples ", *al-mukhtāra* means " very important, significant ".

The unique manuscript of this work was included in the Arabic manuscript Bankipore 2468 (now 2519) in the Khuda Bakhsh Oriental Public Library in Patna (India), composed mainly in 631/632 A.H. (1234 A.D.)[4]. It unites over forty valuable treatises of medieval Islamic mathematicians and astronomers. Among those there are almost all treatises of Ibn Sinān that have come down to us except for *The book on shadow-making instruments* (*Kitāb fī ālāt al-aẓlal*)[5].

Ibn Sinān's treatise *Al-masā'il al-mukhtāra* includes 41 problems on construction with elements of geometrical algebra, using mainly the methods of Euclid's *Elements* and Apollonius' *Conics*. In a few lines of introduction Ibn Sinān says, that he gives mostly the analysis of problems, but leaves synthesis for the interested reader to perform by himself.

The attempt of Ibn Sinān to solve the Apollonius' problem is Problem 40 of the treatise. It should be noted, that Apollonius' name is nowhere in Ibn Sinān's reasoning mentioned.

Let us consider the Ibn Sinān's solution[6]. He examines three cases.

The first case : all three given circles - AB with the center F, CD with the center L and EG with the center M are equal.

FIGURE 1.

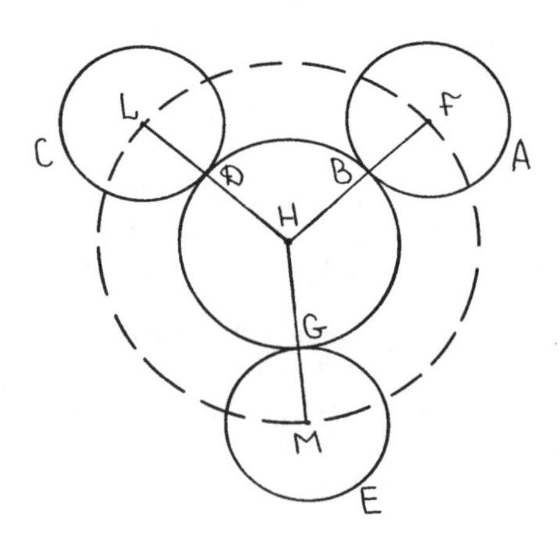

4. On the rearrangement of the Bankipore Manuscript 2468 and the identification of treatises included in it see : J.P. Hogendijk, " Rearranging the Arabic Mathematical and Astronomical Manuscript Bankipore 2468 ", *Journal for the History of Arabic Science*, 6, n° 1-2 (1982), 133-159 ; A.S. Saidan, " The Rasail of al-Bīrūnī and Ibn Sinān. A rearrangement ", *Islamic culture*, 34 (1960), 173-175.

5. All these treatises have been published in Arabic by A.S. Saidan (ed.), see *The works of Ibrāhīm ibn Sinān*, Kuwait, 1983.

6. *Idem*, 242-258.

The analysis of Ibn Sinān is following[7]. He supposes that the circle tangent to three given circles has been built. Let it be the circle BDG with the center H. Points B, D and G are the points of tangency, hence the lines HF, HL and HM are the straight lines[8]. The segments DL, BF, GM are the radii of the equal circles, so they are equal. The segments HB, HG, HD are the radii of one and the same circle. Hence HF, HL, HM are equal and point H is the center of the circle, which passes through the centers of the given circles, *i.e.* through the known in position points L, F and M.

But the point H is simultaneously the center of a circle to be found. So the problem (or more precisely the first case) comes down to the problem of construction a center of the circle described around the known in position triangle LFM and determining the points of tangency B, D and G by joining this center H with the centers of the given circles.

The second case : only two of three given circles are equal. Let it be CD and EG.

FIGURE 2.

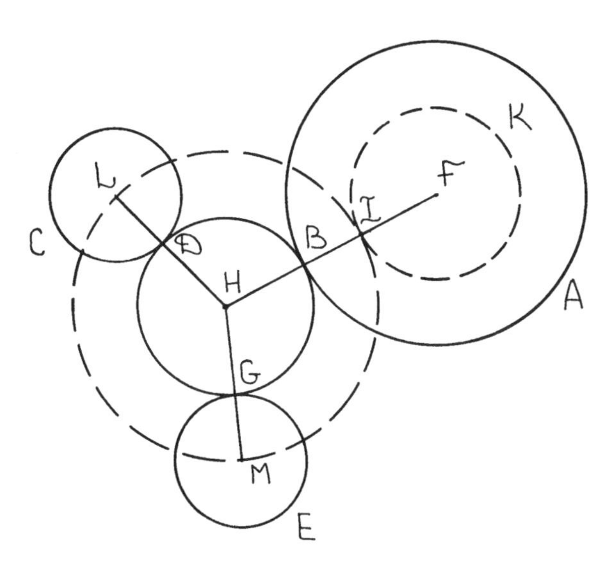

Ibn Sinān supposes that a circle tangent to three given circles CD, EG and AB has been built. Let it be BDG.

7. Here and further we give slightly adapted for the modern reader version of Ibn Sinān's analysis, keeping as close as possible to the medieval Arabic style and terminology.

8. Proposition 12 of book III of book of the Euclid's *Elements* : if two circles touch each other from outside, then the straight line joining their centers passes through the point of their tangency.

Since the circles CD and EG are equal, then LD is equal to GM. Therefore LH is equal to MH.

Ibn Sinān draws segment BI equal to MG, so the size of BI is known. The size and the position of the circle AB are known, therefore the size of its radius BF is known. The segment FI is equal to the difference between the segments BF and BI. The sizes of the segments BF and BI are known. Therefore the size of FI is known. According to the construction the segments HL, HM, HI are equal. The position of the center F of the given circle AB is known, the size of the segment FI is known. So, the size and the position of the circle IK are known. According to the construction HIF is a straight line, therefore the point I is the point of tangency of the known circle IK and the circle with the center H and radii HL, HM, HI.

Thus, the second case reduces to the problem of construction a circle which passes through two given points (in our case L and M) and is tangent to given circle (circle IK), with the center H coinciding with a center of circle to be found.

Then, as in the previous case, Ibn Sinān draws the straight lines joining point H with the centers of the given circles, determines the points of tangency B, D and G. So the circle BDG to be found has been determined. As it has been already shown Ibn Sinān has turned the second case into the problem of constructing a circle, which passes through two given points and is tangent to given circle. It is Problem 38 of the treatise.

The analysis of Ibn Sinān of this problem is following. Let given points be D and E and given circle - AB.

FIGURE 3.

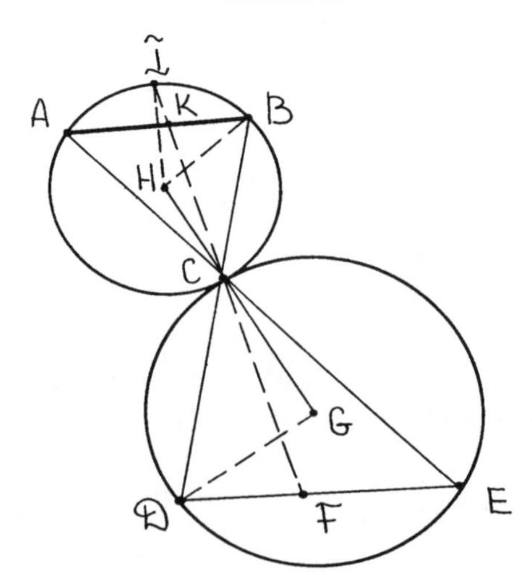

Ibn Sinān supposes that a circle to be found has been built. Let it be DCE.

He draws the straight lines EC and DC and marks off the points A and B - the points of intersection of the straight lines with the circle AB. Then he draws the straight lines DE, AB, BH and DG. The point C is the point of tangency, therefore CHG is the straight line.

Then Ibn Sinān proves the similarity of the triangles BHC and DCG, AHC and ECG and obtains respectively two relations

$$CD/CB = CG/CH \text{ and } CE/CA = CG/CH.$$

Hence, CD/CB = CE/CA and the triangles ABC and DCE are similar. Therefore, the angles BAC and CED are equal. Thus, the straight line AB is parallel to DE[9].

Since the triangles ABC and DCE are similar, then

$$AC/EC = BC/DC.$$

Therefore, according to Euclid's *Elements*[10], Ibn Sinān obtains the relation

$$(AC + EC)/EC = (BC + DC)/DC, \textit{ i.e.},$$

$$AE/EC = BD/DC.$$

Hence, $AE \cdot EC/EC^2 = BD \cdot DC/DC^2$ and $AE \cdot EC/BD \cdot DC = EC^2/DC^2$.

The product AE·EC is known since it is equal to the square of the tangent[11] from the point E (the position of E is known) to the known circle AB. Analogously, the product BD·DC is known. Therefore, the ratio AE·EC/BD·DC is known. Since it is equal to the ratio EC^2/DC^2, the ratio EC^2/DC^2 is known. Hence, the ratio EC/DC is known.

Then Ibn Sinān marks off such point F on DE that DF/FE = EC/DC. Therefore, the angles FCE and DCF are equal[12]. Further he draws the straight line FC and marks off the point I of its intersection with the circle AB. The angles KCB and KCA are equal (since the angles DCF and FCE are equal). Therefore, the arcs BI and AI are equal[13]. Hence, the perpendicular dropped from the point I onto the chord AB will bisect it and pass through the center H[14]. That is the end of Ibn Sinān's reasoning.

9. This result is based on Proposition 27 of book I of the Euclid's *Elements* : if a straight line, that crossed other two straight lines, makes the alternate angles equal to one another, these straight lines will be parallel.

10. Ibn Sinān applies here Definition 14 of book V of the Euclid's *Elements* : if a/b = c/d, then (a + b)/b = (c + d)/d.

11. Proposition 36 of book III of the Euclid's *Elements*.

12. Ibn Sinān applies the property of the bisectrix of an angle of a triangle to divide the side opposite to this angle in proportion to the lengths of the other sides. It is Proposition 3 of book IV of the Euclid's *Elements*.

13. Proposition 26 of book III of the Euclid's *Elements* : in equal circles equal angles rest on equal arcs.

14. This result follows from Proposition 30 of book III of the Euclid's *Elements* (problem of bisecting the arc) and Proposition 3 of the same book : if a straight line is a midpoint perpendicular to some chord in the circle, then it passes through the center of this circle.

The analysis of Ibn Sinān of this auxiliary problem, perhaps vague at first sight, gives nevertheless complete (both cases of tangency of circles — external and internal — have been taken into account) and original method of construction a circle, which passes through two given points and is tangent to the given circle.

So, let us consider our synthesis of this problem corresponding to the Ibn Sinān's analysis. It should be noted, that according to the ancient classification all these problems are plane problems. Therefore, our constructions have to be carried out only by means of a compass and a ruler.

Let given points be D and E and given circle - AB.

FIGURE 4.

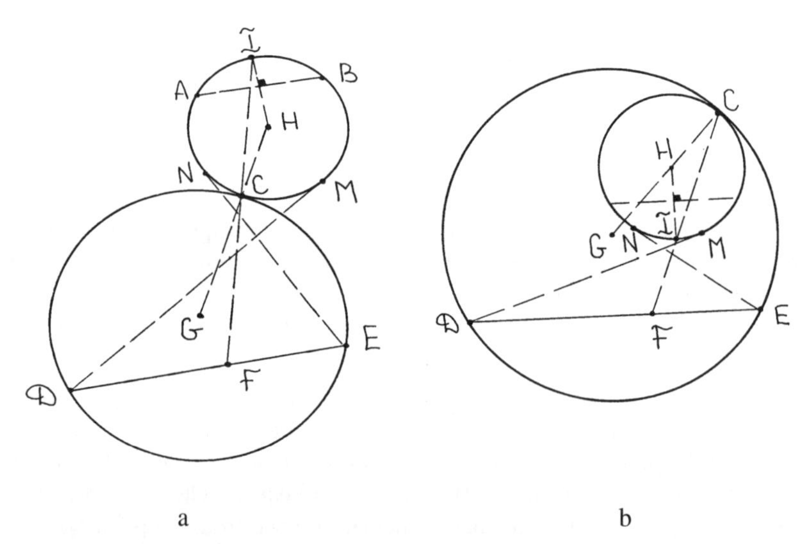

a b

From the given points D and E we draw tangents to the given circle AB (this construction carried out only with the help of a compass and a ruler is described in Proposition 17 of book III of Euclid's *Elements*).

Then, we divide the segment DE in point F so that the ratio of DF to FE is equal to the ratio of the tangent DM to the tangent EN (the construction is described in Proposition 10 of book VI of *Elements*). Further we draw some chord AB parallel to DE (two cases are possible : figures 4a, 4b). Then we draw perpendicular HL at the midpoint of AB, find the point I and join the points I and F. We mark off the point of intersection of the straight line IF and the circle AB - point C. Hence, the point C of tangency has been built.

Thus, three points of the circle to be found is known - D, E and C and we can draw it (the construction of a circle passing through three given points is described in Proposition 5 of book IV of *Elements*).

In addition it is worth noting that Ibn Sinān in this analysis has in fact proved the statement : if two circles are tangent (fig. 3) and are described around two triangles with the vertices at the point of tangency and opposite to this point sides parallel to one another, then the triangles are similar. The matter is, that F. Viète, who was perhaps the first to get the complete plane solution of Apollonius' problem, had considered the case similar to Ibn Sinān's second case and also had reduced it to the problem of construction a circle, which passes through two given points and is tangent to the given circle[15]. Moreover, Viète had proved this auxiliary problem by means of his Lemma III inverse to Ibn Sinān's statement mentioned above : if two similar triangles have the common vertex and parallel sides opposite to this vertex, then the circles described around the triangles are tangent at this vertex).

And, finally, Ibn Sinān's analysis of the third case : all three given circles AB, CD and EG are different.

FIGURE 5.

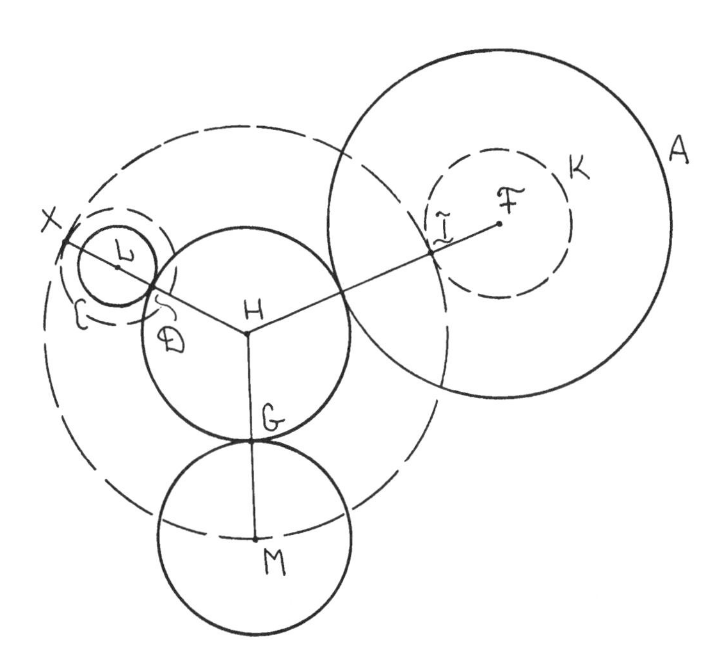

Ibn Sinān supposes that a circle tangent to three given circles has been built. Let it be BDG. He draws BI equal to MG and DX equal to MG.

15. *Apollonius Gallus sive exsuscitata Apollonii Pergaei. De tactionibus geometria*, in F. Viète (ed.), Paris, 1600.

Since the line DL is the radius of the given circle and according to the construction the size of DX is known, then LX is equal to the difference between DX and DL and therefore is known. Thus, the size and the position of the circle LX with the center in the given point L and the radius LX are known.

According to the construction, the size of BI equal to MG is known. The line BF is the radius of the given circle. The segment FI is equal to the difference between BF and BI. Hence, the size of FI is known. Thus, the size and the position of the circle KI with the center in the given point F and the radius FI are known.

The radii HB, HD and HG of one and the same circle are equal. According to the construction IB, MG and DX are equal. Then HI, HM and HX are equal. Hence, the point H (the center of circle to be found) is the center of the circle, which passes through the given point M (the center of the given circle) and is tangent to two circles KI and LX, which have been just determined above.

Thus Ibn Sinān has turned the third case into the problem of construction a circle, which passes through one given point and is tangent to two given circles (Problem 39).

But his analysis of this problem is actually the analysis of its special case, when the ratio DI to HQ (or the difference between DI and HQ) is given (figure 6). Besides, he has considered only one case of the possible disposition of circles.

FIGURE 6.

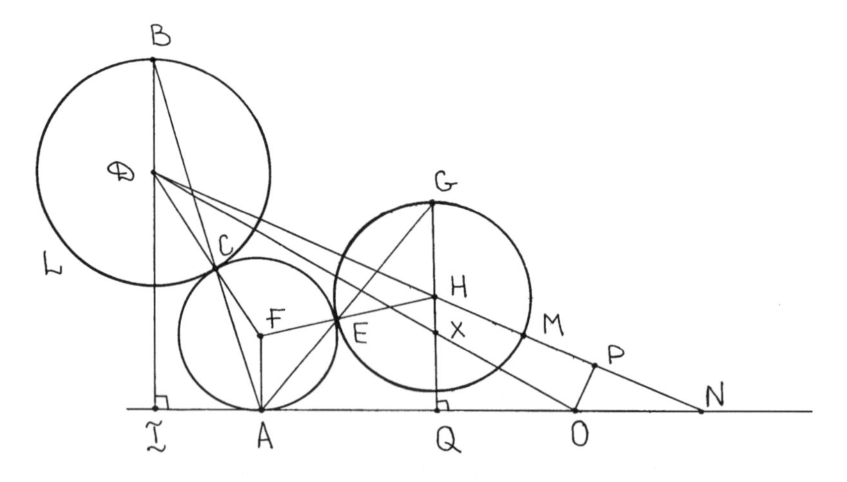

We shall not go into this incomplete solution. But it should be noted that while considering this case Ibn Sinān cites another his treatise *Kitāb fī'l-dawāir al-mutamāssa* (*The book on tangent circles*), which did not survive and, as Ibn

Sinān asserts, contained a lot of similar problems (perhaps it contained also the complete solution of the second auxiliary problem).

As for F. Viète, he had also considered the case similar to Ibn Sinān's third case and had turned it into the same auxiliary problem. But he had succeeded in its solving and had got the complete solution for all eight possible cases of the disposition of circles.

After giving his own solution Ibn Sinān presents the solutions belonged to two geometers of Xth century, Abū Yaḥyā al-Marwazī and Abū'l-'Alā ibn Karnīb. They have reduced the Apollonius' problem to the following : how to find the point D in the given triangle ABC, if the sums

$$AD + DB \text{ and } BD + DC \text{ are known.}$$

But their arguments, perhaps due to insufficient skill of copyist, sometimes are faulty, sometimes are vague, and from our point of view do not lead to the desired result.

Transmission et innovation : l'exemple du miroir parabolique

Roshdi Rashed

Parmi les objets de science, il est une classe particulière où se mêlent intimement une *theoria* et une *technè*. Il ne suffit donc pas, si l'on veut décrire l'un ou l'autre de ces objets, d'évoquer les concepts et leurs connexions réglées, mais il faut en même temps rappeler les procédés techniques nécessaires à sa fabrication.

La tâche de l'historien des sciences se double à l'évidence du métier d'historien des techniques. Mais cette tâche se complique encore dès lors que ces objets appartiennent à un passé lointain. Il y en a en effet qui viennent de si loin que personne n'oserait en fixer les origines. Celles-ci se perdent souvent à l'aube de l'astronomie, de la géométrie et de la statique. C'est le cas de beaucoup d'instruments : les cadrans solaires, les astrolabes, les leviers, les balances, les mésolabes, les miroirs ardents, et bien d'autres instruments mathématiques. Tous ces objets sont l'oeuvre des anciens géomètres et ingénieurs. Tous pourraient relever de ce que d'aucuns nommeraient aujourd'hui les mathématiques appliquées. Tous se présentent comme objets techniques, orientés vers un but pratique et doués d'une utilité sociale. Que cette utilité soit effective ou simplement l'effet de l'imaginaire social, cela importe relativement peu ; elle assigne en tous les cas à cet objet géométrico-technique une finalité qui dépasse la simple connaissance qu'il exprime. L'astrolabe, par exemple, ne se réduit ni à la connaissance en astronomie exigée par sa fabrication, ni aux procédés de l'artisan qui l'a réalisé ; il évoque aussi l'utilité précieuse qu'il offre à l'astronome, et celle que l'imagination de l'astrologue et du médecin lui attribue. C'est cette utilité multiple qui a suscité une demande sociale responsable de l'institution d'une profession, celle des “ astrolabistes ”, reconnue comme telle par les historiens et les biobibliographes des IXe-Xe siècles tout au moins.

Mais astrolabes, miroirs ardents, etc. furent non seulement objets de transmission, mais aussi vecteurs de la propagation du savoir scientifique. Il y a donc toute une réflexion à mener sur cette transmission, ses formes particulières, et sur l'intégration successive de ces instruments aux différentes traditions. C'est cette réflexion que je vais entamer ici, pour le seul miroir parabolique.

Commençons par nous interroger sur ce que pouvait signifier aux yeux d'un ancien grec un miroir ardent. Ce n'est rien d'autre qu'un *organon*, une machine de construction délicate destinée à un usage pratique. Ce but pratique, qu'il soit spéculatif ou effectif, attribué au miroir, a joué un rôle important dans l'incitation à la recherche depuis le second siècle avant l'ère chrétienne au moins, et jusqu'au XVIIIe siècle, en attirant non seulement les géomètres de premier rang, mais aussi toute une foule de mathématiciens de moindre prestige. D'autre part, au cours de cette recherche, le miroir ardent, objet apparemment simple, ne tarde pas à révéler une complexité dont l'élucidation renvoie à plusieurs traditions : géométrique, catoptrique, technique, et, parfois, astronomique. Examinons tout cela dans le cas du miroir ardent parabolique, en passant d'Alexandrie à Byzance, avant de nous rendre à Bagdad, au Caire, en Europe du Sud ..., c'est-à-dire dans les plus fameux centres scientifiques, jusqu'au XVIIe siècle environ.

C'est Dioclès[1] qui nous fournit la première information importante sur le miroir parabolique, au deuxième siècle avant l'ère chrétienne. Il nous rapporte qu'il y a eu toute une tradition de recherche sur ce miroir, conduite dans le milieu de Conon d'Alexandrie, et parmi les correspondants d'Archimède. Des mathématiciens comme Dosithée y ont contribué, mais aussi des astronomes comme Hippodamos. Il nous apprend enfin que ce miroir servait aussi bien à illuminer les temples à l'occasion de certaines fêtes, dans plusieurs villes, qu'à faire fonctionner des gnomons. Dioclès rappelle que l'astronome Hippodamos cherchait un miroir tel que les rayons du soleil se réfléchissent en un seul point, et embrasent en ce point. C'est ce problème que Dosithée a résolu. Dioclès lui-même commence par démontrer la propriété focale de la parabole, en utilisant pour cela les propriétés de la sous-tangente et de la sous-normale. Il montre ensuite que si la surface intérieure du paraboloïde engendré par la rotation d'un arc de la parabole autour de l'axe est réfléchissante, et si l'axe est dirigé vers le soleil, tout rayon solaire tombant sur ce miroir est réfléchi vers le foyer. Pour construire le miroir parabolique, Dioclès passe à la détermination de la parabole génératrice d'un miroir parabolique de révolution, de sorte que les rayons réfléchis se rencontrent en un point situé à une distance donnée du centre du miroir. Il explique ensuite comment faire, à partir de cette construction, un gabarit du miroir cherché. On voit donc que, à l'époque de Dioclès et avec lui, les mathématiciens ont amorcé l'étude de la propriété anaclastique de la parabole, dégagé la propriété foyer-directrice et construit, à l'aide de la géométrie, le gabarit du miroir parabolique. Mais si l'on veut aller au-delà des résultats nus, pour saisir l'esprit de cette recherche, on ne manque pas de noter que celle-ci fait partie de la géométrie des coniques : ce sont les propriétés anaclas-

1. Voir R. Rashed, *Les Catoptriciens grecs : les miroirs ardents*, à paraître aux Belles Lettres.

tiques des courbes qui intéressent le mathématicien ; certes, les retombées optiques ne sont pas négligeables, mais on peut dire qu'elles sont là de surcroît.

De l'impact de l'étude de Dioclès, il ne nous reste aucune trace. Perdu assez tôt, le texte n'est connu que dans une compilation arabe. Mais ce n'est pas le seul texte consacré aux miroirs ardents qui ait été rendu en arabe. On a également traduit en cette langue l'écrit d'Anthémius de Tralles, un traité d'un certain Didyme, perdu en grec. En fait, tout laisse penser qu'il y avait au IXe siècle une recherche active en arabe sur les miroirs ardents et la catoptrique, menée principalement par al-Kindī et Qusṭā ibn Lūqā[2], et que c'est cette recherche qui a suscité un mouvement massif de traduction des écrits grecs traitant de ce sujet. Ainsi, sur quatre textes connus consacrés aux miroirs ardents, les trois les plus importants ont été rendus en arabe, dont deux sont perdus dans leur langue originelle (le quatrième est la manuscrit de Bobbio). Mais, avant de suivre la destinée de cette recherche en arabe, repérons le contexte des autres textes grecs, et leurs perspectives. L'étude de Didyme est en quelque sorte dans la même lignée que celle de Dioclès : la géométrie des coniques. Didyme commence par montrer la propriété focale de la parabole, puis il établit que la distance du sommet de la parabole au foyer est égale au quart du côté droit ; il construit alors par points la parabole, et montre enfin comment façonner le gabarit du miroir. Le cas d'Anthémius de Tralles est différent : le savant byzantin n'appartient pas à la tradition de la géométrie des coniques. Il prend comme point de départ la légende selon laquelle Archimède aurait incendié la flotte de Marcellus lors de l'attaque de Syracuse. Anthémius veut établir la possibilité de cette légende, et répondre à la question suivante : comment faire parvenir un rayon solaire à un point éloigné de nous d'une distance donnée. Il examine alors plusieurs mi-roirs, et conclut sur l'étude du miroir parabolique. Il procède alors à la construction par points et tangentes d'une parabole dont on connaît le foyer et la directrice. À tout prendre, le traité d'Anthémius se situe davantage dans une perspective catoptrique que dans celle de la géométrie des coniques. Or, c'est précisément cette perspective, mais incomparablement renforcée, qui l'emportera à l'heure de la transmission chez les savants arabes, assurant au livre d'Anthémius une pérennité que les autres n'ont pu connaître. En effet, à la différence des autres textes traduits du grec, l'ouvrage d'Anthémius a été généreusement consulté ; il fut même l'objet d'un commentaire critique d'al-Kindī ; Aḥmad ibn 'Īsā l'a fréquemment cité ; 'Uṭārid, au Xe siècle, le reprend intégralement dans son traité. Tous ces faits sont corroborés par Ibn Sahl et par Ibn al-Haytham[3]. Ibn Sahl cite la légende d'Archimède mentionnée par Anthémius, et reprend l'étude du miroir ellipsoïdal, tandis qu'Ibn al-Haytham cite, dans son traité sur le miroir parabolique, Anthémius de Tralles

2. Voir R. Rashed, *Oeuvres philosophiques et scientifiques d'al-Kindī,* vol. I : " L'Optique et la catoptrique ", Leiden, 1997.

3. *Cf.* R. Rashed, *Géométrie et dioptrique au Xe siècle* : *Ibn Sahl, al-Qūhī et Ibn al-Haytham*, Paris, 1993.

comme seul nom aux côtés d'Archimède. Cependant, si on la compare aux textes de Dioclès et de Didyme, l'étude d'Anthémius paraît géométriquement beaucoup moins rigoureuse, et cette faiblesse ne pouvait échapper à des mathématiciens de la classe d'Ibn Sahl ou d'Ibn al-Haytham. Le succès du traité d'Anthémius revenait donc sans doute au changement de perspective qu'il laissait entrevoir en filigrane.

Or, ce changement de perspective au moment de la transmission est étroitement lié à l'émergence d'un nouveau caractère, que l'on n'a pas assez souligné, et qui distingue la recherche sur les miroirs ardents au IXe siècle : celle-ci fait désormais partie de la catoptrique. De cela, la traduction concrète est qu'un seul et même savant s'occupe d'optique ou de catoptrique, en même temps que des miroirs ardents. C'est le cas d'al-Kindī, de Qusṭā ibn Lūqā, comme d'Ibn Sahl au Xe siècle ; c'est aussi le cas d'Ibn al-Haytham ensuite, d'al-Ghundijānī, d'Ibn Ṣāliḥ plus tard. Or, ni Dioclès, ni Didyme, ni l'auteur du fragment Bobbio, ni d'ailleurs Anthémius lui-même, malgré la différence notable déjà relevée, n'ont poursuivi en même temps une recherche optique. C'est à al-Kindī le premier, semble-t-il, que revient d'avoir unifié des champs disparates, modifiant ainsi la perspective. Il a écrit sa fameuse *Optique*, connue sous le titre *De aspectibus* dans la traduction latine de son texte arabe perdu, ainsi que *La rectification des erreurs et des difficultés dues à Euclide dans son livre appelé l'Optique* ; on lui doit aussi plusieurs traités en catoptrique. Il consacre un traité aux miroirs ardents, intitulé *Sur les rayons solaires*, où il déclare vouloir remédier aux insuffisances de l'étude d'Anthémius de Tralles, et la compléter. Ce traité s'achève sur une étude du miroir parabolique. Le contemporain d'al-Kindī, de Qusṭā ibn Lūqā, écrit quant à lui une catoptrique, et, aux dires des biobibliographes, un traité sur les miroirs ardents. Leur successeur du Xe siècle, qui est aussi le compilateur d'al-Kindī, Ibn 'Īsā, regroupe dans un même livre optique, catoptrique, optique météorologique et miroirs ardents. Il est donc clair que la transmission pour la recherche, celle qui s'opère au IXe siècle, n'est nullement une livraison de résultats nus : elle s'accompagne, à l'évidence, d'une rénovation. Déjà présente chez les savants du IXe siècle, elle apparaîtra plus tard avec tout son éclat. Avec Ibn Sahl à la fin du siècle suivant, l'unification opérée par al-Kindī et ses contemporains connaîtra toute son ampleur, et s'achèvera dans la naissance d'un nouveau chapitre de l'optique : l'anaclastique ou la dioptrique.

Ibn Sahl, rappelons-le, est le premier mathématicien connu qui ait élaboré une théorie géométrique des lentilles, et formulé la loi dite de Snell. En fait, il a conçu un chapitre de l'optique, qui porte sur les instruments ardents, miroirs et lentilles. Son point de départ est plus général que celui de ses prédécesseurs : non seulement Ibn Sahl connaissait l'*Optique* de Ptolémée, et donc le cinquième chapitre consacré à la réfraction — ce qu'al-Kindī et Ibn Lūqā ignoraient ; mais il a infléchi la théorie transmise — grâce à l'importance prise par le concept de milieu réfringent, certes, mais aussi parce qu'il a caractérisé ce milieu par un certain rapport constant. Ibn Sahl considère l'embrasement non seulement comme l'effet de la réflexion, mais aussi de la réfraction, relatif à la distance de la source, que cette distance soit finie ou infinie. Le miroir parabolique n'apparaît pas pour lui-même ; ce n'est pas non plus un miroir parmi les autres : il se place à l'intersection de la réflexion et de la distance infinie, c'est-à-dire là où l'embrasement se fait par réflexion, la source étant à l'infini. Le miroir parabolique, tout comme le miroir ellipsoïdal, ont chacun une place particulière dans une étude plus générale qui porte sur les miroirs et les lentilles. Dans son étude du miroir parabolique, Ibn Sahl commence par examiner les propriétés anaclastiques de la parabole avant de procéder au tracé continu de la courbe à l'aide du foyer et de la directrice. Il fabrique à cette fin un appareil mécanique conçu pour le tracé continu de trois courbes coniques.

Au cours de son étude des deux miroirs, parabolique et ellipsoïdal, Ibn Sahl s'attache tout particulièrement à la détermination du plan tangent au point d'impact de la lumière incidente à la surface réfléchissante, ainsi qu'à l'unicité de ce plan. Pourquoi cet intérêt ? Certes, on y retrouve bien la connaissance qu'Ibn Sahl avait de la théorie des coniques, mais aussi sa conception de la réflexion de la lumière. Ibn Sahl veut non seulement s'assurer de l'égalité de l'angle d'incidence et de l'angle de réflexion, mais aussi vérifier que la droite suivant laquelle la lumière parvient au point d'une surface, la droite suivant laquelle cette lumière est réfléchie, et, enfin, la normale menée au plan tangent à la surface en ce point, sont dans un même plan. En fait, pour Ibn Sahl, ce n'est pas la surface réfléchissante qui importe, mais bien ce plan tangent.

Avec Ibn Sahl, l'étude du miroir parabolique fait désormais partie de l'optique géométrique. Et là, nous sommes à la veille d'une grande transformation, celle qu'accomplit Ibn al-Haytham. Cependant, l'étude du miroir parabolique que nous a laissée ce dernier est proche de celle d'Ibn Sahl, mais à une différence près, qui n'est pas négligeable : Ibn al-Haytham souligne le contenu physique des notions géométriques, comme celles de rayon lumineux et de faisceau lumineux. Il commence par établir les propriétés anaclastiques de la parabole, celles du paraboloïde de révolution ensuite, puis il s'emploie à la fabrication du miroir. Il explique comment construire sur des plaques d'acier les gabarits nécessaires à la confection des miroirs. Il distingue deux types de plaques : plaque du sommet de la section, et plaque du milieu de la section. Si donc on construit un miroir ovoïdal on utilise une plaque du premier type, le côté droit de la parabole étant choisi à partir de la distance souhaitée pour l'embrasement ; le côté droit sera égal au quadruple de la distance. Mais si on veut construire un miroir parabolique en forme d'anneau, on détermine la plaque en supposant connues la distance à laquelle on veut embraser et le côté droit de la parabole.

Ainsi se poursuit la recherche sur le miroir parabolique, depuis la transmission des travaux grecs jusqu'à Ibn al-Haytham. Le développement de cette recherche ne se réduit pas, comme on le voit, à un accroissement mécanique des résultats, ni à un perfectionnement linéaire des techniques ; c'est l'histoire de la modification du sens, elle-même effet d'une rénovation des perspectives.

Beaucoup plus que toute autre, l'étude d'Ibn al-Haytham eut un grand impact sur les travaux consacrés au miroir parabolique, que ce soit en arabe ou en latin. Mais on verra brièvement que cet impact ne fut nullement le même dans chaque cas.

Rédigé au Caire avant les années quarante du XIe siècle, le mémoire d'Ibn al-Haytham, en effet, n'a pas seulement été lu en arabe, mais aussi en latin à partir du XIIe siècle. La différence qui, pendant un temps tout au moins, a distingué ces deux lectures, est riche d'enseignement pour une réflexion sur la transmission. En fait, la ligne de clivage passe pour ainsi dire entre optique et géométrie.

Amplement diffusé en arabe, ce mémoire a été commenté à Bagdad, à l'Est comme à l'Ouest islamiques, en tant qu'écrit optique : celui qui s'intéressait à la théorie des coniques avait largement de quoi se satisfaire ailleurs. D'autre part, le seul développement de nous connu de la recherche d'Ibn al-Haytham dans ce mémoire est optico-technique, nullement géométrique. À cela, rien d'étonnant : la tradition de l'optique est déjà bien établie, avec ses noms, ses références et ses problématiques. Quant à la recherche en théorie des coniques, elle est déjà trop avancée pour qu'un mathématicien qui s'y adonne aille chercher des informations dans ce mémoire. Mais il en va tout autrement au XIIe siècle, en latin. La traduction du traité d'Ibn al-Haytham par Gérard de Crémone ne semble pas répondre aux besoins de la recherche en optique, alors surtout que celle-ci n'existait pas en fait. En revanche, cette tradition offrait le premier accès en latin à la géométrie des coniques. Lisons ce qu'écrit Marshall Clagett : " Avant le douzième siècle, la connaissance des sections coniques à l'Ouest était inexistante ", et il poursuit : " Les premières traces d'une quelconque connaissance des sections coniques en Occident résultaient de la traduction latine de deux ouvrages d'Alhazen (Ibn al-Haytham). Le premier était la traduction par Gérard de Crémone du *Liber speculis comburentibus* d'Alhazen... "[4]. Il fallait attendre la traduction du *De aspectibus* d'al-Kindī et celle du *Livre de l'Optique* d'Ibn al-Haytham pour que fût entamée en latin la recherche en optique. Il reste que l'impact de ce mémoire en géométrie des coniques fut grand, comme l'a montré M. Clagett à l'aide de travaux comme le *Speculi almukefi compositio*, anonyme, ou le *Libellus de seccione nukefi*, de Johannes Fusoris.

Comme je viens de le rappeler, le rôle de ce mémoire en arabe était bien différent, du fait même de son intégration dans une tradition de recherche continue, et bien établie depuis déjà deux siècles. Pour illustrer son rôle dans le développement de la recherche future, je m'en tiendrai à un seul exemple, d'un

4. M. Clagett, *Archimedes in the Middle Ages*, vol. 4 : " A Supplement on the Medieval Latin Traditions of Conic Sections ", Part I : " Texts and Analysis ", Philadelphia, 1980, 3.

auteur jusqu'ici inconnu, un certain Ibn Ṣāliḥ[5]. Celui-ci a écrit un volumineux mémoire sur le miroir parabolique, dans lequel il emprunte de nombreux paragraphes au texte d'Ibn al-Haytham.

Ibn Ṣāliḥ commence par des considérations sur la théorie des coniques, en mêlant le langage des *Éléments* d'Euclide à celui des *Coniques* d'Apollonius — on peut à ce propos souligner tout le danger qu'il y a à déterminer la succession des auteurs à l'aide de leur seule langue. Puis, Ibn Ṣāliḥ revient au miroir parabolique. La principale difficulté qu'il pressent concernant ce miroir est d'ordre technique : réussir la courbure d'un miroir parabolique d'une surface assez grande pour augmenter l'embrasement. On comprend qu'un tel miroir est difficile, voire impossible, à fabriquer pour l'artisan de l'époque. L'idée est donc la suivante : perdre un peu en focalisation pour gagner en surface. Or, on se souvient qu'Ibn al-Haytham a étudié le miroir " en forme d'anneau ", dont l'axe est toujours l'axe du paraboloïde, et le point d'embrasement le foyer. On peut dans ce cas choisir l'arc *EB* générateur d'un tel miroir, pour que le foyer *F* soit à une distance arbitrairement choisie du centre du cercle décrit par le point *E*.

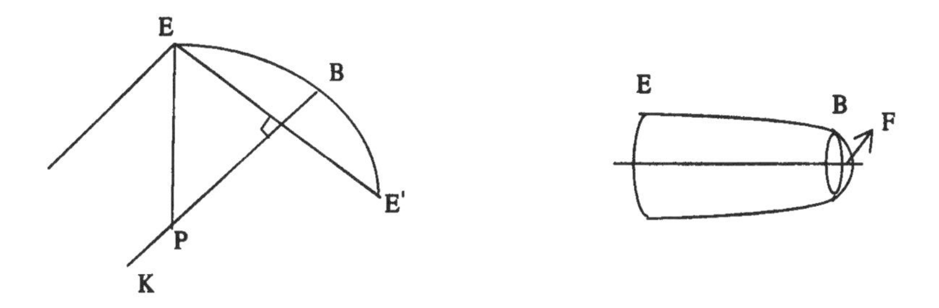

Ibn Ṣāliḥ considère alors un miroir concave engendré par cet arc *EB*, par sa rotation autour de la normale *BK* au point *B* ; *BK* sera donc l'axe du miroir. Tous les rayons incidents parallèles à *BK* et tombant sur le cercle engendré par *E* sont réfléchis vers un point *P* de *BK*. La distance *BP* dépend du point *E* qui détermine le bord du miroir. Un rayon incident parallèle à *BK* tombant en un point quelconque du miroir rencontrera l'axe en un point différent de *P*, point plus proche de *B*. Ibn Ṣāliḥ procède au calcul de ces points. Il n'y a donc pas un point d'embrasement, mais une concentration de rayons réfléchis au voisinage de l'axe *BK*.

Ibn Ṣāliḥ expose ensuite un procédé qui permet d'obtenir un arc de parabole comme section plane d'un tronc de cône de révolution ; il décrit en détail les procédés techniques employés à la fabrication de ce miroir, description très intéressante, mais que nous ne pouvons expliquer ici.

5. Voir notre étude à paraître " Les miroirs ardents d' Ibn Ṣāliḥ ".

L'exemple d'Ibn Ṣāliḥ est clair : si Ibn Ṣāliḥ conçoit ce type de miroir, jamais pensé auparavant, c'est poussé par une contrainte technique. Ni les matériaux, ni les procédés de fabrication de son temps, ne permettaient en effet de construire un miroir parabolique avec une grande courbure. On peut dire que cette dialectique entre la tradition conceptuelle développée depuis al-Kindī et la tradition technique, domine de part en part l'histoire du miroir parabolique, et, plus généralement, celle des miroir ardents, en arabe.

Concluons. Entre grec, arabe et latin, nous venons d'assister à deux types de transmission, qui diffèrent selon les rapports que chacune entretient avec la recherche. La première est la transmission pour la recherche, comme on l'a vue au IXe siècle : cette transmission s'accompagne d'une traduction massive des écrits ; elle est critique et innovatrice. On reprend les anciens acquis pour les intégrer dans de nouvelles traditions conceptuelles et techniques, en formation ; à leur tour, les écrit transmis sont productifs dans ce mouvement d'élaboration d'une nouvelle tradition. La seconde transmission, à l'exemple de celle de l'arabe en latin au XIIe siècle, n'est pas une réponse aux besoins de la recherche ; elle répond plutôt aux exigences de l'apprentissage : la transmission n'est alors ni critique, ni créatrice : elle prend de l'oeuvre transmise ce qu'il est urgent d'apprendre. Du traité optique d'Ibn al-Haytham sur le miroir ardent, c'est seulement la géométrie des coniques que l'on retient ; le contenu optique, pour sa part, restera là en attente d'une recherche future. Cet exemple semble une parfaite illustration : on risquerait de ne rien comprendre à la transmission scientifique, son rôle, sa rapidité…, si l'on n'exhibe pas les rapports qui l'unissent avec la recherche.

IBN AL-KAMMĀD'S ASTRONOMICAL WORK IN IBN AL-HĀ'IM'S *AL-ZĪJ AL-KĀMIL FĪ-L-TAᶜĀLĪM*

I. SOLAR YEAR, TREPIDATION, AND TIMEKEEPING[1]

Emilia CALVO

INTRODUCTION

The purpose of this paper is to analyze some aspects of Ibn al-Kammād's astronomical work as they are described by Ibn al-Hā'im's *al-Zīj al-Kāmil fī-l-Taᶜālīm*. Both astronomers are inscribed in what is called the Zarqāllian tradition, which begins with Ibn al-Zarqālluh in the 11th century.

As is well known, Abū Isḥāq Ibrāhīm ibn Yaḥyā al-Naqqāsh, Ibn al-Zarqālluh, was one of the outstanding astronomers of al-Andalus. He worked in Toledo and Cordoba and died around 1100. Ibn al-Zarqālluh's astronomical work exerted a considerable influence on later astronomers in al-Andalus, North Africa, and in Europe as well[2].

One of the astronomers influenced by Ibn al-Zarqālluh's work — possibly the first — was Ibn al-Kammād, who was probably a disciple of his, and who

1. This paper is part of a research project on *Astronomía teórica y tablas astronómicas* run by the Department of Arabic Studies of the University of Barcelona sponsored by the Spanish *Dirección General de Investigación Científica y Técnica* of the Spanish *Ministerio de Educación y Cultura*.

2. On Ibn al-Zarqālluh *cf.* J. Samsó, *Las Ciencias de los Antiguos en al-Andalus*, Madrid, 1992, 147-152, 166-240 ; J. Samsó, " Azarquiel e Ibn al-Bannā' ", *Relaciones de la Península Ibérica con el Magreb (Siglos XIII-XVI)*, Madrid, 1989, 361-372 ; R. Puig, *Al-Šakkāziyya de Ibn Naqqāš al-Zarqāllūh, Edición, traducción y estudio*, Barcelona, 1986 ; R. Puig, *Los Tratados de Construcción y Uso de la Azafea de Azarquiel*, Madrid, 1987 ; J. Vernet, " Al-Zarqālī ", *Dictionary of Scientific Biography*, vol. XIV, New York, 1976, 592-595 ; E. Calvo, " Ibn al-Zarqālluh ", *Encyclopaedia of the History of Science, Technology and Medicine in Non-Western Cultures*, Dordrecht, 1997, 415-416.

lived during the first half of the twelfth century as can be deduced from later sources (fl. 1116 A.D.)[3].

His complete name was Abū Ja^cfar Aḥmad ibn Yūsuf Ibn al-Kammād (although Ibn Isḥāq, and Ibn al-Hā'im call him Abū-l-ᶜAbbās) and he was the author of several astronomical handbooks. We know the titles of three of them :

- *al-Amad ᶜalā al-Abad (The Valid for the Eternity),*
- *al-Kawr ᶜalā al-Dawr (The Periodic Rotations),*
- *al-Muqtabas (The Compilation — of the other two —).*

None of them seem to survive in their original Arabic version, with the exception of some passages of the second, *al-Kawr*[4], and chapter 28 of the third, *al-Muqtabas*, which was also translated into Latin, in 1262, by Johannes of Dumpno. This translation, which is preserved in the *Biblioteca Nacional* in Madrid (in ms. 10.023), played an important role in the transmission of Ibn al-Zarqālluh's astronomical theories into the Latin World, and their reception by the astronomers of the Renaissance[5]. In spite of their importance, Ibn al-Kammād's tables have been studied only recently[6].

Some elements of Ibn al-Kammād's astronomical work are preserved in later sources. One of them is the *zīj* of Ibn Isḥāq al-Tūnisī, an astronomer who worked in Marrākush in the first half of the 13th century (fl. 1222 A.D.). There we are told, for instance, that Ibn al-Kammād cast a horoscope in Cordoba in the year 510 H./1116-7 A.D. We also find some information on Ibn al-Kammād in Abū-l-Ḥasan ᶜAlī al-Marrākushī's (*ca.* 1262) work on *mīqāt*, entitled *Jāmiᶜ al-Mabādī wa-l-Ghayāt fī ᶜIlm al-Mīqāt*. There, for instance, the author praises the precision and correctness of the tables included in *al-Amad ᶜalā al-Abad* to determine the solar longitude[7].

Another source of information is the work of Ibn al-Hā'im, whose complete name was Abū Muḥammad ᶜAbd al-Ḥaqq al-Gāfiqī al-Ishbīlī Ibn al-Hā'im. He was the author of an astronomical book entitled *al-Zīj al-Kāmil fī-l-Taᶜālīm*, a *zīj* composed in the first years of the 13th century and dedicated to the Caliph Abū ᶜAbd Allāh Muḥammad al-Nāṣir (1199-1213). All we know about Ibn al-

3. For instance Ibn Isḥāq in his *zīj*. *Cf.* A. Mestres, " Maghribī Astronomy in the 13th Century : A description of Manuscript Hyderabad Andra Pradesh State Library 298 ", in J. Casulleras, J. Samsó (eds), *From Baghdad to Barcelona. Studies in the Islamic Exact Sciences in Honour of Prof. Juan Vernet*, vol. I, Barcelona, 1996, 383-443, especially 404.

4. They are preserved in ms. 939,4 El Escorial.

5. On Ibn al-Kammād *cf.* J. Samsó, *Las Ciencias de los Antiguos en al-Andalus*, *op. cit.*, 320-324 and J. Vernet, " Un tractat d'obstetricia astrológica ", *Estudios de Historia de la Ciencia Medieval*, Barcelona, 1979, 273-300.

6. *Cf.* J. Chabás, B.R. Goldstein, " Andalusian Astronomy : al-Zīj al-Muqtabis of Ibn al-Kammād ", *Archive for the History of Exact Sciences*, 48 (1994), 1-41 and B.R. Goldstein, J. Chabás, " Ibn al-Kammād's Star List ", *Centaurus*, 38 (1996), 317-334.

7. *Cf.* J.M. Millàs Vallicrosa, *Estudios sobre Azarquiel*, Madrid, Granada, 1943-1950, 347.

Hā'im's life is that he appears to have worked in North Africa under the Almohad dynasty[8].

IBN AL-HĀ'IM'S *AL-ZĪJ AL-KĀMIL FĪ-L-TAᶜĀLĪM*

Ibn al-Hā'im's *al-Zīj al-Kāmil* is included in MS Arab 285 (Marsh 618) preserved in the Bodleian Library, Oxford. The manuscript is badly damaged.

The last two or three lines of each page are very difficult (and sometimes impossible) to read due to humidity, stains and a number of holes. It is quite a long text (170 pages), comprising an introduction and seven books (*maqālāt*).

In his *zīj* Ibn al-Hā'im gives some historical data on Ibn al-Zarqālluh, who seems to have had a considerable influence on his work. But he also mentions other astronomers, among them Ibn al-Kammād, to whom he devotes several commentaries in many chapters of his *zīj*, and especially in the long introduction to this work. In fact, the introduction of Ibn al-Hā'im's *zīj* is devoted almost entirely to a criticism of several aspects of Ibn al-Kammād's astronomical works. And in some passages he is extremely harsh with his predecessor. Since Ibn al-Kammād's works are not well known, the information given by Ibn al-Hā'im is of great importance.

CRITICISMS IN THE INTRODUCTION TO IBN AL-HĀ'IM'S *ZĪJ*

Ibn al-Hā'im mentions the discrepancies between calculation (*taqwīm*) and observation (*raṣad*) — the differences between the values calculated from the tables and those actually observed — and calls for a revision of the tables, because, after a period of time they become obsolete. This is a basic fact that some astronomers seem to ignore. As an example, Ibn al-Hā'im considers that his own work is valid for 40 years. After this period it should be updated.

One of his criticisms of Ibn al-Kammād is that the latter considered his work valid for ever, or, at least, this can be deduced from the title of one of his works —*al-Amad ᶜalā al-Abad*.

With this example Ibn al-Hā'im demonstrates the divergence between the astronomers who only compile and repeat what they find in earlier works, and the innovative astronomers who make use of their own observations to carry out modifications to the theories inherited.

There are also astronomers who modify and make additions to the works of their predecessors, but introduce new errors. It is therefore necessary to correct these errors and mistakes. As an example[9], Ibn al-Hā'im mentions Abū-l-ᶜAb-

8. On Ibn al-Hā'im *cf.* J. Samsó, *Las Ciencias de los Antiguos en al-Andalus*, *op. cit.*, 320-325 and E. Calvo, " Astronomical Theories Related to the Sun in Ibn al-Hā'im's *al-Zīj al-Kāmil fī-l-Taᶜālīm* ", to be published in *Zeitschrift für Geschichte der Arabisch-Islamischen Wissenschaften*.

9. *Cf.* ms. Bodleian O. 285, Marsh 618 fol. 3 v, 6.

bās al-Kammād as the author of two books : *al-Kawr* ᶜ*alā al-Dawr* and *al-Muqtabas* in which the author made some mistakes, especially in *al-Kawr* ᶜ*alā al-Dawr.*

Ibn al-Hā'im is also very critical of those who, in order to make calculations easier, are content with approximate results ; this is the case of some procedures described by Ibn al-Kammād concerning the determination of eclipses.

In fact, as I have said, he is the first astronomer criticized in the introduction to Ibn al-Hā'im's *zīj*. Here, Ibn al-Hā'im includes 25 corrections to errors he found in *al-Kawr* ᶜ*alā al-Dawr* and *al-Muqtabas*.

In some cases Ibn al-Hā'im gives his corrected option, but, more often, he refers the reader to the corresponding chapter of *al-Zīj al-Kāmil* in which he will find the detailed explanation of the correct procedure[10].

As another example of these statements, he mentions a contemporary astronomer who was the author of a *zīj* called *al-Muntakhab* which included the mean motions taken from Abū Marwān al-Istījjī, the equations [already taken by al-Istījjī from al-Battānī and also some of the errors that can be found in al-Kammād's *al-Kawr* ᶜ*alā al-Dawr*[11]. In this *zīj* the author compiled the mean motion of the sun, the equations, the excess of revolution and some other parameters following the statements of Abū Marwān al-Istījjī, in his *Risāla fī-l-Iqbāl wa-l-Idbār* (Book on the Access and Recess), which, according to Ibn al-Hā'im, was valid for Abū Marwān's time and a short period of time after him. But it is no longer valid for Ibn al-Hā'im's time, since approximately 150 years have already passed. Furthermore, Abū Marwān considered that the solar equation had not suffered any variation since al-Battānī's times.

Finally and, maybe as an example of innovative astronomers, Ibn al-Hā'im mentions Ibn al-Zarqālluh and the *qāḍī* and *faqīh* Abū ᶜAbd Allāh Ibn Bargūth who, from their own observations (*bi-arṣādihimā*), found that the solar equation had changed notably since al-Battānī's times.

Topics criticized in Ibn al-Kammād

After these considerations, Ibn al-Hā'im introduces the mistakes and errors (he says *al-agālīṭ wa-l-awhām*) in Ibn al-Kammād's work, at the request of some contemporary colleagues[12].

The 25 topics concerned in the criticisms appear numbered in *abujad* notation in the margin of the manuscript. Here they are in an abridged table :

10. Some of these chapters, however, seem to be missing in Ibn al-Hā'ims *zīj*. This is the case of chapters related to astrology, eclipses, timekeeping, etc.

11. *Cf.* ms. Bodleian O. 285, Marsh 618 fol. 4 r, 7.

12. This information seems to point to more astronomical activities, in Ibn al-Hā'im's time. Unfortunately, he does not give more details.

1	Solar year	
2	Trepidation	
3-5	Solar theory	
6	Arcs of visibility	
7	Trigonometry	
8	Mediation	
9-10	Timekeeping	
11	Astrology (Equalization of houses)	
12-21	Moon	
	12	Conjunction and opposition
	13-16	Parallax
	17-21	Eclipses
22	Solar year	
23-24	Astrology (excess of revolution, *tasyīr*)	
25	Sun	

The second of them, the one dealing with trepidation, is divided into four aspects (*jiha*, pl. *jihāt*) which could be considered as four different criticisms. This passage of Ibn al-Hā'im's *zīj* was known to later astronomers in North Africa, some of whom mentioned it in their works[13].

Here I will analyze five of these 25 topics, namely numbers 1 and 22, related to the length of the solar year, number 2 related to trepidation, and numbers 9 and 10 related to timekeeping.

The length of the solar year

(#1) The first criticism[14] concerns " the most evident error of that book " (*al-Kawr*), in Ibn al-Hā'im's words[15] : the length of the solar year stated at the beginning of the book (that is to say, *al-Kawr ᶜalā al-Dawr*) does not coincide with the solar mean motion determined in the tables.

In fact, at least three different lengths of the solar year are associated with Ibn al-Kammād[16] :

13. For instance, it is mentioned in a commentary to the *Rawḍat al-azhār fī ᶜilm waqt al-layl wa-l-nahār*, the urŷuza of Ibn al-Yādirī (14th century). The title of the commentary is *Natā'ij al-afkār fī sharh rawdat al-azhār* and its author could be Abū-ᶜAbbās Aḥmad ibn Muḥammad ibn ᶜIsā al-Mawāsī al-Fāsī (d. 1505 A.D.).

14. *Cf.* ms. Bodleian O. 285, Marsh 618 fol. 5 r, 9, 1. 9.

15. He says : *min awḍaḥ jaṭā' waqaᶜa fī ḏalika al-kitāb*.

16. *Cf.* J. Chabás, B.R. Goldstein, *Andalusian Astronomy : al-Zīj al-Muqtabis of Ibn al-Kammād, op. cit.*, 35.

- At the beginning of *al-Muqtabas*[17] Ibn al-Kammād gives the length of the solar year with a value of

I. 365; 15,36,19,34,12 days.

- In another passage of the same book[18] we find a different value for the length of the solar year, namely

II. 365; 15,36,19,35,32 days.

As for Ibn al-Kammād's tables, the recalculated value involved for the solar mean motion is

0;59,8,9,21,15,17,21,17,39,34... °/day[19].

which yields a year-length of

III. 365; 15,36,34,58,16,54,28,11,41,19... days[20].

These two values are very close to the ones given by Ibn al-Hā'im in the fifth chapter of the second book of his *zīj*. There he gives for the length of the anomalistic year (he says *zamān al-ᶜawda fī-l-khārij al-markaz*, " period of revolution in the eccentric circle "[21]) a value of :

IV. 365;15,36,35,0,8 days

This value corresponds to a value for the solar mean motion[22] of

0;59,8,9,21,15 °/day[23]

It seems fairly probable that *al-Kawr* had a table similar to the one in *al-Muqtabas*, or even the same one.

(#22) Related to this, Ibn al-Hā'im criticizes Ibn al-Kammād's use of the period of revolution in the eccentric circle (*zamān al-ᶜawda fī-l-khārij al-markaz*) where he should use the period of revolution in the ecliptic (*zamān al-ᶜawda fī falak al burūj*) to carry out astrological calculations[24] : the excess of revolution, for instance. This means that, according to Ibn al-Hā'im, Ibn al-Kammād is using the anomalistic year instead of the sidereal one for astrological purposes, which is a major error in Ibn al-Hā'im's view.

17. *Cf.* Ms. 10023 Biblioteca Nacional de Madrid, fol. 2v.

18. *Cf.* Ms. 10023 Biblioteca Nacional de Madrid, fol. 65 v.

19. *Cf.* ms. 10023 Biblioteca Nacional de Madrid fol. 28 r. table of the solar mean motions. Chabás and Goldstein give 0;59,8,21,15...°/day. *Cf.* J. Chabás, B.R. Goldstein, *Andalusian Astronomy : al-Zīj al-Muqtabis of Ibn al-Kammād, op. cit.*, 28.

20. Chabás and Goldstein give 365; 15,36,34... days. *Cf.* J. Chabás, B.R. Goldstein, *Andalusian Astronomy : al-Zīj al-Muqtabis of Ibn al-Kammād, op. cit.*, 28.

21. *Cf.* Ms. Bodleian O. 285, Marsh, 618, fol. 31 r. and E. Calvo, " Astronomical Theories Related to the sun in Ibn al-Hā'im's *al-Zīj al-Kāmil fī-l-Taᶜālīm* ", *op. cit.*, § 3.1 and Arabic text § [7].

22. *Cf.* Ms. Bodleian O. 285, Marsh, 618 fol. 32 r. and E. Calvo, " Astronomical Theories Related to the sun in Ibn al-Hā'im's *al-Zīj al-Kāmil fī-l-Taᶜālīm* ", *op. cit.*, § 3.1 and Arabic text § [9].

23. This is a rounded value. There must be a fraction which is neglected. This is why Ibn al-Hā'im's value for the year-length is somewhat greater than Ibn al-Kammād's.

24. *Cf.* ms. Bodleian O. 285, Marsh 618 fol. 71r.

Trepidation

(#2) the second criticism is related to the theory of trepidation. As I have said, it includes four different aspects. Ibn al-Hā'im says that Ibn al-Kammād was wrong on four counts. Some of them have already been mentioned by Samsó in *Las Ciencias de los Antiguos*[25].

(#2.1) The first of the four aspects concerns the theory of trepidation in general as described by Ibn al-Kammād, of which we find here a rather vague description : Ibn al-Hā'im says that at the beginning of the book (*al-Kawr*), in his description of the structure of this motion, Ibn al-Kammād mentions that when the pole of the ecliptic has a mean position in the epicycle which justifies the variation of the obliquity (*dā'irat ikhtilāf al-mayl al-kullī*) then, the beginning of Aries is on the equator and the two points : the equinox and the head of Aries are the same. Ibn al-Hā'im says that, in that case, the motion of the head of Aries in its circle of trepidation (*dā'irat al-iqbāl*) must be similar (*shabīha*) to the motion of the pole of the ecliptic in its circle of the difference (*dā'irat ikhtilāfi-hi*)[26] and that the two periods of revolution must be the same. Ibn al-Hā'im considers these statements erroneous, and says that the correct proofs show that their motions and periods of revolution are completely different [*al-amr fi-himā ᶜala khilāf dhalika*].

The same description as the one criticized by Ibn al-Hā'im can be found in a Castilian translation of Ibn al-Kammād's *al-Kawr* found in Segovia (Spain)[27].

(#2.2) The second question (*al-jiha al-thāniya*) concerns the structure of the table of trepidation. Ibn al-Hā'im says that Ibn al-Kammād compiled a table for this motion for 90° degrees (for the two values : of trepidation and of the variation in the obliquity) and considered it valid for all time (*li-jamiᶜ al-azmān*). He placed six signs in the upper part and the other six in the lower part, just as they appear in the tables of equations. But this is also erroneous,

25. *Cf.* J. Samsó, *Las Ciencias de los Antiguos en al-Andalus, op. cit.*, 322-324. The trepidation tables in *al-Muqtabas* have been analysed in J. Chabás, B.R. Goldstein, *Andalusian Astronomy : al-Zīj al-Muqtabis of Ibn al-Kammād, op. cit.*, 24 & ff. (*Cf.* ms. 10023 Biblioteca Nacional de Madrid, fols 28 & 35) together with the instructions given in chapter 12 of the Canons. The model of trepidation in Ibn al-Hā'im's *al-Zīj al-Kāmil* has been analyzed by M. Comes (in a forthcoming paper). It follows Ibn al-Zarqālluh's third model as described in his book on the motion of the fixed stars.

26. This " circle of the difference " is a small epicycle in which the pole of the ecliptic moves with uniform motion. It is a Zarqāllian model for the justification of the variation in the obliquity of the ecliptic through the centuries. On this model *cf.* J. Samsó, *Las Ciencias de los Antiguos en al-Andalus, op. cit.*, 232-233. This model can also be found in the astronomical books written by the astronomers in the Zarqāllian tradition. On his influence in later astronomers in al-Andalus and the North of Africa *cf.* M. Comes, " Accession and Recession in al-Andalus and the North of Africa ", *From Baghdad to Barcelona*, vol. I, 349-364, especially 352 & ff.

27. *Cf.* Biblioteca Catedral Segovia Ms. 115, ff. 218vb-220vb : Yuçaf Benacomed, *Libro sobre çircunferençcia de moto sacado por tiempo seculo*. I came to know this manuscript thanks to the kindness of M. Comes.

because the pole of the ecliptic does not complete a revolution in that period (*jami*ᶜ *muyūli-hi al-kulliyya*, that is, the return to the same value of the obliquity). The problem of tables that are valid for all time is that a fixed value of the obliquity is usually used for its calculation. But the obliquity is not constant, and must therefore be changed from time to time to adjust its value. Therefore, the tables must be calculated again with the changed value for the obliquity.

It seems clear that, according to Ibn al-Hā'im, Ibn al-Kammād has abandoned the Zarqāllian model in the theory of trepidation. Ibn al-Zarqālluh makes this motion, over an equatorial epicycle, independent from the variation in the obliquity of the ecliptic (which is due to the motion of the pole of the ecliptic over an epicycle, the radius of which is 0;10°). In the Zarqāllian model, it is not possible to calculate the value of trepidation with a table of equations, as it is in the *Liber de Motu*, but Ibn al-Kammād introduces a table of this kind in which, according to Ibn al-Hā'im, the head of Aries and the pole of the ecliptic have the same period of revolution.

(#2.3) The third question (*al-jiha al-thālitha*)[28] gives some more details on Ibn al-Kammād's trepidation table in *al-Kawr* : he (Ibn al-Kammād) put the beginning of the motion of the head of Aries at the beginning of the table [together with the] circle of the equator and stated that the pole was at its mean distance. If this table is completed according to this premise, for 180 degrees, the pole completes the maximum value of the obliquity in this period. However, the beginning of the motion of the head of Aries is placed at one of the two extremes : either the southern or the northern because the pole is at one of its two extremes (the apogee or the perigee). From this position the pole will have attained the maximum obliquity (he says " the complete declination ") at the middle of the revolution and, then, it would be necessary, according to this premise to generalize, this table for all time.

This is a rather obscure passage and requires some comment. My impression is that Ibn al-Hā'im is criticizing a table which is not the one we find in *al-Muqtabas* fol. 28 v. In the lost table of *al-Kawr* the same argument seems to be used to calculate, in two different columns, the equation of trepidation and the value of the obliquity of the ecliptic, $\in$. Ibn al-Hā'im seems to imply that $\in$ oscillates between a minimum and a mean value for a variation of the argument between 1° and 90°. This column in the table probably computed positive or negative differences between the mean value and one of the two extremes.

Besides, according to Ibn al-Hā'im, Ibn al-Kammād uses an argument in which 0° degrees corresponds to the moment at which the head of Aries crosses the equator and, at this precise moment, the pole of the ecliptic will be at a mean distance from the pole of the equator, which implies that the obliq-

28. *Cf.* ms. Bodleian O. 285, Marsh 618 fol. 5 v, 10.

uity will also attain its mean value. This would confirm Goldstein and Chabás' hypothesis that Ibn al-Kammād is using Ibn al-Zarqālluh's second model.

In my opinion, as Ibn al-Kammād's table depends on an argument that goes from 0° to 90°, the column which corresponds to the obliquity will tabulate $\Delta\in$ (difference between the value of $\in$ for a given argument and the mean value of $\in$). Therefore, for a change for the argument from 0° to 180°, $\Delta\in$ will change between 0° and $\in_{max}$, positive or negative, and the actual value of $\in$ will pass from its mean value to one of the two extremes (and will not " attain all its values "). As Ibn al-Hā'im states, $\in$ would only attain all its values in 180° if the argument begins at one of the two extremes ($\in_{max}$ or $\in_{min}$) which correspond to the maximum or minimum distance of the pole of the ecliptic from the pole of the equator.

(#2.4) Finally, the fourth question (*al-jiha al-rābiᶜa*) deals with the values given by Ibn al-Kammād in the table of trepidation. Ibn al-Hā'im says that they are not correct. In fact, he says that the motion established for trepidation in them *(iqbāl wa-idbār)* does not coincide with the values deduced from observation.

He gives two examples : the radix (*aṣl*) used by the observer (he means Ibn al-Zarqālluh) to construct the parameters of this motion implied that the precession (*iqbāl*) in Hipparchus' times had to be 9;29° approximately and that the argument (*khāṣṣa*) of the head of Aries was 292;33° approximately (in *abjad* notation he gives : 9^s 22;32,40°). The precession (*iqbāl*) in Ptolemy was 6;42° and the argument (*khāṣṣa*) of the head of Aries was 319;2° (in *abjad* 10^s 19, 2°).

Ibn al-Hā'im says that if we enter with these two anomalies in the tables of equation of the motion of the head of Aries in this book (*al-Kawr*) we obtain, as for Hipparchus' precession, 9;19°, which is 10 minutes less than the radix given ; and for Ptolemy's precession, 6;57° with a difference of 15 minutes from the radix given. This is what is erroneous and deficient in this motion of (*iqbāl*) in Ibn al-Kammād's tables, according to Ibn al-Hā'im.

The values ascribed to Ibn al-Zarqālluh can be found in the fifth chapter of the first section of his treatise on the motion of the fixed stars[29] and the values ascribed to Ibn al-Kammād can be obtained from the table included in *al-Muqtabas*[30].

As for the Canons, in *al-Muqtabas* this question, the theory of trepidation, is described in canon (porta) 12 and the author refers the reader to his work *al-Amad ᶜalā al-Abad* which, as I have said, is not preserved.

These values are represented in the table below :

29. *Cf.* J.M. Millàs Vallicrosa, *Estudios sobre Azarquiel*, *op. cit.*, 318.

30. *Cf.* Ms. 10023 Biblioteca Nacional de Madrid, fol. 35 v and J. Chabás, B.R. Goldstein, *Andalusian Astronomy : al-Zīj al-Muqtabis of Ibn al-Kammād, op. cit.*, 25, table 8.

Differences in Ibn al-Kammād's tables of trepidation
According to Ibn al-Hā'im

Argument		Precession		
		Ibn al-Zarqālluh	al-Kammād	difference
Hipparchus	9^s 22;32,40°	9;29°	9;19°	0;10°
Ptolemy	10^s 19;2°	6;42°	6;57°	0;15°

From Ibn al-Zarqālluh's *Fixed Stars*

	Argument	Precession
Hipparchus	9^s 22;32,12°	9;28,30°
Ptolemy	10^s 19;1°	6;42,45°

From Ibn al-Kammād's *al-Muqtabas*

Argument	Precession
9;22°	9;21°
9,23°	9;17°
10;19°	6;57°

It seems fairly sure that these values in Ibn al-Kammād's tables are obtained by interpolation.

Timekeeping

(#9-10) Another passage which gives an idea of the differences that may be found between Ibn al-Kammād's *al-Kawr* and *al-Muqtabas* is the one devoted to timekeeping. Ibn al-Hā'im's criticisms numbered 9 and 10 are related to the determination of the altitude of the sun from the hour and vice-versa (the hour from the altitude of the sun). Ibn al-Hā'im does not give the procedure in Ibn al-Kammād but says that it is only correct at the equinoxes. Therefore it could be the approximate formula of Indian origin

I. $\sin t = \sin h / \sin H$

where t is the time elapsed since sunrise in degrees, h is the altitude of the sun at the moment given and H is the meridian altitude of the sun in that day.

In chapter 22 of *al-Muqtabas*' Canons Ibn al-Kammād includes instructions to determine the time of the day from the altitude of the sun using this formula. But afterwards, he explains that this formula is only useful at the equinoxes and, otherwise, another procedure must be used. He then describes the exact formula

II. vers T = vers D - vers D sin h/sin H[31]

where T is the horary angle and D the half length of daylight. Then he explains that the time from sunrise, t, can be obtained making

$$t = D - T.$$

In fol. 47v. of the *Muqtabas* we find a table which gives the half length of daylight as a function of ϕ = 38;30 and $\in$ = 23;33 which derives from the exact formula :

$$\cos D = -\tan(\phi) \tan(\in)$$

in modern notation.

This can be considered as another difference between *al-Kawr* and *al-Muqtabas*.

Concluding remarks

There are other questions and criticisms that will be analyzed in the near future but, as a provisional conclusion, I would like to summarize some of Ibn al-Hā'im's reasons for criticizing Ibn al-Kammād's astronomical work :

- First, the inconsistencies he found in Ibn al-Kammād's text, for instance, in the length of the year ; also, its inappropriate use.

- Second, the lack of accuracy in certain calculations which are not in accordance with the needs of the astronomers.

- Third, that Ibn al-Kammād sought in some instances to derive a model valid for all time, which is erroneous according to Ibn al-Hā'im. From his experience, models and, in consequence, tables have to be corrected periodically and their validity is for a period of no more than 40 years after compilation.

As for Ibn al-Hā'im's work, from this study there are reasons to think that *al-Zīj al-Kāmil* in its present form is a reduced version of the original, which probably included some other topics in detail as astrology, timekeeping and eclipses for instance, since Ibn al-Hā'im refers the reader to them. It may also have contained tables although there are no traces in the only copy extant of his *al-Ẓīj al-Kāmil*[32].

31. This formula is also of Indian origin and was known in the Islamic East from the 9th century onwards.

32. On the possibility that this *zīj* had tables *cf.* M. Abdulrahman, " Wujūd jadāwil fī zīj Ibn al-Hā'im ". English abstract with the title " Ibn al-Hā'im's *zīj* did have numerical tables ", *From Baghdad to Barcelona*, vol. I, 365-381.

As for Ibn al-Kammād's work, it seems that there were several differences between *al-Kawr* and *al-Muqtabas*, although some tables may be similar in the two works.

In conclusion, Ibn al-Hā'im's criticisms are a useful aid to understanding the astronomical work of Ibn al-Kammād who, no doubt, was a major figure in the history of Astronomy in al-Andalus.

Some new Maghribī sources dealing with trepidation[1]

Mercè Comes

Short introduction on trepidation or accession and recession theory[2]

As is well known, trepidation or accession and recession theory seems to have had its origins in pre-Ptolemaic Greek astronomy. It appears to have reached the Muslim world, particularly al-Andalus, from India, where the earliest reference dates from *c.* 550 A.D.[3]. In the Iberian peninsula the theory was highly developed and from Muslim Spain passed on to Latin Europe where it was in vogue until at least Copernicus' time. However in al-Maghrib, as we will see, the theory was already being questioned at the very beginnings of 15th century.

Descriptions of the theory are found in a large number of sources and are based on the belief in a backward and forward motion of the equinoctial points along an arc of 8°, due to which the value of the precession of the equinoxes was sometimes negative and sometimes positive. It seems that the origins of these 8° and of the theory of a linear zig-zag function are to be found in ancient Babylon.

1. This paper has been written as part of a research programme on *Astronomical Theory and Tables in al-Andalus and al-Maghrib between the twelfth and thirteenth centuries* sponsored by the *Dirección General de Investigación Científica y Técnica* of the Spanish *Ministerio de Educación y Ciencia.*

2. A whole account of the most recent bibliography on that subject is to be found in J. Samsó, *Las Ciencias de los Antiguos en al-Andalus*, Madrid, 1992, 219-25. As well as in the following papers : J. Ragep, " Al-Battānī, Cosmology and the History of Trepidation in Islam " ; R. Mercier, " Accession and Recession : Reconstruction of the Parameters " and M. Comes, " Accession and Recession Theory in al-Andalus and the North of Africa ", in J. Casulleras, J. Samsó (eds), *From Baghdad to Barcelona. Studies in the Islamic Exact Sciences in Honour of Prof. Juan Vernet*, Barcelona, 1996, 267-364.

3. D. Pingree, " Precession and Trepidation in Indian Astronomy before A.D. 1200 ", *Journal for the History of Astronomy*, III (1972), 27-35.

These descriptions were at the beginning only qualitative. We find a description of this kind for the first time in Theon's commentary to the *Handy Tables*[4] ; he attributes the theory to the " ancient astrologers " (*hoi palaioi ton apostelesmatikon*). A term that will be reproduced in Arabic works as " Ahl al-ṭalismāt ". Theon's words were often reproduced, in one way or another, by Arabic sources, sometimes, though, only to be rejected as is the case of al-Battānī or al-Bīrūnī.

The first time we find this theory described by means of geometrical models is in Ibrahim b. Sinān's *Kitāb ḥarakat al-shams*. According to al-Bīrūnī's *Athar al-Bakiyya*, the best explanations are found in the *zīj al-ṣafā'iḥ* by al-Khāzin (d. b. 961-971) and in the book by Ibrahim b. Sinān (d. 946). However, their models show no parameters at all[5].

In al-Andalus, the theory was more successful, probably due to the enormous influence that Indian astronomy had in the early days of the introduction of this science in the Iberian peninsula.

In Ṣāᶜid al-Andalusī's *Ṭabaqat al-umam* we learn that the theory was introduced in al-Andalus through the *Kitāb naẓm al-ᶜikd* by Ibn al-Adamī (d. 920)[6]. Some authors, according to Ibn al-Hā'im (fl. 1205), including Ṣāᶜid himself, Abū Marwān al-Istijī (fl. 1045), *etc*., studied it, but Azarquiel (d. 1100) was the one who devised his three famous geometrical models, the third of which was in use in al-Andalus and North of Africa until the 15th century as we will see[7]. In Europe the theory will have an even longer life, reaching Copernicus and Longomontanus, the student of Tycho Brahe ; so it endured until the mid-17th century[8]. The end of that century would bring Newton and the definitive demise of the theory.

The aim of this paper is, then, to give an account of some new Maghribī texts dealing with trepidation which may shed some light on the evolution of this theory.

4. See A. Tihon, *Le* " Petit Commentaire " *de Théon d'Alexandrie aux tables faciles de Ptolémée*, Vaticano, 1978, 236-237 and 319.

5. For the *Kitāb ḥarakāt al-shams*, see A.S. Saᶜīdān, *Rasā'il ibn Sinān*, Kuwait, 1983 ; for *al-zīj al-ṣafā'iḥ* I have used ms. 314 from Srinagar Research Library, a photocopy of which we owe to the generosity of Prof. David A. King.

6. Sāᶜid al-Andalusī, *Kitāb ṭabaqāt al-umam*, in H. Bū ᶜAlwān (ed.), Beirut, 1985, 146-147.

7. Ibn al-Hā'im *Kitāb al-kāmil*, ms. 285 (Marsh 618) Bodleian Library II, 2, ff. 3v, 8v, 28r and 59r.

8. See K.P. Moesgaard, " Tychonian observations, perfect numbers and the date of creation : Longomontanus's solar and precessional theories ", *Journal for the History of Astronomy*, VI (1975), 84-99 and " The 1717 Egyptian Years and the Copernican Theory of Precession ", *Centaurus*, XIII (1968), 120-138 ; see also W. Hartner, " Trepidation and Planetary theories ", *Oriens. Occidens*, II, 277-282.

Texts analyzed[9]

a) *Kitāb al-adwār fī tasyīr al-anwār* (821/1418-19). Written by the Moroccan astrologer Abū ᶜAbd Allāh al-Baqqār (fl. beginning of 15th c.)[10]. It consists of 5 short chapters, the first of which, entitled *Fī bayān ḥarakat al-falak al-kuliyya*, deals with trepidation motion and in it the theory's beginnings, life and death are expounded. The author draws on Azarquiel and Ibn Isḥāq, especially the latter. When he describes the motion and the relative positions of the Head of Aries and the equinox, he reproduces Ibn Isḥāq's text almost word for word[11]. The other 4 chapters deal with astrological items such as projection of rays, tasyīrāt, etc.

b) A short fragment due to Ibn ᶜAbd Rabbihi al-Ḥafīd (530-602h/1135-1206) answering Ibn Rushd al-Ḥafīd's *Fī mas'ala falakiyya*[12]. The author, though born in al-Andalus, travelled to Marrakesh *c.* 550h where he seems to have lived and died. In this fragment the trepidation theory is used to predict catastrophic events and especially to refute the astrologers' opinion that the end of the world was near.

c) Four commentaries (from now on c1, c2 and c3 and c4) to the *Rawḍat al-azhār fī ᶜilm waqt al-layl wa-l-nahār*[13] (794/1391-92), the *urjūza* on time-keeping written in Fes by Abū Zayd ᶜAbd al-Raḥmān al-Lakhmī al-Jādirī (777-818/1375-1416)[14]. First one (c1), anonymous (1257/1842) and second one (c2) entitled *Qatf al-anwār min Rawḍat al-azhār* by Abū Zayd ᶜAbd al-Raḥmān b. ᶜUmar b. Aḥmad al-Sūsī al-Jazūlī al-Buᶜaqīlī, also called Ibn al-Muftī (d. 1020/1611)[15], and probably based on the commentary by M. b. Aḥmad b. al-Ḥabbāk al-Tilimsānī d. 867/1462-63)[16]. Third one (c3) is commentary titled *Natā'ij al-afkār fī sharḥ rawḍat al-azhār.* There are three copies : two anonymous copies dated 1183/1770 ; *annus praesens* 920/1515[17] and another copy not dated, but

9. Prof. Julio Samsó and myself have analized these new sources. The information gathered was presented at the *XXth International Congress of History of Science*, Liège, *July 1997.*

10. Ms. Escorial 418. See also J. Vernet, " Tradición e innovación en la ciencia medieval ", *Estudios sobre Historia de la Ciencia Medieval*, Barcelona, 1979, 188-189.

11. Al-Baqqār, f. 238v. Ibn Isḥāq's *zīj* is extant in ms. Hyderabad Andra Pradesh State Library n° 298, fol. 92. We owe a photocopy of this manuscript to the kindness of Prof. D. King.

12. Microfilm of the ms. in Rabat National Library 47. There is an edition based on a different ms. in M. Ben Cherifa's *Ibn ᶜAbd Rabbihi al-Ḥafīd. Fuṣūl min sīra mansibba*, Beirut, 1992, 196-197.

13. Ms. 80 in the Maktabat al-zāwiyya al-ḥamzawiyya (Ayt Ayache), 203-220.

14. D. Lamrabet, *Introduction à l'Histoire des Mathématiques Maghrébines*, Rabat, 1994, n° 432.

15. Both in ms. Cairo K 7584. See D.A. King, *A Survey of the Scientific Manuscripts in the Egiptian National Library*, Winona Lake, Indiana, 1986, 142 (n. F43) and *Fihris al-makhṭūṭāt al-ᶜilmiyya al-maḥfūẓa bi-Dār al-Kutub al-Miṣriyya*, I, Cairo, 1981, 352.

16. D. Lamrabet, *Introduction à l'Histoire des Mathématiques Maghrébines, op. cit.*, n° 445.

17. In mss. Cairo K 4311 and London British Library Or 411 (D. King, *Survey*, 139, F 26). For c2 I will use ms K 4311 unless otherwise expresed.

probable written around the end of the 16th century, in which the name of the author appears as Abū Zayd ᶜAbd al-Raḥmān al-Jānātī al-Nafāwī[18]. Most of the data found in c3 are to be found more expanded in the *Kanz al-asrār wa-natā'ij al-afkār fī sharḥ rawḍat al-azhār* (c4) by Abū'l-ᶜAbbās al-Māwāsī al-Fāsī (d. 911/1505)[19]. These commentaries contain some interesting remarks indicating that the trepidation theory was no longer applicable. If we compare the four commentaries we can see that in order of knowledge of their authors and also in order of interest the first must be c3 and c4, followed by c2 and finally c1, in which almost nothing new is found.

INFORMATION FOUND IN THESE SOURCES

Two types of information related to accession and recession theory can be found in these sources :

1) Non-scientific information, that is to say, folk-tales, traditions, etc., mainly on the primitive myths of creation ; the preservation of science and astrological predictions.

2) Scientific information, by which I mean basically astronomical data derived from observation or calculations, and information on the various authors who have dealt with this theory.

Non scientific information is found mainly in b and c1. I would like to mention several subjects.

The creation myth

In the anonymous commentary to the *urjūza* by al-Jādirī (c.1, f. 87r) we find the assertion that the two points called *nuqṭat al-ṭabīᶜiyya wa-nuqṭat al-dhātiyya* (that is the tropical and sidereal points) coincide at the moment of Creation, when Time, supposedly, started its motion. This idea seems to be somehow original, but was probably stated by someone more influential in Europe, because it will be found even in Longomontanus, as late as the early 17th c., who supports the idea, because, according to him, at the very time when Heaven and Earth came into being, amongst other celestial phenomena related to the sun and the obliquity of the ecliptic, the oscillatory term of precession set out from its neutral mean position[20]. This seems to have something

18. Maktaba ḥamzawiyya (Ayt Ayache) 80, 228-334.

19. A microfilm of the Rabat Ḥasaniyya ms. 2151 was available to me after the submision of this paper at the Liege Congress. As the ms. is fairly long (234 pp.), I have took it into account just to verify some data. A full account on trepidation, some paragraphs of which reproduce almost word by word ms. K4311, ff. 4r-5v, appears in pp. 35-36. Pp. 37-40 are devoted to the problem of the obliquity of the ecliptic.

20. See K.P. Moesgaard, " Tychonian observations, perfect numbers and the date of creation : Longomontanus' solar and precessional theories ", *op. cit.*, 84.

to do with the old Indian concept of world year, supported by Abū Maᶜshar, according to which all the planets would coincide in the Head of Aries at the moment of the Creation[21], but, in our case, related to the trepidation motion.

According to Ibn ᶜAbd Rabbihi al-Ḥafīd (b, f. 28r) in the Sindhind book (he follows here al-Bakrī's Egyptian section of the *Kitāb al-masālik wa-l-mamālik*) on which Almagest and others are based on (*sic*) it is stated that the first rotation of the sun took place from the beginning of Aries and the last one will be after 4.320.000.000 years[22], a figure that once again recalls Abū Maᶜshar, for whom the life of the world was 4.320.000.000 solar years, dating from the first day on which the planets started to move from the beginning of Aries. Abraham Bar Ḥiyya in his *Meguil. lat hamegal-lè*[23] also affirms that at the moment of the Creation the Sun was at the beginning of the sign of Aries and started its motion at this moment, a motion that will take 4.320.000.000 years. And Rheticus relates the beginning of Creation and the end of the world to the position of the centre of the sun eccentric on his small circle. The figure 4.320.000.000 years, transmitted by Abū Maᶜshar, seems to have an Indian origin for we can find it in the *Paitāmahasiddhānta* (*c*. 400-450) as well as in the *Sindhind*.

It is worth mentioning here something which, although is not new, appears repeatedly in these texts and will solve an apparent problem of terminology. In all these manuscripts, as well as in others treatises like those of al-Marrākushī, Ibn al-Raqqām, Ibn Isḥāq, Ibn Abī 'l-Shukr al-Maghribī, etc., we find the terms *al-mabdā' al-ṭabiᶜī* and *al-mabdā' al-dhātī* to distinguish between tropical and sidereal calculations, sometimes as in the case of al-Baqqār accompanied by a full and extensive explanation of the exact meaning of these terms. That is *al-mabdā' al-ṭabiᶜī* for positions calculated from the equinox and *al-mabdā' al-dhātī* for positions calculated from the Head of Aries. This terminology, found in most of the texts dealing with accession and recession movement, appears also in certain star tables in which the title *jadwāl al-kawākib min al-mabdā' al-dhātī* has often been misunderstood[24].

The myth of the preservation of science

Among the old nations there are a great many legends, revolving around the myth of the Flood and aiming to explain the survival of an antediluvian culture or knowledge, often considered as superhuman, in the days which followed the cataclysm.

21. See E.S. Kennedy, " The world-year of the Persians " and " The world year concept in Islamic Astrology ", *Studies in the Islamic Exact Sciences*, Beirut, 1983, 338-350 and 351-371.

22. 4.420.000.000 years in the manuscript.

23. See Guttman and Millás, *Llibre revelador* (*Meguil. lat Hamegal-lè*), Barcelona, 1929, 19.

24. See for instance J. Chabás, B.R. Goldstein, " Andalusian Astronomy : *al-Zīj al-Muqtabis* of Ibn al-Kammād ", *Archive for History of Exact Sciences*, 48 (1994), 1-41.

The Babylonians talked of the tablets on which the knowledge of the seven sages was recorded ; Plato mentions the writings of the antediluvian Egyptians at the beginning of his *Timaeus* ; according to a legend found in some Byzantine sources, Seth constructed two towers, one of stone and another of clay, for the same purpose ; Abū Maᶜshar introduces a similar legend to accord the authority of the antediluvian sages to his system of thousands devised to interpret history by astrological means. He claims to have found a manuscript of his book in a building called Sārawīya in Jay, built to preserve science from a flood ; according to al-Bīrūnī[25], Tahmūrath chose Ispahān as the site where all the scientific books will be preserved from a flood in a building similar to the two pyramids built in Egypt[26].

References to the Hermetic legend[27] — the preservation of science by Hermes I in the upper Egypt before the arrival of the flood — can be found in al-Suyūṭī[28], the *Tabula Smaragdina*[29], Ibn Juljul, Ṣāᶜid al-Andalusī and the King Alfonso's *Primera Crónica General*[30], among others.

Ibn ᶜAbd Rabbihi al-Ḥafīd (b, ff. 27v-28r) offers us an account of the old myth of the Pyramids as keepers of ancient science related also in this case to the trepidation motion. According to him, the Ancients observed that when the stars had completed their path a sacrificial fire descended from the heavens.

Then, when the king of Egypt was told that the star *qalb al-asad* (Regulus) had already completed two thirds of its advance and once the remaining third was accomplished the celestial vault will disintegrate, he ordered all the knowledge of the ancients to be written in books and these books kept in the Pyramids.

Obviously, the text refers here to the accession motion which was about to be accomplished and reflected the tradition according to which the change in direction, that is from accession to recession or vice versa, is sign of great changes, mostly disasters and cataclysms.

Here, the use of *qalb al-asad* (Regulus) for the time of the Flood is of course an anachronism because this star in fact was used for the first time as a star reference near the ecliptic for sidereal co-ordinates in Ptolemy's *Handy Tables* and *Canobic Inscription*.

25. Al-Bīrūnī, *The Chronology of Ancients Nations*, E. Sachau (transl.), London, 1879, 27-28.

26. See at this respect D. Pingree, *The Thousands of Abū Maᶜshar*, London, 1968, 15.

27. On this see R.P.O.P. Festugière, *La Révélation d'Hermes Trismégiste*, I. *L'Astrologie et les Sciences occultes. Avec un appendice sur l'Hermétisme Arabe par M. Louis Massignon*, Paris, 1981 and D. Pingree, *The Thousands of Abū Maᶜshar, op. cit.*, 14.

28. L. Nemoy, " The Treatise on the Egyptian Pyramids ", *Isis*, 30 (1939), 21-22 and 29.

29. J. Ruska, *Tabula Smaragdina*, Heidelberg, 1926, 61-64.

30. Alfonso X, *Primera Crónica General*, I, 12b-13a. See J. Samsó, " Astrology, Pre-Islamic Spain and the Conquest of al-Andalus ", *Islamic Astronomy and Medieval Spain*, II, Aldershot, 1994, 87.

Recurrent world events related to the moment of change of direction

It is well known that celestial phenomena related to catastrophic events had often attracted the attention of astrologers. In this case, the change of direction was seen as a sign of bad omens in general.

Apart from the view already stated, Ibn ᶜAbd Rabbihi (b, ff. 27v-28r) presents other opinions on world disasters and their causes also related to trepidation motion. According to him, some people stated that when *qalb al-asad* reached the degree 16 of the sign of Leo (some ms. state the end of 6°), a flood wiped out the world leaving no animal behind. After that, the Earth recovered its warm as established from the moment of the creation. This again seems to associate the beginning of the motion with the moment of the creation. In fact the Ikhwān al-Ṣafā' apply this theory to Ptolemy's precession motion. According to them every 36.000 years, that is a whole revolution of the precession at 1°/100 years, the Earth would suffer great changes[31].

If we consider that, according of the star tables calculated for the beginning of the accession and recession movement, *qalb al-asad* was at that moment at Leo 9;08°, when this star reached the 16° of the same sign, the change of direction would be about to start. However, the amplitude would be only of 7°. The text edited by Ben Cherifa is different at that point and states that this had happened when the star reached the end of Leo 6°. That could means that *qalb al-asad* should cover almost 7° of Leo (16°-9;08°=6;52°) before starting its motion in the opposite direction.

Maybe, for several reasons, we should consider a combination of both statements : not 16° nor the end of 6°, but the end of 16°. Mainly because the author immediately insists that al-Bakrī in his *Kitāb al-masālik wa-l-mamālik* states that this is not true, because *qalb al-asad* was observed to be at Leo 13;30° in year 214h (829/39) and then in his time, year 461 (1068/69), it should be at " the end " of 16th degree, and consequently this would mean great disturbances, but were none. Besides, this would imply an amplitude of about 8° (*c.* 7;52°) in consonance with the opinion of the Ahl al-ṭalismāt.

There are a couple of interesting points in this statement. First of all, we know that this value (13;30°) correspond to some of the *mumtaḥan* tables for year 211 and 214, which according to Ibn Yūnus was the year in which Ḥabash made his observations[32]. If we consider the approximately 250 years of difference at a rate of 1°/66 years, the value of constant precession given in the *mumtaḥan* tables and used by most of the Arabic astronomers of the age, we have to add to 13;30° something more that 3° which will take us to the end of degree 16°. (If we use 1°/80 years, which is the value implied in Theon's the-

31. *Rasā'il Ikhwān al-Ṣafā'*, ms. Escorial, n° 928, fol. 141v.

32. Ibn Yūnis, *al-zīj al-Ḥākimī*. See Caussin de Perceval, *Notices et Extraits des Manuscrits de la Bibliothèque Nationale*, VII, Paris, 1803, 58.

ory the difference would be of more than 4° and hence the value of the star would reach the middle of degree 17°)[33]. This means that al-Bakrī used *al-mumtaḥan* tables in al-Andalus in year 1068. The second point is the date 461h. as *annus praesens* of al-Bakrī, which confirms Lévi-Provençal's hypothesis that the lack of any mention to the Almoravids' entry in al-Andalus suggests that al-Bakrī finished the book in year 460/1068[34].

According to al-Nāẓir, the author's source, the sidereal position (*al-mawḍi*ᶜ *al-dhāti*) of *qalb al-asad* was Leo 9;08°. This is indeed the value found in the tables calculated for the beginning of the accession and recession motion (*jadwāl al-kawākib min al-mabdā' al-dhātī*), as for instance the star table in the *zīj al-Muqtabis* of Ibn al-Kammād or the one in Ibn al-Bannā's *Minhāj* or al-Marrākushī's *Jāmi*ᶜ *al-mabādi*ᶜ *wa-l-ghāyāt*[35]. Furthermore, the distance between *qalb al-asad*'s maximum accession and recession should be 19;58° (the arc covered, according to him, should go from Leo 9;08° to Leo 19° and then to Cancro 29;09°), which means a P_{max} of 9;59° (19;58°/2), that is the one found in the tables of Ibn al-Kammād, al-Marrākushī, Peter the Ceremonious and Ibn ᶜAzzūz al-Qusunṭīnī, one of the different branches of astronomers using Azarquiel's trepidation models[36]. However a correction should be applied, because 29;09°-9;08°=9;59° ; while 19°-9;08°=9;52°. So that these 19° should be corrected to 19;07°. That the other values are correct can be internally proved : 9;59°-9;08°=0;51° ; 30°-0;51°=29;09°.

Furthermore, according to Ibn ᶜAbd Rabbihi, Ibn Rushd informed him in year 558 that within approximately 60 years the accession motion would reach its end and the recession motion will start. Our author, however, was not entirely convinced, saying that this statement must be submitted to observations and calculations because if it were true there would be great disturbances.

Here, Ibn Rushd is probably mixing two different pieces of information that point to the same end. On the one hand, the old idea of trepidation covering an arc of 8° every 800 years. That means that 1° would take 80 years, so the direction would change every 640 years. In this case 584 plus 60 will give us year 644, which implies that he considers the beginning of trepidation to coincide with the beginning of the Hijra, a mistake found in more than one source. In fact al-Marrākushī himself entitle his longitude star table as for the beginning of Hijra, when he is using the same increase on Ptolemaic longitudes used in the tables of Ibn al-Kammād and Ibn al-Bannā' calculated for the beginning of

33. According to Ibn Bargūt (c. 1050) for his time *qalb al-asad* was in Leo 16;20°. See Millás, *Estudios sobre Azarquiel*, Madrid, Granada, 1943-1950, 318-319, where a list of observations of *qalb al-asad* is showed.

34. See M. Ben Cherifa, *Ibn ᶜAbd Rabbihi al-Ḥafīd. Fuṣūl min sīra mansibba, op. cit.*, 196-197 ; and Lévi-Provençal, *Encyclopédie de l'Islam*, I, Nouvelle édition, 161.

35. M. Comes, " Deux échos andalus a Ibn al-Bannā' de Marrākush ", *Le Patrimoine andalous dans la culture arabe et espagnole*, Tunis, 1991, 81-94.

36. See M. Comes, " Accession and Recession Theory in al-Andalus and the North of Africa ", *op. cit.*, 349-364.

the trepidation motion, some 40 years before the Hijra, as al-Marrākushī himself states, but called the Prophet's epoch in most of the sources, as in text c1 (f. 87r)[37]. In fact, different sources date the Prophet's epoch between 556 and 580[38]. This means that it would have coincide with the birth of the Prophet although some authors confuse this epoch with the beginning of the Hijra. On the other hand, it is stated in several sources that the astrologers thought that the destruction of the world would take place near the year 582h because for that time all the planets will coincide in the sign of Libra.

In the *al-Dhayl wa-l-takmīla* by M. b. M. al-Marrākushī[39] it is said that ᶜAbd Allah al-Anṣārī confirmed a statement found in an Egyptian book based on Indian sources according to which after year 580 all the planets will coincide in Libra and this will imply the destruction of the world. Al-Anṣārī goes farther and says that this conjunction will take place the 29th of *jumādā al-akhira* of 582h, which corresponds to 16th September 1186. In fact, it is true that the date coincide with the moment in which the Sun entered Libra and the rest of the planets were between 180° and 190°, that is at the very beginning of the sign of Libra[40]. Of course this must be understood according to the ancient notion that the Universe is cyclically created and destroyed and that the successive cataclysms coincide with grand conjunctions of the planets at the zero point of the ecliptic, an idea already found in Babylonian texts[41]. In this case, the fact that Libra was the sign chosen may be related to the models of trepidation in which the moving Heads of Aries and Libra rotate around their twin small epicycles. However, it has nothing to do with Andalusian trepidation models because using them the change of direction would take place around year 939h, that is some 300 years later. Berosus, however, already considered the possibility that the conjunction took place in a specifically sign, mainly solstices. According to him, when all the planets coincide in Cancer

37. From radix positions and mean motion tables in Azarquiel's trepidation tables we can deduce a zero position for the Head of Aries for a date comprised between January 1st 581 and March 20th 581. See H. Mielgo, " A Method of Analysis for Mean Motion Astronomical Tables ", in J. Casulleras, J. Samsó, *From Baghdad to Barcelona. Studies in the Islamic Exact Sciences in Honour of Prof. Juan Vernet,* Barcelona, 1996, 178.

38. According to the *Brahmasphutasiddhānta*, the solar sidereal and tropical longitudes coincided in year 580. See J. Samsó, " Trepidation in al-Andalus in the 11th century ", *Islamic Astronomy and Medieval Spain*, VIII, Aldershot, 1944, 22. On the the Saturn and Jupiter conjunction indicating the rise of Islam (year 571), see Kennedy's and Van der Warden " The world-year of the Persians ", 343 and Kennedy's " The World-Year concept in Islamic Astrology ", 360. Both reprinted in E.S. Kennedy, " Colleagues and Former Students ", *Studies in the Islamic Exact Sciences*, Beirut, 1983.

39. M. b. M. al-Marrākushī, *al-Dhayl wa-l-takmila,* IV, in Aḥsān ᶜAbbās (ed.), Beirut, 211.

40. See *Solar and Planetary Longitudes for Years -2500 to 2000 prepared by W.D. Stahlman and O. Gingerich*, Madison, 1963, 451.

41. E.S. Kennedy, " The World-year of the Persians ", *op. cit.,* 351 ; see also D. Pingree, *The Thousands of Abū Maᶜshar, op. cit.*, 27-28.

(summer solstice) the world will be destroyed by fire and when in Capricorn (winter solstice) by water[42].

Obviously, all these traditions must be connected to the astrological predictions and the world year notion in which astronomical events, be they conjunctions of planets or in this case changes in the direction of a motion, imply world changes, negative or positive as is the case of the world destruction or creation.

Scientific information

As far as scientific information is concerned, we can find in these texts several interesting points.

First of all almost all the texts show the beginning of the end of the theory of accession and recession as a consequence of the doubts aroused by the new observations.

It becomes clear to the authors that the estimations of the accession values carried out for their times are not correct and that there is a disagreement between calculations and observations. The maximum accession cannot be around 10° as stated by previous astronomers because observations prove that for their times accession has reached a much higher value of about 13° or 14°. Azarquiel had had the same doubts when he realizes that the values of maximum accession or recession stated by the previous astronomers were amply surpassed, as he states in the introduction of his *Book on the fixed stars*[43]. He, then, instead of refuting this motion, looks for possible explanations of these irregularities.

Al-Baqqār in his *Kitāb al-adwār fī tasyīr al-anwār* (a, ff. 240v-241r) introduces the following calculations to prove that old models and parameters are no longer valid :

According to our author using Ibn al-Raqqām's *zījes* we would obtain for year 680 an *iqbāl* of 9;26°. This value, recalculated using Ibn al-Raqqām's tables, proved to be correct.

On the other hand, he states that Ibn al-Raqqām places *qalb al-asad* for the year of his star table (680) at Leo 19;16°. This again is correct because Ibn al-Raqqām's star tables are different in his *zījes*[44], but in *al-zīj al-Qawīm* and in *al-zīj al-Mustawfī*, where the title of the star table shows the date of year 680[45],

42. See also D. Pingree, " Astronomy and Astrology in India and Iran ", *Isis*, 54, 2 (1963), 229-246.

43. J.M. Millás, *Estudios sobre Azarquiel*, *op. cit.*, 274-281.

44. *Al-zīj al-Qawīm*, Rabat, General Library, ms. 260 and *al-zīj al-Shāmil*, Istambul, Kandilli Museum, ms 249. A ms. of *al-zīj al-Mustawfī*, Rabat, General Library, ms. 2461, has been lately available to me.

45. This year corresponds also to Ibn Isḥāq's recension and to al-Marrākushī's star tables and examples dealing with trepidation.

the position of *qalb al-asad* is exactly Leo 19;16° (surprisingly enough in *al-zīj al-Shāmil* the star table is apparently for the year 887 but *qalb al-asad* is placed at Leo 19;20°, then with an increase of only 4' for 207 years. This prove, in my opinion, that this table was probably for Ibn al-Raqqām's time and probably composed by him, but that some errors were introduced by the copyist, amongst them the date in the title, which should be 687, not 887, and could help to determine that *al-Qawīm* and *al-Mustawfī* were earlier than *al-Shāmil*. Once recalculated, in effect, the table in *al-Qawīm* and *al-Mustawfī*, in which 16;46° are added to Ptolemy's longitudes, corresponds to year 680, while the one in *al-Shāmil,* showing an increase on Ptolemy's longitudes of 16;50°, corresponds to year 687.

Then, al-Baqqār substract the *iqbāl* from the position of *qalb al-asad* and obtains a position radix for the beginning of the accession motion (*al-mabdā' al-dhātī*) of 9;50°. This, according to al-Baqqār, who once more is right, differs from the opinion of the ancients according to whom the position at the beginning of the motion was 9;08°.

Al-Baqqār, then, concludes that the *iqbāl* for the end of year 680 should be of 10;08° and not of 9;26°, the *iqbāl* given by the tables.

To confirm this he introduces the horoscope cast for Ibn al-Raqqām's son, in which for year 685 the *iqbāl* considered was 10°, which would correspond, according to Ibn Isḥāq's tables, to year 821, that is more or less the epoch of Ibn al-Baqqār and approximately 140 year after Ibn al-Raqqām's time.

After seeing this, the author calculates the *iqbāl* for the year of the book 821 H./1418 A.D. and finds it near the limit values described, according to him, by Azarquiel (10°) and Ibn Isḥāq (10;10°). These values, however, do not correspond to Azarquiel and Ibn Isḥāq (10;24°), but are the values deduced by al-Baqqār using the tables of Azarquiel and Ibn Isḥāq. In general the values range from *c.* 9;45° to *c.* 10;55°[46].

He then compares it with the observed position of the sun and finds a difference of 2°, from which he concludes that the *iqbāl* must be 12°.

This arouses his doubts, because this value surpasses the maximum *iqbāl* accepted by the old astronomers on which he has been relaying until now. The author, then, introduces the possibility of the existence of a different motion different.

The anonymous commentator of the *urjūza* on timekeeping by al-Jādirī (c1, f. 84r) simply states that the *iqbāl* for al-Jādirī was 10° but for his time it had reached 13°.

al-Jazūlī (c2, f. 119r), however, considers that although al-Jādirī states that *iqbāl* for his time was 10°, it should be around 12°. He thinks that by al-Jādirī's

46. See M. Comes, " Accession and Recession Theory in al-Andalus and the North of Africa ", *op. cit.*, 359-360.

time the difference may have been too small to be apprcciated. He adds that P_{max} for Ibn Isḥāq al-Tūnisī and Ibn al-Bannā' was 10;24°, which is true, but by his time trepidation has reached 13° and even 14° according to some, which would invalidate the above mentioned P_{max}.

In the *Natā'ij al-afkār* (c3, f. 5r), the author goes a little farther, giving a series of data to prove the theory that the old values and tables are no longer useful. The data are the following :

Author	Year	Trep. value
al-Jādirī (following Ibn Isḥāq P_{max} = 10;24°)		10°
According to Ibn Abī l'Shukr's observations (Damascus) must be		12°
ᶜAlī b. Yūnis al-Balansī (Egypt)[47]	732 h.	*c*. 13°
The author's time	920 h.	14°

So, the author of the *Natā'ij al-afkār* concludes that the maximum accession cannot be 10;24° as stated by the branch following Ibn Isḥāq's school, such as Ibn al-Raqqām, Ibn al-Bannā' and the *Tashīl al-ᶜamal wa-l-ᶜibārat*, a *zīj* ascribed to Ibn Qunfud al-Qusunṭīnī, the three astronomers most followed in his time in al-Maghrīb. His explanation is that there must be another motion to add to the accession and recession movement.

Secondly, and as we have already seen, some of these texts also announce the end of the accession and recession theory on the basis of the observations made in Damascus by Yahyà b. Abī l-Shukr al-Maghribī (d. 1.283) and expounded in his *zīj* entitled *Tāj al-azyāj*[48], which influence in the Maghrib has recently been described by J. Samsó[49].

In the *zījes* of some eastern astronomers who carried out observations in Damascus and refute accession and recession theory, mainly in Ibn Abī 'l-Shukr al-Maghrībī's *Tāj al-azyāj*, al-Baqqār (a, f. 241r), finds an increase in the

47. Further on he is quoted as " al-shaykh al-fāḍil Abū al-Ḥasan ᶜAlī b. Yūnis al-Balansī al-Hākimī ". There seems to be a confusion with the well known Ibn Yūnus.

48. There are two manuscripts, the wellknown Escorial ar. 932 and the ms. kept in the Arabic Department of the University of Barcelona. On the last one see J. Samsó, " An outline of the history of Maghribī " zījes " from the end of the 13th century " to appear in the *Journal for the History of Astronomy*.

49. Part of the information is to be found in J. Samsó, " An outline of the history of Maghribī " zījes " from the end of the 13th century ", *op. cit.*

position of the tropical sun for the end of year 810 with respect to the calculations based on the observations carried out by Ibn Isḥāq of *c.* 12;16,45°, after taking into account the difference in longitude between the two cities (not stated, but probably Damascus and Fes).

This quantity, which must be the distance between the two beginnings (tropical and sidereal) for the given moment that corresponds to the author's epoch, coincides with the observations and this convinced our author that Ibn Abī 'l-Shukr's rejection of trepidation motion was correct.

All these findings, of course, aroused his doubts. According to him, on the one hand, the observations did not account for another motion which could be added to trepidation ; and, on the other, the limits established for this motion (*c.* 10°) had been amply surpassed. He, then, suggests that for all the astrological questions such as projection of rays, *tasyīrāt, al-nīmūdarāt*, the making of talismans, etc., it is necessary to take into account that for his time the difference between tropical and sidereal positions amounts to these 12° and some minutes.

Al-Jazūlī (c2, f. 120r) mentions the name of Ibn Abī-l-Shukr and his new observations but only in relation to the obliquity of the ecliptic.

However, as we have already seen, for the time of al-Jādirī he mentions the value of 12° attributed in the *Natā'ij* to Ibn Abī-l-Shukr's new observations in Damascus.

Towards the end of the book, al-Jazūlī (148v) comments on some verses by al-Jādirī in which it is stated that the latitudes of the fixed stars never changed while 1° every 66 years must be added to or substracted from their longitudes, depending on whether the sphere is in accession or recession.

He states that there are different opinions on the subject, because some say that 1° must be added every 80 years and others every 100 years.

This then implies a misinterpretation of the theory of trepidation to which he attributes the values of precession.

The anonymous commentator (c1, 108r) insists also on the fact that the accession or recession value to be added is 1°/66 years.

The author of the *Natā'ij al-afkār* (c3, 30v-31v), after concluding that new observations and calculations based on the old tables do not agree, tests Ibn Abī 'l-Shukr precession and obtains a value for his time of 13;40°, a results that agrees with observations. So, in the rest of the treatise, he uses this new value as the value of precession for his time (920H).

We should also note that the author of the *Natā'ij al-afkār,* once he has declared his preference for Ibn Abī 'l-Shukr's precession, also introduces a list of values for precession which he calls *iqbāl wa-idbār*, that is to say " accession and recession ", adding that the followers of trepidation consider this motion not to be uniform (*Ḥaraka ghayr mushawiyya*) in accordance with al-Battānī's and Azarquiel's opinion, once again mixing concepts.

Author[50]	Precession value/year
al-Jādirī	1°/66 Arabic years
Ptolemy and the Ancients	1°/100
Ancient Babylonians	1°/80
Al-Battānī	1°/66
The rest of the oriental astronomers and ᶜAbd al-ᶜAzīz al-Raqqām (In fact this is the parameter of Ibn Abī 'l-Shukr al-Maghribī)	1°/72 Py = *c*. 74 Ay

So, at the end of year 920h the motion from the beginning of Hijra would be of 12;35°.

Then he adds the correct values for 1, 30, 60 and 90 lunar years, extracted, it seems, from a table for lunar years. These values coincide with the ones found in the table for precession *jadwāl ḥarakat al-falak al-mukawkab fī 'l-falak al-aqsā al-ghayr mukawkab* in Ibn Abī 'l-Shukr's *Tāj al-azyāj*, once the Arabic years are converted in Persian years. In fact Ibn Abī 'l-Shukr usually gives his values in Persian years, although the tables in the extant manuscripts, which are written in the Maghrib, are for Arabic years. I have also found some other precession table based on that of Ibn Abī 'l-Shukr. For instance, in a Maghribī astrological manuscript for year 785[51] and in Peter the Ceremonious' *Barcelona Tables*[52].

As far as the later is concerned, Chabbas concludes that the table corresponds to a precessional constant to be added to a periodic term of trepidation. However, three facts point to its independence, confirming Millás hypothesis that Jacob Corsuno, the compiler of the *Barcelona Tables*, was just trying to present both possibilities : trepidation and precession. First of all, Chabas himself admits that using the suggested combination, results are too high and not at all acceptable. Secondly, the table underlies the constant precessional value found in Ibn Abī 'l-Shukr's *Taj al-azyāj* precession table. And thirdly, the presence of the Jewish Jacob Corsuno in other texts in which this value is proposed for a constant precession[53].

50. I have used the ms. British Library Or 411 to correct the errors and fill the blanks found in ms. K4311.

51. *Risālat al-sayb fī ᶜamal al-jayb* by ᶜAbd al-ᶜAzīz b. Masᶜūd, ms. Escorial ar. 918.

52. Although this table does not appear in all the manuscripts, we find it in almost three : The Catalan ms. 21 of Biblioteca Mata (Ripoll) and the Hebrew ones, the 132 of the Wien Nationalbibliothek and the 10263 of the Bibliothèque Nationale of Paris and 356 of the Biblioteca Vaticana. See Millás, *Las Tablas Astronómicas del Rey Don Pedro el Ceremonioso*, Madrid, Barcelona, 1962.

53. J. Samsó informs me that recently a ms. containing a copy of the Almagest written by Jacob Corsuno in 14th century Zaragoza has appeared.

All this acceptation is not surprising because of the skill and precision with which, in the words of Prof. George Saliba[54], al-Maghribī conducted his observations being able to determine planetary positions to within one degree of their true positions as ascertained from modern computations.

In fact, his last *zīj* contains the only parameters that are definitely determined by observation and not copied from early sources, as is the case with many other *zījes*.

At that respect, we cannot forget that al-Maghribī worked at the Marāgha Observatory where a large-scale program to update the Ptolemaic parameters, comparable only to that of Copernicus, was undertaken.

Thirdly, some of our authors also question the value of the obliquity of the ecliptic, related since the very beginnings to the accession and recession theory and report different measurements. Ibn al-Muftī (c2, 120r) offers the following values :

Author	Obliquity
Ibn Isḥāq & Ibn al-Bannā' (actual value for Ibn Isḥāq 23;32,30°)	23;33°
Ibn Jubayr (?)	23;32°
Ibn Abī-l'Shukr (Used also by al-Ṭūsī, Ibn al-Shāṭir and al-Kāshī)[55]	23;30°

As far as the obliquity of the ecliptic is concerned, the author of the *Natā'ij al-afkār* (c3, 13v) does not reach any conclusion but follows the same procedure he used in the case of trepidation to show that the old parameters are no longer valid.

He gives the following series of data to prove his theory[56] :

Author	Obliquity
al-Muṣannif = al-Jādirī (following Ibn Abī 'l-Shukr)	23;30°
The Ancients	*c.* 24°

54. See his book *A History of Arabic Astronomy. Planetary Theories during the Golden Age of Islam*, New York, 1994.

55. In *Ilkhānī zīj ; al-zīj al-jadīd* and *zīj-i-qānī* respectively. See E.S. Kennedy, " A Survey of Islamic Astronomical Tables ", *Transactions of the American Philosophical Society*, 46, 2 (1956) 161, 163-164.

56. For another list of obliquities see F.J. Ragep, *Nasīr al-Dīn al-Ṭūsī's Memoir on Astronomy (al-tadhkira fī ᶜilm al-hay'a)*, II, New York, Berlin, Heidelberg, 1993, 394.

Author	Obliquity
Ptolemy's *Almagest*	23;51°
Mumtaḥan's author (for al-Ma'mūn)	23;33°
Ibn al-Bannā' (following al-mumtaḥan/*sic*)	23;33°
Abū ᶜAbd Allāh Ibn al-Raqqām's *zīj al-mustawfī*[57] (following al-mumtaḥan/*sic*)	23;33°
al-Battānī	23;35°
ᶜAbd a-ᶜAzīz al-Rassām (following al-Battānī)	23;35°
Ibn Qunfud al-Qusunṭīnī[58] (following al-Battānī)	23;35°
The Imam Ibn Shāṭir (following al-Battānī) (fl. Damascus, 1350) (used together with 23;30°)	23;31°
Ibn ᶜAbd Allāh al-Raqqām b. Abī 'l-ᶜAbbās al-shahīr bi-l-Mizzī (fl. Damascus *c*. 1350)	23;31°
Observation made by a man *min Ahl Miknāsah* (according to Abū Isḥāq) 602/1205-6 [Correspond to Ibn Isḥāq value[59]]	23;32,30°
Ibn Hilāl[60] (Sibta) [597-678/1199-1279]	23;32,30°
Al-Ḥakīm al-Mirrikh[61] [master of Ibn al-Bannā'] (Marrākush) 704/1304-5. (In Ḥassaniyya 2151, appears as Al-wazīr ᶜAbd al-Raḥmān Ibn ᶜUmar al-maᶜrūf bi-Ibn Khamis al-Ṣūfī, but in a marginal note is identified as al-Mirrikh)	23;26,57°
Ibn al-Tarjumān [699/1299-1300] end 7th c.h.	23;26°

57. D. Lamrabet, *Introduction à l'Histoire des Mathématiques Maghrébines, op. cit.*, n° 134. In this *zīj,* the value for the obliquity of the ecliptic is 23;32,40°. Same value often found in Ibn al-Raqqām, although in some of his tables he sets the obliquity at 23;32,30°.

58. D. Lamrabet, *Introduction à l'Histoire des Mathématiques Maghrébines, op. cit.*, n° 425.

59. On the use of this parameter by Alphons the Xth and Sanjūfīnī, see M. Comes, " A propos de l'influence d'al-Zarqallūh en Afrique du Nord : l'Apogée solaire et l'obliquité de l'écliptique dans le zīdj d'Ibn Isḥāq ", *Actas del II Coloquio Hispano-Marroquí de Ciencias Históricas. Historia, Ciencia y Sociedad*, Madrid, 1992, 151-153.

60. D. Lamrabet, *Introduction à l'Histoire des Mathématiques Maghrébines, op. cit.*, n° 393.

61. D. Lamrabet, *Introduction à l'Histoire des Mathématiques Maghrébines, op. cit*. In Ibn al-Bannā's entry, 80.

According to our author[62] this list of obliquities is surprising because some scholars had stated that the obliquity varies between two values which are 23;51° and 23;35° (*sic*), which, of course, refers to Azarquiel's model of the obliquity of the ecliptic which he combines with the accession and recession third and definitive model and considers a maximum obliquity for Ptolemy's times and a minimum for his owns.

However, these figures are not exact because according to Azarquiel the obliquity will vary between 23;53° and 23;33° and according to Ibn Isḥāq, which seems to be our author's old source, between 23;53,30° and 23;32,30°. Be that as it may, the fact that according to observations obliquity was still decreasing at the end of 7th c. Hijra, having reached 23;26°, that is more than 5 minutes below the lower limit, invalidate the old models and parameters.

Maybe this list should be related to the table found in Ibn Isḥāq's *zīj* in which there is a list of 24 observed (*sic*) obliquities of the ecliptic from 5th century B.C. to the beginning of the 13th century[63].

Finally, in some of these texts, especially in al-Baqqār's, we find a list of the astronomers who were well known in their times for having contributed, in one way or another, to the birth and development of the theory.

In the *Kitāb al-adwār fī tasyīr al-anwār* by al-Baqqār (a, 239r) we find the fullest account on the birth, development and death of the theory. The origins, according to our author have to be found in :

- Madhhab al-hind,

- Hermes,

- Aṣḥāb al-ṭalismāt (8°/800 years),

- *Kitāb al-filaḥa al-nabaṭīyya*[64] (9°/900 years)[65] and the *Ṭīqānā* (or *Ṭabqānā*) also by Ibn Waḥshīya, in which the author, according to al-Baqqār, corrected this motion,

- Thāwūn al-Iskandarānī (where a combination of precession and trepidation seems to be implied),

- *Zīj* Ḥanash (Obviously referring to Ḥabash), probably to his first *zīj* based on the *Sindhind* according to Ṣāᶜid's *Ṭabaqāt*.

Curiously enough, the confusion between Ḥabash and Ḥanash is not new in Andalusian sources. We find it in the treatise on the use of the astrolabe of Ibn al-Samḥ, where certain methods applied to the astrolabe are attributed to

62. In a marginal note 23;28,30° appears as a value for the obliquity of the ecliptic based on a motion of 48"/year and calculated for 1263.

63. See M. Comes, " A propos de l'influence d'al-Zarqallūh en Afrique du Nord : l'Apogée solaire et l'obliquité de l'écliptique dans le zīdj d'Ibn Isḥāq ", *op. cit.*, 147-159.

64. Ibn Washiyya, *L'Agriculture Nabatéenne*, I, in T. Fahd (ed.), Damascus, 1993, 276.

65. A mixture of both is to be found in Pedro Alfonso (8°/900 years). See Millás, " La aportación astronómica de Pedro Alfonso ", *Sefarad*, III, 73.

Ḥanash. M. Viladrich in her study of the text identifies this Ḥanash with Ḥabash al-Ḥāsib, who also wrote a book on this instrument. All the references to Ḥanash/Ḥabash were eliminated in the alfonsine treatise on the astrolabio redondo based on the above mentioned text by Ibn al-Samḥ[66].

J. Samsó has also shown that the same mistake appears in a late British Museum Arabic manuscript which ascribes to Ḥanash a kind of astrolabe called *āfāqī*, which can be used for different horizons, thus being a predecessor of Azarquiel's *ṣafīḥa*[67].

I have also found it in one of the manuscript of the *Ṭabaqāt al-ūmmam* by Ṣāᶜid al-Andalusī. In the copy kept at the Chester Beatty Library (3950) in Dublin we read the whole name Aḥmad b. ᶜAbd Allah al-Baghdādī al-maᶜrūf bi-Ḥanash, when talking about Aḥmad b. ᶜAbd Allāh al-Marwazī, Ḥabash al-Ḥāsib. Ṣaᶜid states that he is one of al-Ma'mūn's astronomers who wrote three *zījes*, in the first of which, based on the *Sindhind*, he corrected the longitudes of the stars using the accession and recession motion following al-Iskandārānī.

The confusion is easy to understand. First of all because of the similarity of the names, Hanash b. ᶜAbd Allah and Ḥabash b. ᶜAbd Allah, taking into account that the difference between a *bā'* and a *nūn* in Arabic script depends only on the point being below or above the same graphical trait. A point that often is not written in the manuscripts. Secondly because our Ḥanash is Ḥanash b. ᶜAbd Allah al Ṣanᶜānī (from Ṣanᶜā', near Damascus), one of the *tābiᶜūn*, credited with specific knowledge in the field of astronomy and astrology at the time of the Islamic conquest of Spain. Moreover, both flourished around 835, and while Ḥabash was a renowned astronomer in the East, Ḥanash had a high reputation as astrologer and talismanist in al-Andalus. On the other hand, when astrological and astronomical texts reached al-Andalus from the East, Ḥabash al-Ḥāsib was not well known in al-Andalus while Ḥanash was well known as an astronomer as well, having determined the azimuth of the qibla for the *jāmiᶜ* mosque of Zaragoza. His association with talismans also links him to the subject of trepidation[68].

Amongst the authors who express doubts about the accession and recession theory, al-Baqqār mentions the following :

- *Zīj* and *Kitāb al-qirānāt* by Jaᶜfar b.M. al-Balkhī al-maᶜrūf bi-Abī 'l-Maᶜshar. [According to al-Hāshimī, Abū 'l-Maᶜshar in his *zīj al-hazārāt* talks about trepidation but does not agree with the models of Theon, Yaḥyā b. Abī

66. M. Viladrich, *El " Kitāb al-ᶜamal bi-l-asṭurlāb " (Llibre de l'us de l'astrolabi) d'Ibn al-Samḥ*, Barcelona, 1986, 70 and " Una nueva evidencia de materiales árabes en la astronomía alfonsí ", in M. Comes, R. Puig, J. Samsó (eds), *De Astronomia Alphonsi Regis*, Barcelona, 1987, 105-116, esp. 114.

67. See J. Samsó, " Astrology, Pre-Islamic Spain and the Conquest of al-Andalus ", *Islamic Astronomy and Medieval Spain*, II, Aldershot, 1994, 89-90.

68. On Ḥanash see M. Marín, " Ṣaḥāba et tābi'ūn dans al-Andalus : histoire et légende ", *Studia Islamica,* 54 (1981), 25-36.

Manṣūr, al-Fazārī, Ḥabash or *al-Arkand*][69] Azarquiel, mentions also the *zīj* by Abū Maᶜshar, but does not talk about the *Kitāb al-qirānāt*.

- Muḥammad b. Jābir al-Battānī, who states that his doubts were aroused because the stars move from the spring equinox travelling different distances in equal times (which in fact corresponds to al-Battānī's *zīj* and indeed Azarquiel attributes the very same words to al-Battānī. Azarquiel's models just try to solve this problem)[70].

Amongst the authors that reject this theory, he mentions the following, although, curiously enough, Azarquiel says nothing about them.

- al-Hamdhānī in his *Kitāb sarā'ir al-ḥikma*[71],

- *Zīj al-ᶜAlūmīn* (Also found in Azarquiel's introduction to his *Book on the fixed stars*) [Maybe Ptolemy's *Handy Tables* ? The fact that in the rest of the book references to the *mumtahan tables* are frequent, and that in this text the only reference to this al-ᶜAlūmī is that its precession of 1°/100 years invalidates J.M. Millás' suggestion that we should interpret it as the *Tabulae Probatae*. Pointing also to this invalidation, although for other reasons, J. Ragep suggests that it could be another of Ḥabash's *zījes*[72]],

- Ptolemy,

- *Ahl al-mumtahan.*

Finally, al-Baqqār states that Azarquiel, after having observed the sun for 25 years, was the first one to give a geometrical and mathematical model to explain the accession and recession motion.

As we have seen, all of this is found also in other authors. Azarquiel himself in the introduction to his *Book on the fixed stars* offers a complete account in which this part of the text by al-Baqqār seems to draw on, in most parts it is reproduced almost word for word. The only additions by al-Baqqār are the authors who rejected the theory, as well as Ibn Waḥshiya, Hermes, the Aṣḥāb al-ṭalismāt and Ḥanash. As we can see, the authors using trepidation just for astrological means.

Ibn ᶜEzra (12th c.) in his *Book on the Foundations of the Astronomical Tables* offers a similar account in which, however, there is something rather strange. He distinguishes between those who accept the accession and recession motion, that is to say the oldest authors and those who discussed this theory, that is to say the authors of the *mumtaḥan* tables and their followers except Abencine. In fact, Ibn Sīnā in the *Shifā'* seems to describe the trepidation

69. D. Pingree, *The Thousands of Abū Maᶜshar, op. cit.*, 56.

70. See F.J. Ragep, " Al-Battānī, Cosmology and the History of Trepidation in Islam ", *op. cit.*, 271-273.

71. *Cf.* D. King, *Mathematical Astronomy in Medieval Yemen. A Biobibliographical Survey*, Malibu, 1983, 19-20.

72. See J.M. Millás, *Estudios sobre Azarquiel, op. cit.*, 276 and F.J. Ragep, " Al-Battānī, Cosmology and the History of Trepidation in Islam ", *op. cit.*, 281.

model of Ibrāhīm b. Ṣinān, without mentioning his name[73]. Amongst the ones who accept the theory, he distinguished between those who attribute this northward and southward motion to the motion of the poles, mainly the talismanists, and those who attribute it to the rotation of two circles in the head of Aries and Libra. All of them, however, according to our author agree in an amplitude of 8°. Except Azarquiel who considers this amplitude to be 10;40°. As far as we know, 10;40° cannot be attributed to Azarquiel for who P_{max}=10;24°. But P_{max}=10;40° is the figure found in Ibn al-Raqqām's *zīj al-Qawīm* (680h/1281-82) and in Abū-l-Ḥasan al-Qusunṭīnī (14th c.). Interestingly enough Ibn ᶜEzra wrote in the middle of 12th c. and the manuscripts of his *Book on the Foundations of the Astronomical Tables* used by Millás, where I found this statement, are from the 12th and 13th centuries so that we should probably consider the possibility of attribute this value to an author between Azarquiel and Ibn al-Raqqām. Another possibility is to think that he is referring to the *Liber de motu* value which is 10;45° attributing its authorship to Azarquiel, which is not new in Andalusian sources, specially Hebrew (for example in Isaac Israeli's *Yesod ᶜOlam*[74]).

Texts c3 and c4 are also very interesting from the point of view of the astronomers quoted, as we have seen when dealing with precession and obliquity of the ecliptic values.

CONCLUSIONS

We find in these texts the theory of trepidation applied to astrological predictions, based on the world-year concept according to which astronomical events imply important world disturbances.

It is obvious that we also find here an evaluation of the old trepidation parameters and models, criticized on the basis of new observations, but not reworked. Our authors simply determine the difference and add it, or suggest to do so, to the values when needed but they never try to re-examine the models or their theoretical basis. In this regard, they are very far from astronomers like Ibn Abī'l-Shukr, on whose observations they base their rejection of the old parameters and models, because they are merely interested in application, that is to say in the use of astronomy but not in its development. Nonetheless their accounts are useful to understand why and how a theory in use for centuries came to its end.

Besides, this is an small contribution to establishing the connections between al-Andalus and al-Maghrib traditions in astrology and astronomy, as

73. See F.J. Ragep, *Nasīr al-Dīn al-Ṭūsī's Memoir on Astronomy...*, II, *op. cit.*, 394.

74. On this attribution see also the bibliography in note 2. It is also worth mentioning that Ibn al-Ḥa'im in his *al-zīj al-Kāmil* attributes to Azarquiel besides the wellknown radius of the Head of Aries epicycle of 4;07,58°, another of 4;19,31 very similar to the 4;18,43 of the *Liber*.

well as the influence that Ibn Abī 'l-Shukr al-Maghribī's observations made in Damascus had in al-Maghrib.

Eclipses and comets in the Rawḍ al-Qirṭās of Ibn Abī Zar^{c}[1]

Mònica Rius

Introduction

Some Arabic historians bring in their chronicles astronomical data observed in past times because these phenomena used to cause a great social impact : they were sign of bad omen and they made the population hurry toward the mosques. We do not have to undervalue these information because it brings an interesting material to complement the astronomical and chronological works. In this sense, I pretend to continue the task initiated by Vernet and Stephenson[2]. The second one has been studying the eclipses from long time but he has not included Ibn Abī Zar^{c} as one of his sources.

Sources of Data

Authors

The current study compiles and analyses the meteorological phenomena cited in the *al-Anīs al-Muṭrib bi-Rawḍ al-Qirṭās fī Akhbār Mulūk al-Magrib wa-Tārīkh Madīnat Fās*, usually known as *Qirṭās*, written in A.D. 1326 by Ibn

1. The present paper is a result of research undertaken within a programme on " Astronomical Theory and Tables in al-Andalus in the 12th-14th centuries " sponsored by the *Dirección General de Investigación Científica y Técnica* of the Spanish *Ministerio de Educación y Cultura*.

2. J. Vernet dedicated, in 1981, an article to this topic and in which he picked up and enlarged the carried out work by Hammer-Purgstall and Francisco Codera. *Cf.* J. Vernet, " Algunos fenómenos astronómicos observados bajo los omeyas españoles ", *Revista del Instituto Egipcio de Estudios Islámicos*, 21 (1981-1982), 23-30 [reprint in *De ^{c}Abd al-Raḥmān I a Isabel II,* Barcelona, 1989, 251-258] ; Hammer-Purgstall, " Sur les étoiles filantes ", *Journal Asiatique*, 1 (1837), 391-393 ; F. Codera, " Datos acerca de cometas en dos historiadores árabes ", *Boletín de la Real Academia de la Historia*, 56 (1910), 364-370 ; F.R. Stephenson, *Historical Eclipses and Earth's Rotation*, Cambridge, 1997.

Abī Zar^c[3]. We know almost nothing about this historian, only that he lived in the Fez of the Banū Marīn. Even, we don't know the name of his works for a fact : it has been said that he had written the *Mafākhir al-Barbar*, also some Arabic scholars say that the *Dhakhīra Sannīya*[4], written in A.D. 1310, is one of his works. But, on the other hand, and against what some Western scholars had thought, the *Annales Regum Mauritaniae* is not anonymous but Ibn Abi Zar^{c}'s *Qirṭās*.

The *Qirṭās* covers five centuries of history of the Maghrib, from the eighth to the fourteenth century (A.D.) and Ibn Abī Zar^{c} is very methodical in his book : after each dynasty he introduces a chapter on remarkable records as death of important personalities, famines, floods, plagues, comets and eclipses.

Al-Nāṣirī al-Salāwī (19th A.D.) adds in his *Kitāb al-Istiqṣā'*[5] all the records quoted by this *Fāsī* historian so I have compared both together with the celestial phenomena recorded in the *Dhakhīra Sannīya.*

Geographical coordinates and Sources

Ibn Abī Zar^{c} is from Fez but it is clear that his sources should not necessarily be from the Maghrib. For the ancient times, he includes records from Andalusian sources like Ibn Ḥayyān (A.D. 987-1076) so the observations should have been made in Cordova. Then, I have computed the data locating the observer in two localities : Cordova and Fez[6]. The time difference between them is not very important, a few minutes, but the different latitude is enough to change the visibility of the same event in both cities[7].

CALENDRICAL REMARKS

We have to keep in mind that difference between first visibility and beginning of month added to differences in the beginning of civil and astronomical day could make the date has lack of precision. On the other hand, there are often confusions in the name of months. Finally, the day of the week, when is specified, could be easily confused, but it is, at the same time, useful to fix the

3. Alī b. Abī Zar^{c}, *al-Anīs al-Muṭrib bi-Rawḍ al-Qirṭās fī Ajbār Mulūk al-Magrib wa-Tārīj Madīnat Fās*, Rabat, 1396H./1976 ; Ibn Abī Zar^{c}, *Rawḍ al-qirṭās*, translation and notes by A. Huici Miranda, Valencia, 1964 (2 vols).

4. Ibn Abī Zar^{c}, *al-Dajīra al-Sannīya fī Tārīj al-Dawla al-Marīnīya*, Rabat, 1392H./1972.

5. Al-Nāṣirī al-Salāwī, *Kitāb al-Istiqṣā'*, Rabat, 1373H./1954, edition by Ga^{c}far & Muḥammad al-Nāṣirī.

6. E.S. & M.H. Kennedy, *Geographical Co-ordinates of Localities from Islamic Sources*, Frankfurt, 1987, 118 and 95.

7. In order to contrast and verify the astronomical data I have used the computer astronomical program *Canon of Eclipses* (C.D. Eagle, Willmann-Bell, Inc., 1992), the *EZCosmos*, Astrosoft Inc., 1992 and the *Canon der Finsternisse* of Theodor R. Oppolzer (reprinted in New York, 1962), in the date which corresponds to each case. All three give Universal Time (Greenwich) and that means that the local time for Fez and Cordova would respectively be 20 and 18 minutes earlier (because the longitude of Fez is 5° and Cordova 4° 46' W of Greenwich).

correct date (in the sense that sometimes the weekday allows us to correct the month or the year).

We should remember that, since the Muslim calendar is lunar, all solar eclipses will take place at the end of the month (28/29, when the Moon is new), while all lunar eclipses will occur in the middle of the month (13/15, in the full Moon).

ECLIPSES

The main characteristic of eclipses (in front of comets or aurora borealis) is that they can be exactly computed. Although we must not expect scientific exactitude in historical reports, they offer a far from negligible amount of information : for example, they refer to the magnitude. They are usually total eclipses so the expressions are like " the Sun (or Moon) was totally eclipsed ", " in that year there was a great eclipse ", or " all the disk was hidden ". Only in one record (n° 11) it is said that the Sun was eclipsed in its two thirds. Time and duration are frequently expressed, for example, in relation to the rise or setting of the eclipsed item, to a prayer *(i.e.* " after that prayer ") or to the *zawāl*. Ibn Abī Zarᶜ, as a historian, repeats the record and doesn't care about visibility : obviously it is not the same, in this sense (and specially if it is a solar eclipse) to be in Baghdad or in Fez, but it is nevertheless true that all the records could be observed in the Maghrib.

Solar eclipses

The seven solar eclipses have been located although, some of them, had small errors of date or hour. In those cases in which an eclipse did not take place in the date established by our source, the weekday (which, unfortunately, is not included in all the records) helps to establish the correct date.

Record n° 3 : Wednesday, 29 of Shawwāl of the 289 A.H. (October 6th, 902 A.D.) That day there was no eclipse. A similar account is given by Ibn Ḥayyān in the *Muqtabis* for Wednesday 29 of Shawwāl but ten years after (299 A.H., June 17th, 912 A.D.). Indeed, there was a total eclipse of Sun that day which begun in the late afternoon and continued until the sunset. Immediately after the maximum phase of the eclipse the Sun was partially seen again during, approximately, 20 minutes. It is curious, that Ibn Abī Zarᶜ dramatised the story adding the fact that people made a double Maghrib prayer. Perhaps he took this record from another Andalusian source or simply thought that after a " second sunset " there would have been a " second Maghrib prayer ".

Record n° 4 : 28 Rajab 355 A.H./July 20th, 966 A.D. There was a partial solar eclipse (magnitude 0'70 for Cordova and 0'65 for Fez) from 5; 04 p.m. to 6; 55 p.m. but this does not agree with the hour of the text. The 29 of Rajab of the following year (356 A.H./July 10th, 967 A.D.) there was a total solar eclipse and the Sun rose eclipsed in Cordova. The second one is, then, the correct date.

Record n° 5 : As Ibn Abī Zarc says, there is a confusion in dates, but the correct one is that given by Ibn Muzayn[8] (and not the one offered by Ibn Abī-l-Fayyād[9]). The text seems to indicate that the comet was eclipsed by the Sun. There may be a confusion between two different reports : the first one refer to the solar eclipse (which occurred in October 21st 990 A.D./28 Rajab 380 A.H.) and the second one would be the report on a comet.

Record n° 6 : I have not found eclipses in the year 382 A.H. The nearest one corresponds to the 28th Jumadà II of the following year (383 A.H./August 20th A.D. 993, Sunday). Al-Maqrizī (A.D. 1367 A.H./1442 A.D.) reported this eclipse also. We have here another mistake in the date.

Record n° 9 : The text gives the year and the day but omits the month. However, in year 471 A.H. (1078-1079 A.D.), the 28th was a Monday in Rabīc I and Dhū-l-Ḥijja but a solar eclipse took place in Dhū-l-Ḥijja (July first, 1079 A.D.). Indeed, it happened at noon (*zawāl*) and it was a total eclipse (magnitude 0.98 for Fez and 0.97 for Cordova). This record is included also in the anonymous maghiribian chronicle *Mafājir al-Barbar* (13th Century)[10].

Record n° 11 : This and the next one are events contemporary to Ibn Abī Zarc. In spite of that, there is a mistake in the year (author's fault or copyist' ?). The 29th Rajab 693 A.H. (June 25 of 1294 A.D.) was a Friday (not Sunday) and even though there was a solar eclipse, it happened near midnight. I believe that the correct date is 29th of Rajab of the year before (July 5th, 1293 A.D.). That day was a Sunday, and there was a solar eclipse (magnitude 0'68 for Fez) although not at noon (*zawāl*) but at dawn. On the other hand, there is a remarkable thing, that is the author explains that the imam prayed with the people the Eclipse Prayer[11].

Record n° 12 : Indeed, on Tuesday, November 8th of 1295 A.D. (28 Dhū-l-Ḥijja 694 A.H.) there was a total solar eclipse (magnitude 1 for Fez) after midday.

Lunar eclipses

Our source includes only two reports of lunar eclipses (it is odd because they are more frequent, but, at the same time, they are less impressing). Since a lunar eclipse occurs exactly at full Moon the eclipsed Moon always rises very

8. s. XI. *Cf.* Pons Boigues, *Historiadores y geógrafos arabigo-españoles*, Madrid, 1898, 171.

9. Andalusian chronicler (d. 459 A.H./1066 A.D.). One of his works is the loosed *Kitāb al-cibār*. *cf.* Camilo Álvarez de Morales, " Aproximación a la figura de Ibn Abī-l-Fayyād y su obra histórica ", *Cuadernos de Historia del Islam*, 9 (1978-1979), 29-127.

10. Anonymous, *Mafājir al-Barbar*, Rabat, 1934, 55.

11. In the *Ṣalāt al-kusūf* there is no *adhān*, nor *iqāma* and nor *juṭba*. In the *ḥadīths* it seems that there is no difference between *kusūf* (solar eclipse) and *khusūf* (lunar eclipse). However, Ibn Abī Zayd (10th century) says that the Eclipse Prayer is only optionally with the lunar eclipse.

close to sunset or sets very close to sunrise. Although the Arabic term of a lunar eclipse is *khusuf*, Ibn Abī Zar^c uses the verb : *kusifa al-qamar*. Like in the case of the solar eclipses, the erroneous dates could be derived of miss readings or anomalous copies of the works. I wonder if such mistakes should make us look more carefully at dates given by historians, especially when related to other topics which cannot be computed.

Record n° 1 : The text does not give the day or the month, only the year, but there was a total eclipse (*kusifa al-qamar kullu-hu*, magnitude 1,206) during the night of November 4th 868 A.D. (14 Dhu-l-Qa'da 254 A.H.) from 1; 46 a.m. till moonset. There is a problem with the hour of the beginning of the eclipse. Probably with *awā'il al-layl* the author means that the Moon was eclipsed practically during all night.

Record n° 4 : There was no lunar eclipse the 14th of Rajab of 355 A.H. (July 6th, 966 A.D.). The nearest in date is the total eclipse (magnitude 1'20) of 13 Sha^{c}bān of the same year (August 3rd, 966 A.D.) from 11; 19 p.m. till moonset. It seems, then, that it is another mistake in the name of the month.

Comets

Since we could not compute the reliability of the records which explain that a brilliant object has been seen, classifying comets, meteors or other phenomena become a hard task. On the other hand, Arabic astronomers did not pay especial attention to this topic and historical chronicles acquire, if possible, greater importance. In any case, and in order to compute them, I have compared the records with Baldet's lists of comets[12].

Vocabulary

There is quite a number of Arabic terms which design this kind of celestial objects (*shihāb, kawkab waqqād* ; *najm abū-l-dhawā'ib* ; *nayyirāt*), even though it is not probable that Ibn Abī Zar^c, or his sources, make a rigorous use of them.

Record n° 4 : Tuesday, July 10th of 966 A.D. (18 Rajab of 355 A.H.). Possibly a meteor. The Arabic term used here is *shihāb*. Its interesting the comparison with the *laylat al-qadr* and, moreover, the mention that it was like the big column (*al-^{c}amūd al-^{c}aẓīm*), probably the column in the top of the minaret.

Record n° 5 : As we have seen in the Solar eclipse-section, the correct date is 380 A.H. Indeed, 23rd of Rajab of 380 A.H. (in the following year, the 23rd of Rajab is a Monday) was a Thursday and corresponds to October 16th of the 990 A.D. This phenomenon appears cited, for this date, in the Italian chronicle of

12. M.F. Baldet, " Liste générale des comètes de l'origine à 1948 ", *Annuaire pour l'an 1950 publié par le Bureau des Longitudes*, n° 433 (1950).

Varignana (*ca.* 1425)[13] : according to this chronicle it appeared first in the E and days later in the W. Also Baldet mentions comet number 508 for this year. Passing now to other matter, the text says that the comet is *ka-l-sawma'a al-'aẓima* (the great minaret). The determination makes us think that it is compared with a concrete minaret, possibly that of the Great Mosque of Cordova.

Record n° 8 : With such a short description is difficult to say if it is, in fact, a comet. Baldet however includes one for that year (n° 564).

Record n° 10 : Tuesday 12th of Sha'ban of 661 A.H. corresponds to June 21st of the 1263 A.D. It is comet number 656 of Baldet's list. In the *Dhakīra Sannīya* the comet was seen in the year 652 A.H. (1254-55 A.D.).

Supernovae

Record n° 7 : This record is an explanatory example of the corruption of an original account. In the Rabat's edition we have only one event for year 406 A.H. (1015/1016 A.D.) whereas Tornberg's edition (*Annales Regum Mauritaniae*)[14] gives us two events : one comet for the year 394 A.H./and another for 396 A.H. This version is clearly better because it agrees with the date of the 396 A.H./1006 A.D. supernova. Goldstein determined that the (second) observed phenomenon was the apparition of a supernova[15] and he collected different contemporary sources that refer the amazing event (obviously all of them classify it as a comet). Baldet includes a comet for the year 396 A.H. (n° 519) and denied that it was a nova. The text has, also, the enormous interest of alluding to the recurrence of comets already seen in the past. It has been said that, in an Arabic context, comets interested astrologers more than astronomers. It is true that, in general and following Aristotle[16], these phenomena were located in the sublunar sphere, that is to say, they were considered meteorological phenomena. Some astrologers very well known in Occident like Māshā' Allāh (d. *ca.* 199 A.H./815 A.D.), Ibn Abī Rijāl (d. *ca.* 431 A.H./1040 A.D.) and Ibn Riḍwān (d. 452 A.H./1061 A.D.) dealt with comets and gave several classifications according to their colour or form[17]. But it was Albumasar

13. R.R. Newton, *Medieval Chronicles and the Rotation of the Earth*, London, 1972, 680.

14. *Annales regum Mauritaniae..., ab Abu-l Hasan Ali Ben Abd Allāh Ibn Abi Zer' Fesano...*, in Carl Johann Tornberg (ed.), Upsala, 1839-1843 (2 vols). (Arabic text : I, 74 ; Latin translation : I, 99).

15. B.R. Goldstein, " Evidence for a Supernova of A.D. 1006 ", *The Astronomical Journal*, 70 (1965) 105-114 ; " The 1006 Supernova in Far Eastern Sources ", *The Astronomical Journal*, 70 (1965), 748-753. Goldstein identified it as the NGC 5882. Later, it was said that this NGC was not the corresponding SNR (SuperNova Remnant), but PKS 1459-41 (radioactive source). In 1976 a tenuous filamentous nebulous was detected in this area. *Cf.* D.H. Clark, F.R. Stephenson, *The Historical Supernovae*, Oxford, 1977, 114-139 ; R. Burnham Jr., *Burnham's Celestial Handbook*, New York, 1978, 1117-1122.

16. Aristotle, *Meteorologica*, I, 4.

17. L. Thorndike, " Anonymous Work in Sixteen Chapters Composed in Spain about A.D. 1238 ", *Latin Treatises on Comets. Between 1238 and 1368*, Chicago, 1950, 9-61.

(d. 272 A.H./886 A.D.)[18] the first Arabic author who said, like Seneca[19], that comets transcend the sphere of the Moon and, therefore, they can have an orbit[20]. He was not the only one who followed this road : other astrologers established norms in order to determine the position of the comets[21]. Anyhow, it is surprising to see Ibn Abī Zarᶜ, an historian, sharing this idea.

Aurora Borealis

Last but not least, there is a description of an Aurora Borealis. As the source is probably from Cordova (φ 37.53 N), we have here evidence that such phenomena could be seen in low latitudes. The lower is the latitude, the possibility of seeing it becomes smaller, but we can offer some interesting exceptions : Marbella (φ 36.31 N) in year 1582[22] ; Barcelona (φ 41.25 N) in 1938[23] ; or Lima (φ 12.06 S) in 1989[24].

Since we have found no translation for the term " Aurora Borealis ", Professor Vernet and I propose that *ḥumra* be considered as the Arabic specific term.

Conclusion

The study of these phenomena brings us a set of interesting data. For example, it allows us to establish the sources of Ibn Abī Zarᶜ for the former periods (Ibn Ḥayyān, Ibn Muzayn, Ibn Abī-l-Fayyaḍ, Ibn al-Athīr). We can also find later historians who include his records (Aḥmad al-Nāṣirī al-Salāwī).

Finally, two remarks. The first one moves in the field of the historiography, we can infer from the records of phenomena that can be computed if the historian is actually accurate and if he works with a rigorous method. The second one, with regard to the history of science, is that we have seen that the data offered by the historians complete the material included in the astronomical works.

18. L. Thorndike, " Albumasar in Sadan ", ISIS, 45 (1954), 22-32 ; W. Hartner, " Tycho Brahe et Albumasar ", *Oriens-Occidens*, I (1968), 496-507.

19. Seneca, *Naturales Quaestiones. Liber septimus (sextus). De cometis.*

20. Eclipses and comets are related to *al-Kayd*. This mythical star (related to the periodicity of eclipses) together with *al-Wardī*, are considered by Ibn al-Raqqām (s. XIV), as *ḏ āt al-ḏ awā'ib*. *Cf.* Ibn al-Raqqām, fol. 66, ms. 260 Bibliothèque Générale de Rabat ; Willy Hartner, " Le problème de la planète Kaïd ", *Oriens-Occidens*, 1 (1968), 268-286.

21. E.S. Kennedy, [" Comets in Islamic Astronomy and Astrology ", *Studies in the Islamic Exact Sciences*, Beirut, 1983 (reprint of *Journal of Near Eastern Studies*, 16 (1957) 311-318)] studies *zijs* of the movement of *al-Kayd* and other comets. Ibn Hibintā (s. IX), for example, says that there are seven stars of this kind en the sphere of the sun.

22. Ms. 1252, fol. 19, Biblioteca Nacional, Madrid. I thank to Prof. Vernet and Prof. Codina their help in the identification of this phenomenon.

23. Academia de Ciencias y Artes, Observatorio Fabra, *Estadística de nubes y neblinas*, p. n° 13953, Barcelona, 1938.

24. *Journal of British Astronomical Association*, 105 (1995), 179-181.

Appendix : Ibn Abī Zar^c- Rawḍ al-Qirṭās

١- (p. 96) وفي سنة أربع و خمسين ومئتين كسف القمر كله من أوائل الليل حتى أصبح ولم ينجل

1.- Year 254 A.H. (January 1st to December 19th, 868 A.D.) : the Moon was eclipsed totally from beginning of the night until dawn, and it didn't shine.

٢- (p. 97) وفي سنة ستّ وستّين ومئتين كانت بالسماء حمرة عظيمة من أوّل الليل الى آخره ولم يعهد قبل ذلك مثله وذلك في ليلة السبت لتسع بقين من صفر من السنة المذكورة (١١ أكتوبر ٨٧٩ م).

2.- Year 266 A.H. (August 23rd, 879 A.D. to August 11th, 880 A.D.) : there was in the sky a great red splendour, from beginning to the end of the night ; such a thing had never been before ; it happened the night of Saturday 19th of Safar of the aforementioned year (January 25, 880 A.D.)

٣- (p. 97) وفي سنة تسع وثمانين ومئتين كان الكسوف العظيم للشمس. كسفت الشمس كلها وذلك في يوم الاربعاء التاسع والعشرين من شوال من السنة المذكورة (الثلاثاء ٦ أكتوبر ٩٠٢ م) وذلك بعد صلاة العصر. فبدر أكثر الناس بالاذان في المساجد للمغرب فغاب القرص كله وظهرت النجوم ثم انجلت بعد ذلك وعادت مضيئة قدر ثلث أو نصف ساعة ثم غربت وأعاد الناس الاذان والاقامة والصلاة.

3.- Year 289 A.H. (December 16th, 901 A.D. to December 4th, 902 A.D.). There was a great solar eclipse ; all the Sun was eclipsed on Wednesday, 29th of Shawwal (June 18th, 912 A.D.) of the aforementioned year, after the Afternoon Prayer (*'asr*) and many people went to the mosques when was announced the *maghrib* ; all disk was hidden, and the stars appeared ; then, the Sun reappeared and the light returned during 20 minutes or half an hour, at the end of which the Sun set, and call of the *mu'adhdhin* and the prayers were repeated.

٤- (p. 100) وفي سنة خمس وخمسين وثلاثمائة (...) وفي ليلة الثلاثاء الثامن عشر من شهر رجب الفرد منها (١٠ يوليوز ٩٦٦ م) ظهر في البحر شهاب ثاقب مائل كالعمود العظيم أضاء الليل لسطوع نوره وشُبِّهت بليلة القدر وقارب ضوءها ضوء النهار وفي هذا الشهر كسفت الشمس والقمر. كسف القمر ليلة أربع عشرة منه، وطلعت الشمس مكسوفة في اليوم الثامن والعشرين منه.

4.- Year 355 A.H. (966 A.D.). A brilliant and inclined meteor, like the big column, appeared in the sea during the night of Tuesday 18th Rajab (July 10th). It illuminated the darkness with the splendour of its light and it looked like the night of *al-qadr*. Its light was almost the light of the day. In this month, the Sun and the Moon were eclipsed ; the lunar eclipse was in the night of the 14th (July 6), and the Sun rose eclipsed the 28th (July 20).

٥- (p. 114) وفي سنة أحدا وثمانين وثلاثمائة (...) وفيها ظهر نجم في السماء وذلك في ليلة الخميس الثالث والعشرين من شهر رجب من العام المذكور (٥ أكتوبر ٩٩١ م) كان هذا النجم في رأي العين كالصومعة العظيمة طلع من جهة المشرق وتهافت جريا من بين المغرب والجوف وتطاير منه شرر عظيم فزع الناس منه ودعوا الله تعالى في صرف مكروهه عنهم وكُسف بالشمس في آخر هذا الشهر، قاله ابن الفياض في كتاب القبس، وقال ابن مزين: كان ذلك فى سنة ثمانين وثلاثمائة.

5.- Year 318 A.H. (March 20th, 991 A.D. to March 8th, 992 A.D.) […] during the night of 23rd Rajab of that year (October 5th, 991 A.D.), a star which looked like the great minaret, visible to the naked eye, appeared in the sky. It rose in the East, it moved fast between the West and the North, and it discharged big sparks ; the people feared it, and prayed to God to be delivered of its evil. The Sun was eclipsed at the end of the same month according to Ibn al-Fayyad in his book *al-Qabas* although Ibn Muzayn says that this happened in year 380.

٦- (p. 116) وفي سنة اثنين وثمانين وثلاثمائة (...) وفيها كان الكسوف الذي أذهب القرص كله

6.- Year 382 A.H. (March, 992 A.D. to February, 993 A.D.) […] an eclipse that hid all the solar disk took place.

٧- (p. 117) وفي سنة ست وأربعمائة طلع الكوكب الوقاد في السماء وكان عظيم الجرم كثير الضياء يطلع في الافق الشرقي قال بعض المنجمين ان ذلك النجم يعرف بالمضيء من ذوات الاذناب وهو نجم هائل المنظر مفرط الضياء شديد الاضطراب والحركة له ذوائب أربع محددة الاطراف وهو أحد النيرات الاثني عشر التي ذكرها الاوائل ورصدها علماؤهم في المدة الطويلة وزعموا أنه لا يظهر منها كوكب الا لقضية يحدثها في العالم والله أعلم بغيبه وكان ابتداء ظهوره في أول شعبان من سنة ست و أربعمائة المذكورة طلع أول ظهوره قبل وقت المغرب ثم تقهقر الى أن طلع في الليل وأقام مدة من ستة أشهر ثم غاب وكان بهاذه السنة رياح كثيرة وبروق خاطفة ورعود قاصفة دون مطر.

7.- Year 406 A.H. (June 21st, 1015 A.D. to June 9th, 1016 A.D.) a brilliant star appeared in the sky. It had a great body and was very splendiferous, it rose in the East, and one of the astrologers says that this star is known as the shining one of those who have tails (*Niazak*). This star has a terrible look and an extraordinary shine, it shaken and strongly moved fast, it had four pointed tails and it was one of the twelve stars that were mentioned by the ancients.

The sages observed them during a lot of time and they believed that these stars do not appear unless an event that affects the whole world is going to take place. God knows their secret. It began to appear the first of Sha'ban of the cited year 406 (January 14th, 1016 A.D.) ; it rose in its first apparition before the evening ; then, it was delayed until it rose at night and it stayed during six months, then disappeared.

This year there was a lot of wind, of violent lightning's and of noisy thunders without rain.

٨- (p. 167) و في شهر ذي الحجة من سنة سبع وستين ظهر النجم المعكف بالمغرب

8.- […] In the month of Dhu-l-hijja of the year 467 A.H. (July 25th to August 22nd, 1172 A.D.) a star of tails appeared in the West.

٩- (p. 168) ... وفي سنة احدا [كذا] وسبعين وأربعمائة كسفت الشمس يوم الاثنين عند الزوال في اليوم الثامن والعشرين وهو كسوف الشمس العظيم الذي لم يعهد قبله مثله.

9.- Year 471 A.H. (July 14th, 1078 A.D. to July 3rd, 1079 A.D.) : the Sun was eclipsed at noon of Monday 28th, and it was a great eclipse. No one had ever seen one alike.

١٠- (p. 402) وفي سنة احدا [كذا] وستّين وستّمائة (...) وفيها كان ظهور النجم أبي الذوائب وذلك يوم الثلاثاء الثاني عشر من شعبان من السنة المذكورة وبقي يطلع كل ليلة في وقت السحر مدة شهرين.

10.- This same year (661 A.H./1263 A.D.) appeared a comet on Tuesday 12 Sha'ban (June 23rd, 1263 A.D.) and it appeared every night, near dawn, during two months.

١١- (p. 409) وفي سنة ثلاث وتسعين وستمائة (...) وفيها كسفت الشمس فغاب ثلثا قرصها وذلك يوم الاحد قرب الزوال في التاسع والعشرين من رجب وصلا بالناس صلاة الكسوف الخطيب محمد بن أيوب (أبي الصبر [كذا]) بجامع القرويين حتى انجلت.

11.- Year 693 A.H. (1294 A.D.) […] the Sun was eclipsed in its two thirds on Sunday, near noon of the 29 Rajab (June 24th, 1294 A.D.).

The preacher Abu 'Abd Allah b. Abi-l-Sabr made the eclipse prayer with the people in the mosque of al-Qarawiyyin until the light returned.

١٢- (p. 409) ثم دخلت سنة أربع وتسعين وستمائة (...) وفيها كسف بالشمس الكسوف العظيم الذي غاب القرص كله ورجع النهار ليلا كما يكون بين العشائين بدت نيرات النجوم وعظم الامر لو لا ما تدارك الله سبحانه بسرعة الانجلاء وذلك بعد صلاة الظهر من يوم الثلاثاء الثامن والعشرين من ذي الحجة من سنة أربع وتسعين المذكورة.

12.- Year [694 A.H.] […] a great solar eclipse took place. All the solar disk was eclipsed and the day became night, like between the two twilight's ; the stars began to shine and the situation would have become worse if God had not helped to recover the light.

This happened after the Noon Prayer or the 28th Dhu-l-hijja (November 8th, 1295 A.D.) of the aforementioned year 694.

SOLAR ECLIPSES

Qirṭās date	Computed date	Magnitude	Duration (UT)		
			Begins	Maximum	Ends
Wednesday 29 Shawwāl 289 AH	Wednesday 29 Šhawwāl **299** AH (June 18th 912 AD)	0.98	6 h 20' p.m.	7 h 13 ' p.m.	8 h 25' p.m.
28 Rajab 355 AH	Tuesday 28 Rajab **356** AH (July 10th 967 AD)	0.71	4 h 25' a.m.	5 h 29' a.m.	6 h 32' a.m.
28 Rajab 380 AH	Thursday 28 Rajab 380 AH (October 21 990 AD)	0.53	7 h 58' a.m.	10 h 33' a.m.	11 h 09' a.m.
year 382 AH	Sunday 28 Jumādà II **383** AH (August 20th 993 AD)	0.82	5 h 4' a.m.	6 h 10' a.m.	7 h 16' a.m.
Monday 28 471 AH	Monday 28 **Dhū-l-Ḥijja** 471 AH (July 1st 1079 AD)	0.97	0 h 17' p.m.	1 h 37' p.m.	2 h 56' p.m.
Sunday 29 Rajab 693 AH	Sunday 29 Rajab **692** AH (July 5th 1293 AD)	0.68	7 h 27' a.m.	8 h 42' a.m.	9 h 57' a.m.
28 Dhū-l-Ḥijja 694 AH	Monday 28 Dhū-l-Ḥijja 694 AH (November 8th 1295 AD)	1	0 h 36' p.m.	3 h 15' p.m.	1 h 56' p.m.

LUNAR ECLIPSES

Record *Qirṭās*	Oppolzer num.	*Qirṭās* date	Computed date	Magnitude	Max. eclipse (U. T.)
1	3206	*year 254 AH*	14 *Ḏū-l-Qa^cda* 254 AH (Thursday, Nov 4th 868 AD)	1.206 (total)	4 h 28' a.m.
4	3358	*14 Rajab 355 AH*	13 *Ša^cbān* 355 AH (Wednesday, Aug. 4th 966 AD)	1.201 (total)	2 h 20' a.m.

COMETS

Qirṭās date	Duration	Arabic term	Description
Tuesday, 18 Rajab 355 AH (July 10th 966 AD)	1 night	*shihāb*	Brilliant, inclined meteor, big column; like the *laylat al-qadr*.
Thursday 23 Rajab 380 AH (Oct. 16th 990 AD)	1 night	*najm*	Star like the great minaret, big sparks.
Dhū-l-ḥiŷŷa 467 AH (July/August 1075 AD)	--	*najm muᶜakkaf*	Star of tails.
Tuesday 12 Shaᶜbān 661 AH (June 23rd 1263 AD)	2 months	*Abī-l-dhawā'ib*	Star of tails

AURORA BOREALIS

Qirṭās date	Description	Duration
Saturday 19 Ṣafar 266 AH (August 23rd 879 AD)	*Great red splendour*	*All night*

SUPERNOVAE

Qirṭās date	Computed Date	Duration	Description
1 Shaᶜbān 406 AH (Rabat edition) 1 Shaᶜbān 396 AH (Tornberg edition)	year 396 AH (1006 AD)	6 months	*dhawāt al-dhawā'ib*

Contradictions in Taghwim recent past and present

Mashallah Ali-Ahyaie

Introduction

The Islamic calendar, especially for ritual purposes is based on lunar cycle. Each of the twelve lunar months of the lunar calendar, or lunar Higri, begins the day after the sighting of the youngest lunar crescent is declared. As such the lunar months are commenced and ended in the evening of the first sighting of the youngest lunar crescent.

The process of the first sighting of the new Moon run into confusion and controversy in certain cases, because both physical factors and human errors in sightings are involved in this regard. The controversy usually culminates in the months of Ramazan (fasting month) and Zil-Hajjah (the month of the pilgrimage to Mecca or Hajj ceremony).

To cope with this problem, Fiqh (Religious Jurisprudence) authorities and Muslim scholars have done a great deal of research which made available various guide lines and schools of thoughts, however, the discrepancies in lunar calendar are still existing, as it is proved in this article.

Ficq guide lines

Fiqh authorities over centuries have adapted two major doctrines in fixing the first day of the lunar months, namely, Ekhtilaf al-Mataal'i (difference in horizons), which allows each locality to make its own crescent sightings and Ettehaad al-Matteal'i (unity of the horizons)[1], which accept other countries' sighting of the new Moon.

1. To be more specific, horizon in Arabic and Persian is Ofoq/Ofogh. Mataal'i is the plural form of Matla' which means the rising place of a star or a planet on the horizon. Mataal'i is also used in relation to right ascension.

However, both doctrines are valid and respected under Islamic laws, but the first doctrine is in concordance with the following sayings (Haddith) of the Prophet Mohammad (peace be upon him) : " Fast, when she (the new Moon) appears, and cease fasting when she reappears ".

In addition to the above mentioned two doctrines, there are apparently other doctrines in practice nowadays, in certain Islamic countries, in which as an example, the lunar months are started when the Moon's conjunction with the Sun takes place (no matter at what time), as it is revealed in this study. Furthermore, in contradiction to those who only rely on the first sighting of the youngest lunar crescent, whom are referred to as Ashab al-Roayah (Friends of Sighting), there are others who rely on numbers and tables, whom are named as Ashab al-Adad (Friends of Numbers).

In this connection, Al-Biruni (11th century) criticizes a sect who have unreliable astronomical tables and calculations, by means of which they compute their months, and derive the knowledge of their fast days[2].

Muslim astronomer's achievements

The prediction of the first visibility of planets before sunrise or after sunset has roots in history. In this respect, some criteria are included in Ptolemy's works from the second century A.D. Ptolemy did not mention the first visibility of the lunar crescent but the subject was fully examined by Muslim astronomers soon after the dawn of Islam in the 7th century.

The moonset-lag time (the difference in the time of moonset and sunset) as a criterion for the determination of lunar visibility used by Hindus, was adapted by Muslim astronomers.

The lunar crescent visibility theory of Muslim scholars is included in many extant *Zijes*[3] (Astronomical Handbooks).

Crescent visibility guide line

Sighting any far-away object depends on two major physical parameters among other ones. The first lunar crescent visibility is only possible after some time is passed from the conjunction of the Moon with the Sun. After the conjunction, the Moon elongates from the Sun in the eastward direction. This results more angular distance between two bodies in the sky which gives the opportunity to the Moon remaining above western horizon after sunset for a period of time, when the darkness gradually falls, which this enables to sight the faint crescent with the naked eyes for the first time after the conjunction.

2. *The Chronology of Ancient Nations*, an English version of Arabic text of the Athar ul-Bakiya, of Al-Biruni, translated by C. Edward Sachau, Frankfurt, 1969, 76-81.

3. E.S. Kennedy, *A Survey of Islamic Astronomical Tables*, Philadelphia.

This means that longer period of the moonset-lag time gives more opportunity to the evening sky above the western horizon, getting darker which in turn gives a better contrast to the faint new Moon in relation to the surrounding sky.

The contrast parameter was probably indirectly known to Muslim astronomers, because for the prediction of the crescent visibility, they used to calculate a parameter called Bod-e-Moaddal (the adjusted distance or adjusted celestial longitude) which is almost less than the moonset-lag time within of about half a degree or two minutes of time (one revolution of the earth or 360 degrees equal to 24 hours).

Regarding to the dimension, it could easily be proved that the phase of the Moon, *i.e.*, the ratio of the visible portion of the Moon's disk to the whole illuminated disk is almost equal to (1-cos e)/2, in which " e " is the elongation of the moon (difference between the celestial longitudes of the Moon and the Sun). This means that by the increase in the Moon's elongation, the Moon phase or the crescent dimensions increase as well.

This explains why the Muslim astronomers, for the prediction of the lunar crescent visibility, used to calculate another parameter, called Bod-e-Seva (separated distance or separated celestial longitudes), vis-à-vis Bod-e-Moaddal.

Bod-e-Seva is the Moon's elongation from the Sun at the time of sunset on the twenty-ninth day of the Islamic lunar months.

Based on the experiences available in mid-latitudes, the minimum set limit for the above mentioned two factors (Bod-e-Moaddal and Bod-e-Seva) is almost about ten degrees (40 minutes of time), when the Moon has positive celestial latitude, *i.e.*, being to the north of the ecliptic. Both factors have to meet the set value together, otherwise, the crescent visibility is not predicted or at least it is uncertain. In the meantime, the new Moon's altitude at the time of sunset is an important parameter as well which is considered.

So far, the author has not found any reference in that the exact Moon's location in the western evening sky or its orbit after the sunset, had been determined in the *Zijes*.

This drawback was a major problem of the interested crescent viewers, since they had to search the entire western evening sky for the faint crescent, soon after the sunset, in a limited period of time, which even today in many cases, the results run into false conclusions.

The lack of fixing the new Moon's orbit in the evening sky after sunset is surprising, because the Muslim astronomers were well aware of fixing the ecliptic, since they used to calculate the complement of the angle between the horizon and the ecliptic which is called Arz-e-Ighlim-e-Roayat (the latitude of the visible climate), in the process of the calculation of Bod-e-Moaddal.

Furthermore, fixing the celestial equator should have been not a difficult task because the angle which it makes with the horizon is the complement of the geographical latitude of the locality.

To cope with the above mentioned drawback, it was tried in the series of articles, written by the author in the recent years, to fix the Moon's orbit during the moonset-lag time period, as it is illustrated in the case studied in this article.

CONTRADICTIONS IN TAGHWIM IN THE RECENT PAST

The travel accounts of the past, written by certain authorities, are very good sources to verify the authenticity of the many young crescent sightings claimed or recorded. One of these diaries which was written by S.M. Taleghani in 1952 when he made Hajj Pilgrimage, over four decades ago, is a good example in this regard[4].

Taleghani left Tehran on Sunday evening, Aug. 10, 1952 for Hajj Pilgrimage and came back to Tehran on Wednesday, Oct. 8, 1952, almost after two months.

In that particular year, there was two days difference on the date of the onset of Zil-Hajjah 1371 (A.H.) between the official calendars of Iran and that of Saudia Arabia.

This difference was a major concern for the Iranians performing Hajj ceremony in that year, as Taleghani have described here and there in his book. With other fellow travellers, he tried to sight the new Moon, on their way, in Beirut, Lebanon, on Friday evening of Aug. 22, 1952, but it was not feasible due to hazy weather at the seaside.

While on board from Beirut to Jidda, probably on Saturday, Aug. 23, 1952, he saw the new Moon but doubted whether it was the second, third or fourth night of the month of Zil-Hajjah.

While in Mecca he wrote that it was with much surprise that it was suddenly announced that the new Moon had already been sighted on Wednesday evening, Aug. 20, 1952 and the month of Zil-Hajjah 1371 (A.H.) was started on Thursday, Aug. 20, 1952 in Saudi Arabia.

This was hard to be believed by the Iranian present there, since according to their official calendar, the month of Zil-Hajjah 1371 (A.H.) had to be started two days later on Saturday, Aug. 23, 1952.

THE COMPARISON OF TAGHWIM

It is now a proved fact that any crescent visibility claim in the neighboring countries has a great impact on folks. Especially the one from Saudi Arabia is

4. S.M. Taleghani, *Hajj*, Tehran. A detailed astronomical survey of this book is carried out by the author and will be appeared in Persian, in " Farhang ", in a special issue of *History of Science Journal* published by the Institute for Humanities and Cultural Studies, Tehran, Iran.

taken for a grant, no matter right or wrong because of the holiness of Mecca and Medina.

To cope with this problem, it is wise that any prediction of the lunar crescent visibility in different countries to be observed vis-à-vis the one in Saudi Arabia (Mecca). As such, in this study the beginning of Zil-Hajjah 1371 (A.H.) in Iran is compared with the corresponding date in Mecca and Rabat in Morocco.

Rabat is also chosen because it is located to the extreme west in the Islamic countries of North Africa, almost with a small difference in geographical latitude in comparison to Tehran (1 degree and 39 minutes) (see Table 1).

Besides, the standard time in vogue in Morocco is based on the Greenwich meantime, *i.e.*, Universal Time. So that, the standard time difference between Iran and Morocco is 4.5 hours in spring and summer but 3.5 hours in autumn and winter (see Table 1).

The difference in longitude between Tehran and Rabat is quite high equal to 58 degrees and 18 minutes (3 hours, 53 minutes and 13 seconds) which is almost four hours (see Table 1). As such, if it is predicted that the lunar crescent could not be sighted in Rabat, then the case for Tehran should not be argued. It should be noted that the approximate equality of the latitudes of the two places compared is very important in this comparison and the place being located to the extreme west is not the only requirement.

The Conjunction of the Moon with the Sun on Wednesday, Aug. 20, 1952

The Moon is said to be in conjunction with the Sun, when it could be observed from the Earth, almost in the same direction as the Sun.

Twelve Moon's conjunctions take place in every Islamic lunar year, one around the end of each lunar month.

Astronomically, at conjunction the elongation of the Moon (the difference between the celestial longitudes of the Sun and the Moon) is zero. At conjunction, the angular distance of the Moon and the Sun is only few degrees and it is equal to the celestial latitude of the Moon which its maximum limit could be about ±5 degrees and 8 minutes (the obliquity of the lunar orbit to the ecliptic), which its minimum magnitude could be zero. Generally speaking when the celestial latitude of the Moon at the time of conjunction is close to zero, a partial or total eclipse may occur.

The conjunction of the Moon with the Sun around the end of Zil-Qaadeh 1371 (A.H.) took place on Wednesday, Aug. 20, 1952 (Zil-Qaadeh 28, 1371 (A.H.) according to the then Iranian calendar) at 03:20:13 p.m. U.T.

This was at 06:50:13 p.m. in Iran, 06:20:13 p.m. in Saudia Arabia and 03:20:13 p.m. in Morocco, according to the standard times of today in these countries.

MOONSET-LAG TIME ON WEDNESDAY, AUG. 20, 1952

The calculated moonset and sunset times in Rabat, Mecca and Tehran are listed in Table 1 for Wednesday, Aug. 20, 1952. As seen the Moon had set before the sunset in the three cities. So that it was absolutely impossible the new Moon to be sighted in the evening sky on Wednesday, Aug. 20, 1952. As a result the onset of Zil-Hajjah 1371 (A.H.) could have not been on Thursday, Aug. 20, 1952 in Saudi Arabia and even in the region, as it was then claimed.

LOCATION OF THE LUNAR CRESCENT ON WEDNESDAY, AUG. 20, 1952

However, it had been absolutely impossible to sight the new Moon in the evening sky on Wednesday, Aug. 20, 1952, but for the sake of clarification of the case, the estimated apparent position of the lunar crescent in the evening sky on this day is outlined in Fig. 1 for Mecca and Tehran and in Fig. 2 for Rabat.

Also in Table 2 the corresponding measures of the arcs of great circles and spherical angles shown in Fig. 1 and 2 are tabulated.

Since the astronomical corrections are not taken into account, so that, the topographical results could be in error about one degree or so. One may also note, however, the Moon had set before the sunset in Rabat on Wednesday, Aug. 20, 1952, as shown in Table 1, but in Fig. 2 it is above the horizon at the apparent sunset. The reason for this difference is that, the astronomical corrections are taken into consideration in the data given in Table 1 while it is not considered in Fig. 2.

In Fig. 1 and 2, two main great circles of reference, namely celestial equator and ecliptic are shown and fixed for this date.

The spherical angle FWT, which the celestial equator makes with the horizon on a locality is always equal to co-latitude (90 degrees minus geographical latitude). Point F is the first point of Libra which is the intersection of the ecliptic and the celestial equator at autumnal equinox. The plane of the ecliptic was then inclined at an angle of about 23 degrees and 26.73 minutes to the plane of equator in 1952 (mean obliquity of the ecliptic). In Fig. 1, angle (WFS) is the obliquity of the ecliptic.

The ecliptic cuts the eastern and western horizon at two points at any time, called ascending and descending points respectively which are named as *Talia* and *Ghareb* in Islamic *Zijes*. Ascending and descending points are diametrically opposite to each other on the horizon and their locations are not fixed like cardinal points, but are changing as time goes on.

Point S, the center of the Sun, is located on the descending point at the time of sunset, because the Sun is always positioned on the ecliptic. The measure of the angle of the ecliptic to the horizon (angle FST) also changes with time. The complement of this angle is called the latitude of visible climate (Arz-e-Eqlim-e-Royat) in the Islamic *Zijes*. Arcs (MM') and (SS') are the declination of the Moon and the Sun respectively. Arc (MH) is the altitude of the Moon at the time of sunset. Arc (SM") is the elongation of the Moon at the time of sunset which is called Bod-e-Seva in Islamic *Zijes* when it is calculated on the twenty-ninth day of the Islamic lunar months. Arc (MM") is the Moon latitude.

In the same manner, in Fig. 3 in connection to Table 3, the position of the lunar crescent is determined, at the time of setting for Mecca and Tehran.

It is quite evident from the measures given in Tables 2 and 3 that the Moon was so close to the Sun on Wednesday, Aug. 20, 1952 that it was absolutely impossible the new Moon had been sighted in Mecca, Rabat and Tehran.

The location of the moonset for Rabat is illustrated in Fig. 4 in reference to Table 3 as well.

Solar eclipse on Wednesday, Aug. 20, 1952

As noted earlier, solar eclipse generally occurs at the time of conjunction when the celestial latitude of the Moon is close to zero. In Table 2, the Moon's latitude is arc (MM") which its measure was almost close to zero on Wednesday, Aug. 20, 1952 at the time of sunset. This implies that a solar eclipse was to take place or had taken place on this date.

In fact, an annular eclipse of the sun had taken place the same day on Wednesday, Aug. 20, 1952 at 3:20 p.m. U.T. almost 27 minutes before the sunset in Mecca, which this eclipse had not been observed in Mecca, of course.

The duration of this eclipse was 6 minutes and 40 seconds which the path of annular eclipse was experienced in South America in Peru, Argentina and Bolivia. The northern limit of the eclipse was in South Africa and its southern limit was in Antarctica.

It is surprising how the new Moon could had been sighted almost half an hour after the solar eclipse on Wednesday, Aug. 20, 1952 in Mecca. As a result, the claim of sighting the new Moon on this date was not a true one at all.

However, Taleghani has written that when people were not certain about the claim, it was announced by some authorities there that they are particularly much concerned about the correct onset of the months of Ramazan and Zil-Hajjah, and it is not announced until at least fifty reports of the lunar crescent sightings are received from different places, which this statement was not in accordance to the claim.

MOONSET-LAG TIME ON THURSDAY, AUG. 21, 1952

The calculated moonset-lag times for Mecca, Rabat and Tehran are listed in Table 4 on Thursday, Aug. 21, 1952 (Zil-Qaadeh 29, 1371 (A.H.) in Iran) which were 26,21 and 16 minutes respectively which these periods were not long enough for the appearance of the lunar crescent, to be able to be sighted by naked eyes as well.

LOCATION OF THE LUNAR CRESCENT ON THURSDAY AUG. 21, 1952

Like the case on Wednesday, Aug. 20, 1952, the estimated position of the lunar crescent is determined by the calculation of the arc of great circles and spherical angles on Thursday, Aug. 21, 1952, as tabulated in Tables 5 and 6 in relation to Fig. 5 and 6 respectively.

Summarized in Fig. 7, in connection to Table 7, the location of the lunar crescent and her apparent orbit (arc MM1), in the evening sky, in relation to two fixed points on the horizon, namely, geographical west (W) and the location of the sunset of the date (S), are outlined. Figures shown in Table 7, in the last column is phase of the Moon.

A layman, by the use of this Figure and the corresponding Table, could easily locate the position of the lunar crescent. However, the lunar crescent had reached to a certain altitude in the three cities and arc (SM") which is the difference between the celestial longitudes of the Sun and the Moon, was greater than 10 degrees, but since the periods of the moonset-lag time were quite short, so that the possibility of the then sighting of the new Moon, by naked eyes, on Thursday evening, Aug. 21, 1952 had been very low and almost not possible.

This particular case had been one of the good cases to verify the guide lines set by Moslem scholars which are based on experience, regarding the fulfilment of the both Bod-e-Moaddal and Bod-e-Seva together to reach to 10 degrees, as mentioned earlier.

MOONSET-LAG TIME ON FRIDAY, AUG. 22, 1952

The calculated moonset-lag times for Mecca, Rabat and Tehran, on Friday, Aug. 22, 1952 (Zil-Qaadeh 30, 1371 (A.H.) in Iran) are listed in Table 8. On this date the moonset lag times had been long enough (from 41 to 57 minutes) which are more than the set criteria of about 40 minutes, mentioned earlier, so that, the crescent had been easily sighted, since enough time was available until a fair darkness was fallen while the Moon had not yet set and it was still above the horizon.

LOCATION OF THE LUNAR CRESCENT ON FRIDAY, AUG. 22, 1952

Like the other days, the estimated position of the lunar crescent is determined and illustrated in Fig. 8 in relation to Table 9, Fig. 9 and 10 in connection to Table 10 and Fig. 11 and 12 in reference to Table 11 respectively for the three cities concerned, on Friday, Aug. 22, 1952.

As a result from the data given in these Tables, on this date, the new Moon with a high altitude (arc MH) had elongated enough from the Sun which had resulted a fairly thick crescent, so that, it had then been clearly appeared in the evening sky and sighted by naked eyes, of course.

Since the lunar crescent had a broad width at too high an altitude, this might had then led to the confusion by people as such that the earliest visibility might had taken place the evening before, *i.e.*, on Thursday, Aug. 21, 1952.

In other words, the lunar crescent might had then been mistakenly considered as the one on the second or even third night, as Taleghani had doubted while on board from Beirut to Jidda, probably on Saturday, Aug. 23, 1952, as mentioned earlier. Of course, the lunar crescent on Saturday had been much thicker. It is now clear that it was the first of Zil-Hajjah, 1371 (A.H.) while Taleghani sighted the new Moon on board (second-night sighting of the new Moon).

The possibility of sighting the first night lunar crescent at too high an altitude and broad width is not rare sometimes when the conjunction takes place one or two days earlier.

CONTRADICTIONS IN TAGHWIM IN THE PRESENT

Cases of contradiction in lunar calendar in the recent years are quite ample and the subject is studied and the results are published elsewhere, by the author for various lunar months.

But since the contradiction in the onset of the month of Zil-Hajjah in the recent past was discussed earlier, so that, a few cases of contradiction in the onset of Zil-Hajjah in the recent years are mentioned here :

In 1415 (A.H.), while it had been predicted that a unanimous onset of Zil-Hajjah was to take place in Saudi Arabia and Iran, on the same day[5], on Monday, May 1, 1995, but it was surprisingly claimed that the earliest lunar crescent sighting had been possible on Saturday evening, April 29, 1995, almost before the conjunction of the Moon with the Sun which took place at 20:36 hours the same day according to standard time in Saudi Arabia. However, after a few days, the onset of Zil-Hajjah was altered in this country to the correct date predicted.

5. Mashallah Ali-Ahyaie, " Unanimous Eid-Ui-Azha, 1415 (A.H.) ", *Tehran Times*, April 19, 20 and 22, 1995, Tehran, Iran.

In 1416 (A.H.) too, a unanimous onset of Zil-Hajjah was predicted on Saturday, April 20, 1996 which was two days later than the one predicted in Saudi Arabia[6] in the official calendar which was on Thursday April 18, 1996.

However, the onset of Zil-Hajjah 1416 (A.H.) was altered by one day and shifted to Friday, April 19, 1996 during the Hajj ceremony, but it was surprisingly resumed according to the official calendar afterwards.

In 1417 (A.H.), it was claimed in Saudi Arabia that the new Moon at the end of Zil-Qaadeh had been sighted on Monday, April 7, 1997 while the moonset-lag time on this date was about 7 minutes and 19 seconds only which this claim could not be accepted. This claim resulted that the Zil-Hajjah first was on Tuesday, April 8, 1997 in Saudia Arabia. However, a unanimous onset of Zil-Hajjah 1417 (A.H.) had been predicted on Wednesday, April 9, 1997 in Saudi Arabia and Iran[7].

CONCLUSION

Cases of contradictions in lunar calendar in the past and the recent years are quite ample in the Islamic countries. One of the main reasons attributed to these discrepancies is that a unanimous guide line is not applied while setting the onset of the lunar months in various countries. Even in certain cases the claim of the first new Moon sighting could be rejected by facts and figures right away. More and more educational programmes are required to be performed in this regard.

FIGURES

Table (1)-Moonset - Lag Time, on Wednesday, Aug. 20, 1952

City	Country	Longitude				Latitude				Standard Time (±)GMT, hours	Sunset		Moonset		Moonset-Lag Time	
		°	'	"		°	'	"			h	m	h	m	h	m
Rabat	**Moracco**	6	51	-	W	34	2	-	N	O	19	8	19	5	-	-3
Mecca	**Saudi Arabia**	39	49	46	E	21	25	19	N	+ 3	18	47	18	41	-	-6
Tehran	**Iran**	51	27	-	E	35	41	-	N	+3.5 *	18	48	18	39	-	-9

*** + 3.5 Hours in autumn and winter, but + 4.5 hours in spring & summer nowadays.**

Table 1.

6. Mashallah Ali-Ahyaie, " Locating New Moon at Close of Zil-Qaadeh, 1416 (A.H.) ", *Tehran Times*, April 16-17, 1996, Tehran, Iran.

7. Mashallah Ali-Ahyaie, " Locating New Moon at Close of Zil-Qaadeh, 1417 (A.H.) ", *Tehran Times*, April 7, 1997, Tehran, Iran.

FIGURE 1.

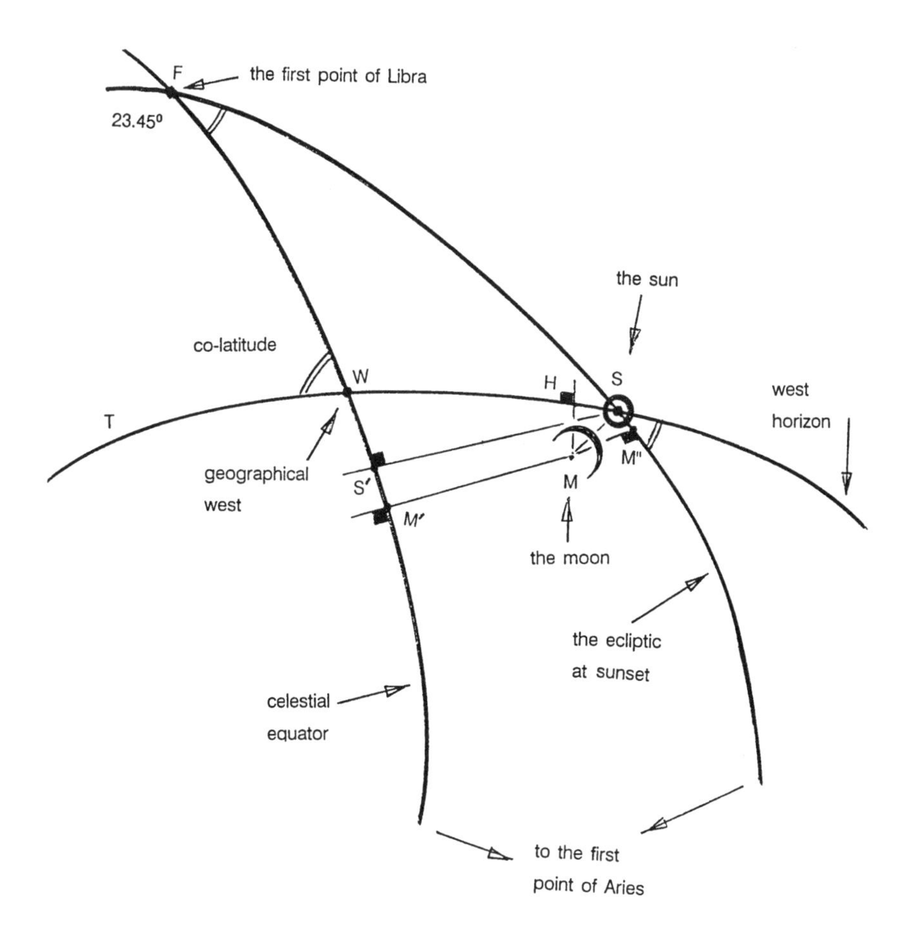

Position of the Moon at sunset, on Wednesday Aug. 20, 1952 in Mecca and Tehran. See Table 2. Not drawn to scale.

FIGURE 2.

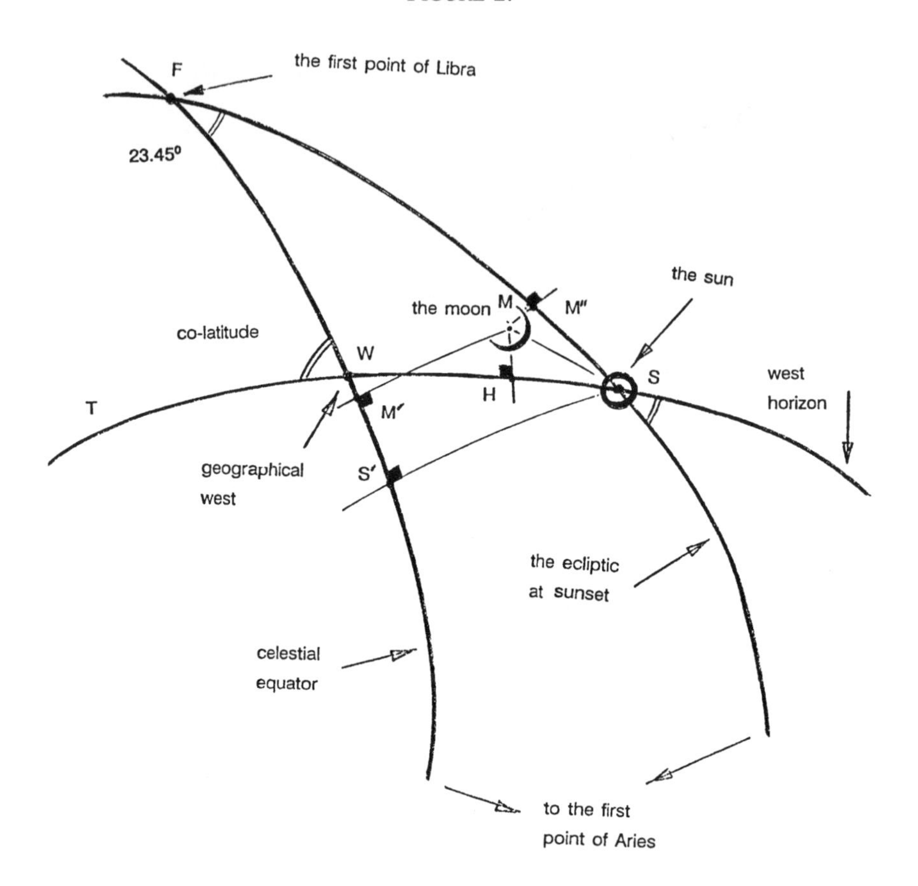

Position of the Moon at sunset, on Wednesday, Aug. 20, 1952 in Rabat. See Table 2. Not drawn to scale.

TABLE 2.

Table (2)-Position of the Moon at sunsent onWednesday, Aug. 20, 1952, see Fig. (1) & (2)

City	Country	arc SW	angle FWT	angle FST	arc MS	arc MH	angle MSW	arc SH	arc FS	arc FM''	arc SM''	arc MM''	arc MM'	arc SS'
		°	°	°	°	°	°	°	°	°	°	°	°	°
Rabat	Moracco	14.88	55.97	34.93	1.84	0.38	11.89	1.8	32.33	30.64	1.69	-0.72	11.01	12.29
Mecca	Saudi Arabia	13.27	68.58	47.95	0.6	-0.25	24.68	0.54	32.46	32.28	0.18	-0.57	11.72	12.34
Tehran	Iran	15.25	54.32	33.24	0.55	-0.48	61.83	0.26	32.48	32.53	0.05	-0.55	11.83	12.34

FIGURE 3.

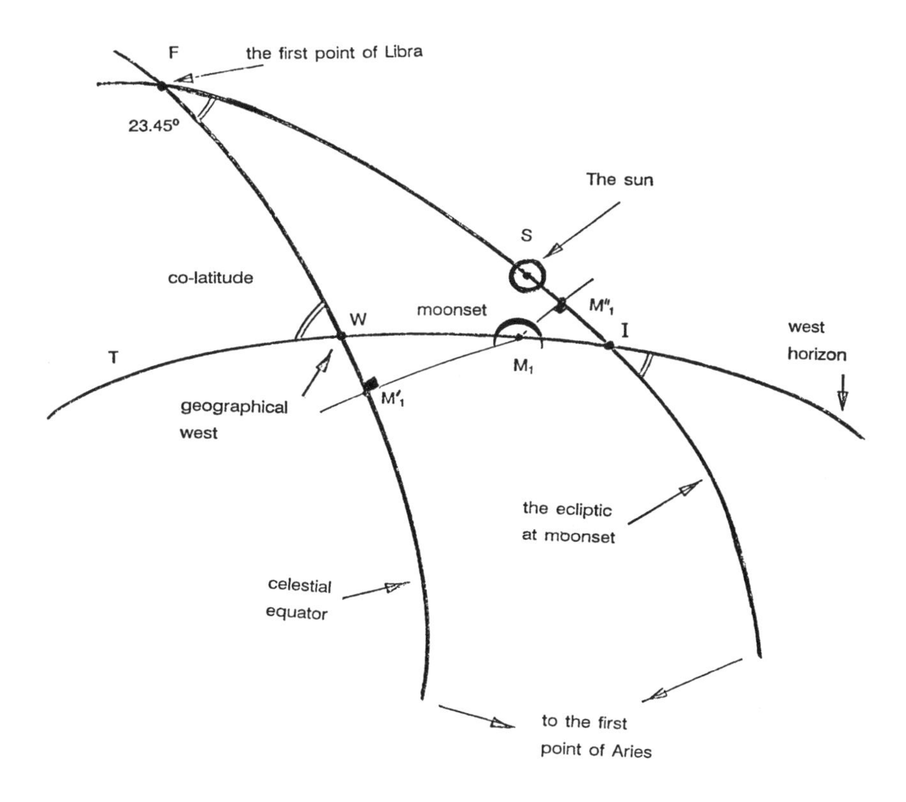

The position of moonset, on Wednesday, Aug. 20 1952 in Mecca and Tehran. See Table 3. Not drawn to scale.

TABLE 3.

Table (3)-Location of moonset on Wednesday, Aug. 20,1952, see Fig (3)&(4)									
City	Country	angle FWT	angle FIT	arc M_1W	arc SM_1	arc M_1I	arc M_1M_1'	arc M_1M_1''	arc SW
		°	°	°	°	°	°	°	°
Rabat	Moracco	55.97	34.84	13.32	1.56	1.26	11.01	-0.72	13.32
Mecca	Saudi Arabia	68.58	48.02	12.61	-	0.76	11.73	-0.57	12.61
Tehran	Iran	54.32	33.39	14.63	-	0.99	11.84	-0.54	14.63

The location of the moonset for Rabat is illustrated in Fig. 4 in reference to Table 3 as well.

FIGURE 4.

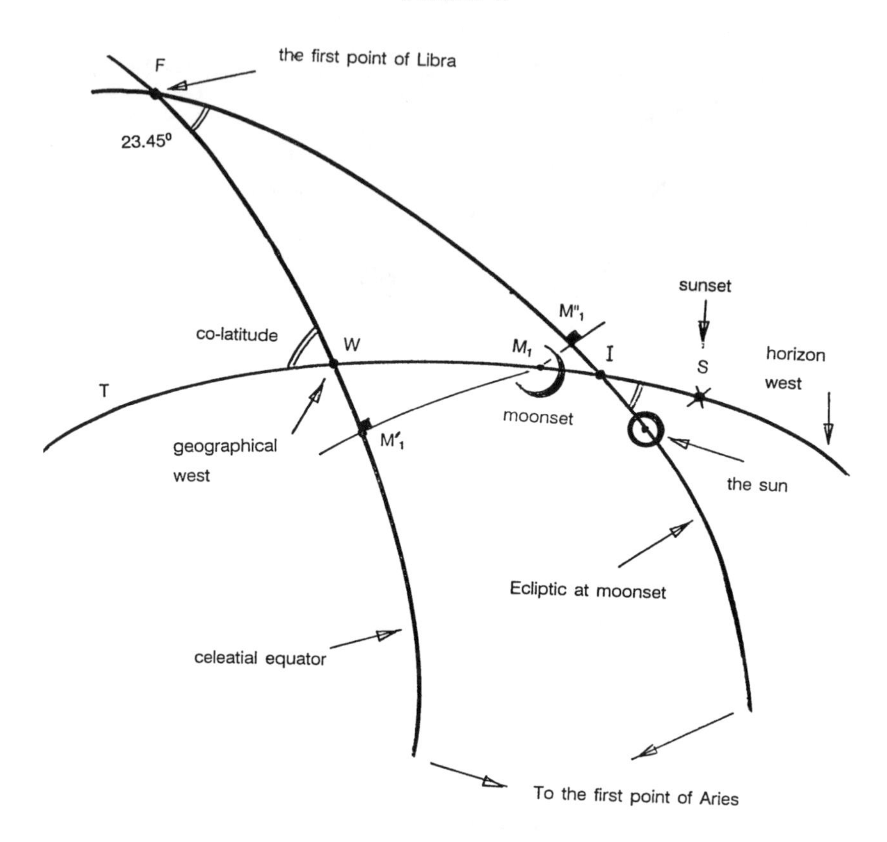

The position of moonset, on Wednesday, Aug. 20, 1952 in Rabat. See Table 3. Not drawn to scale.

TABLE 4.

Table (4) - Moonset - Lag Time, on Thursday, Aug. 21, 1952							
City	Country	Moonset		Sunset		Moonset - Lag Time	
		h	m	h	m	h	m
Mecca	Saudi Arabia	19	12	18	46	-	26
Rabat	Moracco	19	29	19	8	-	21
Tehran	Iran	19	3	18	47	-	16

TABLE 5.

Table (5) - Position of the moon on Thursday, Aug. 21,1952, see Fig. (5)

City	Country	arc	angle	angle	arc	arc	angle	arc	arc	arc	arc	arc	arc	arc
		SW	FWT	FST	MS	MH	MSW	SH	FS	FM"	SM"	MM"	MM'	SS'
		o	o	o	o	o	o	o	o	o	o	o	o	o
Mecca	Saudi Arabia	12.09	68.58	47.78	11.17	7.05	39.29	8.69	31.5	20.45	11.05	-1.64	6.47	12.0
Rabat	Moracco	14.47	55.97	34.79	12.7	5.66	26.65	11.39	31.37	18.79	12.58	-1.78	5.71	11.95
Tehran	Iran	14.84	54.32	33.1	10.94	4.5	24.54	9.97	31.52	20.7	10.82	-1.62	6.59	12.01

FIGURE 5.

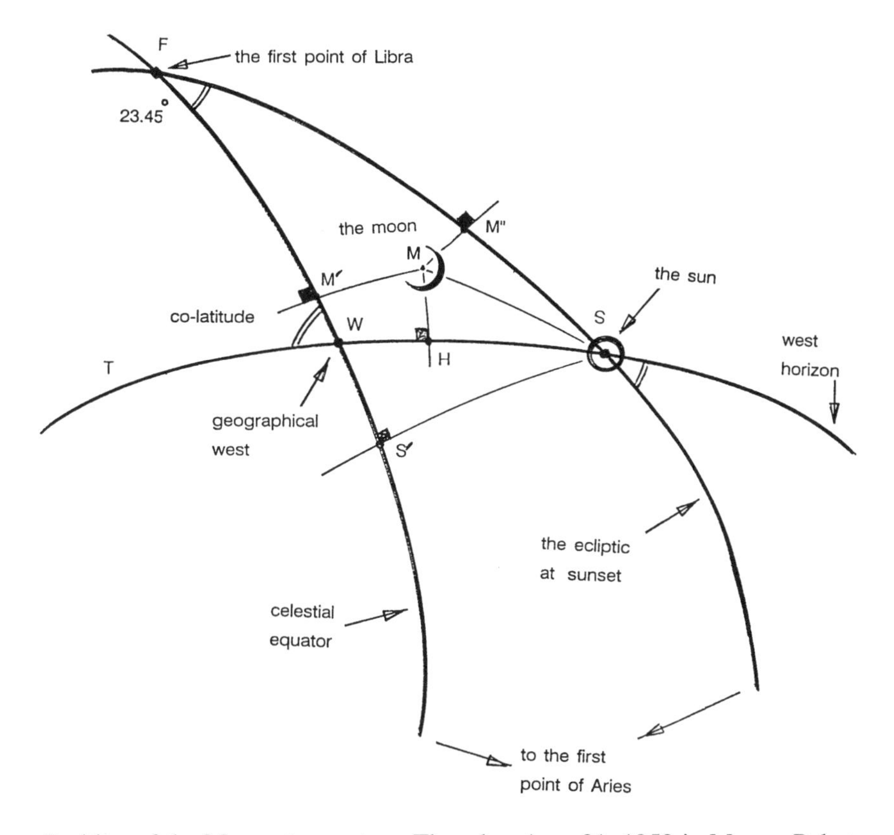

Position of the Moon at sunset, on Thursday, Aug. 21, 1952 in Mecca, Rabat and Tehran. See Table 5. Not drawn to scale.

FIGURE 6.

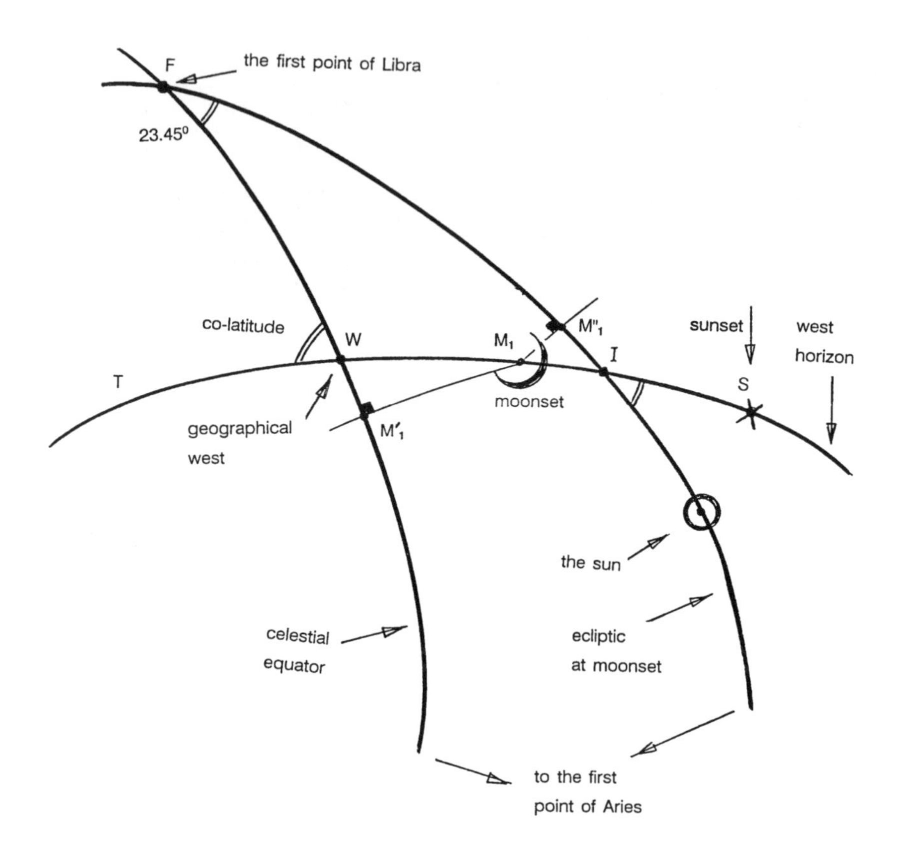

The position of moonset on Thursday, Aug. 21, 1952. See Table 6. Not drawn to scale.

TABLE 6.

Table (6)-Location of moonset on Thursday, Aug. 21,1952, see Fig (6)

City	Country	angle FWT	angle FIT	arc M_1W	arc SM_1	arc M_1I	arc $M_1M'_1$	arc $M_1M''_1$	arc SW
		°	°	°	°	°	°	°	°
Mecca	Saudi Arabia	68.58	46.39	6.83	6.08	2.3	6.35	-1.66	12.9
Rabat	Moracco	55.97	33.55	6.77	7.7	3.26	5.6	-1.8	14.47
Tehran	Iran	54.32	32.06	8.01	6.83	3.08	6.5	-1.64	14.84

FIGURE 7.

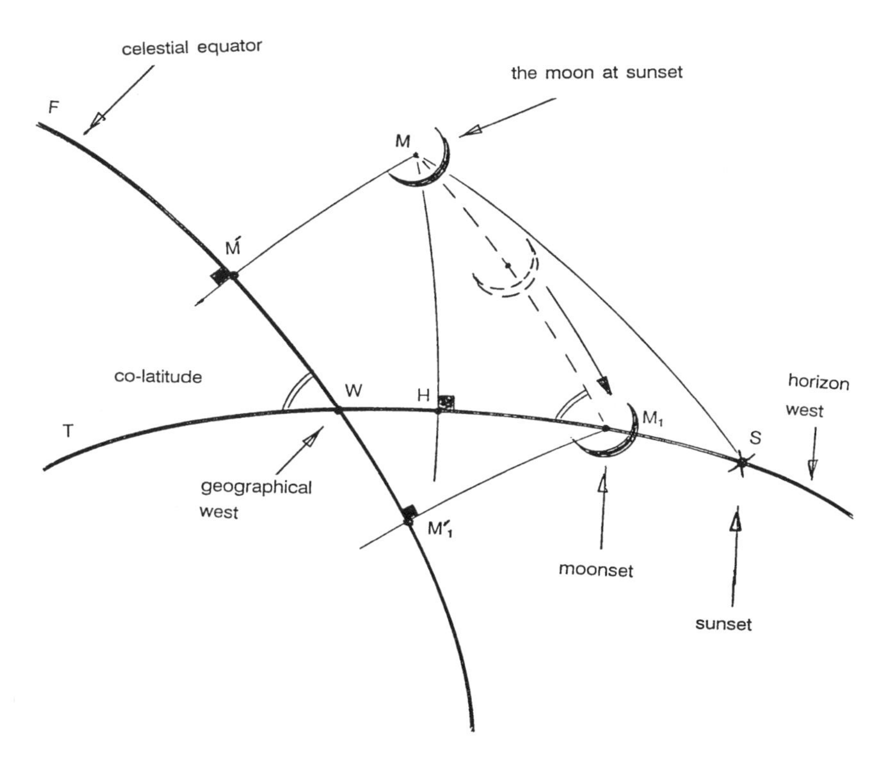

The location of the moonset in reference to celestial equator, the geographical west (W) and sunset (S), in Mecca, Rabat and Tehran, on Thursday, Aug. 21, 1952. See Table 7. Not drawn to scale.

TABLE 7.

Table (7)-Location of moonset in reference to celestial equator, geographical west (W) and sunsent (S) on Thursday, Aug. 21,1952, see Fig (7)

City	Country	angle FWT	angle MM_1W	arc SW	arc M_1S	arc MH	arc M_1W	arc SH	arc MM'	arc M_1M_1'	arc MS	arc MM_1	arc WH	Frac. Ill.*
		°	°	°	°	°	°	°	°	°	°	°	°	°
Mecca	Saudi Arabia	68.58	69.72	12.9	6.08	7.05	6.83	8.69	6.47	6.35	11.17	7.52	4.21	0.95
Robat	Morocco	55.97	57.05	14.47	7.7	5.66	6.77	11.39	5.71	5.6	12.7	6.75	3.08	1.22
Tehran	Iran	54.32	55.22	14.84	6.83	4.52	8.01	9.97	6.59	6.5	10.94	5.51	4.86	0.91

* Frac. Ill. = Fraction Illuminated/phase of the moon.

TABLE 8.

Table (8) - Moonset - Lag Time, on Friday, Aug. 22, 1952

City	Country	Moonset		Sunset		Moonset - Lag Time	
		h	m	h	m	h	m
Mecca	Saudi Arabia	19	43	18	46	-	57
Rabat	Moracco	19	53	19	6	-	47
Tehran	Iran	19	26	18	4	-	41

FIGURE 8.

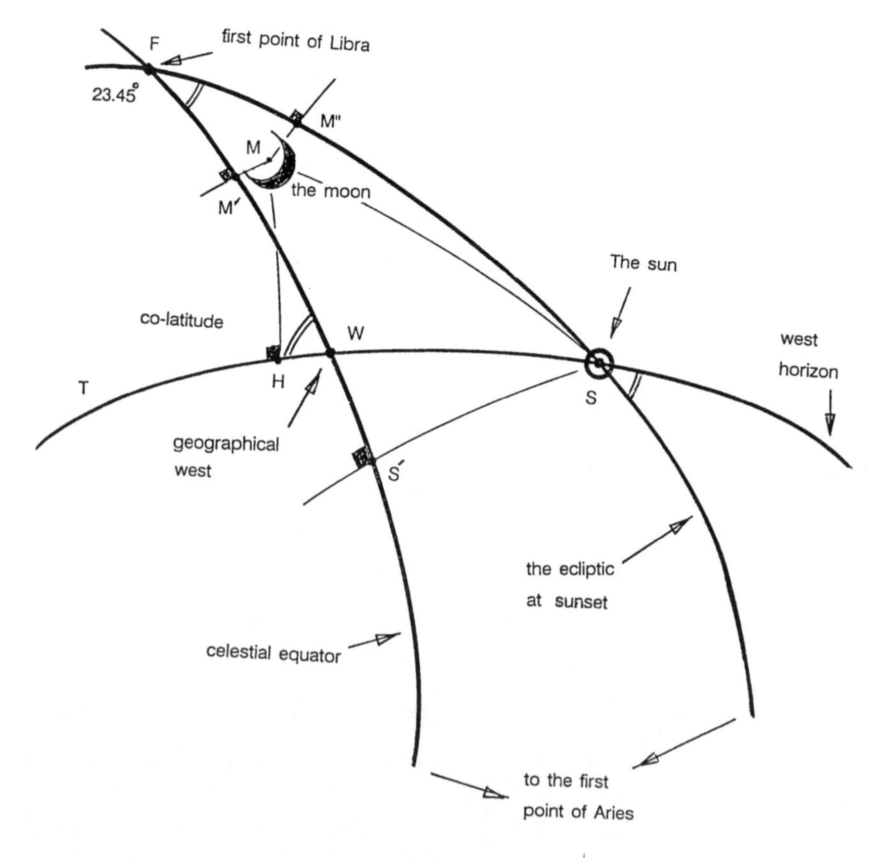

Position of the Moon at sunset, on Friday, Aug. 22, 1952, in Mecca, Rabat and Tehran. See Table 9. Not drawn to scale.

TABLE 9.

Table (9) - Position of the Moon at sunset on Friday, Aug. 22, 1952, see Fig. (8)

City	Country	arc SW	angle FWT	angle FST	arc MS	arc MH	angle MSW	arc SH	arc FS	arc FM"	arc SM"	arc MM"	arc MM'	arc SS'
		°	°	°	°	°	°	°	°	°	°	°	°	°
Mecca	Saudi Arabia	12.54	68.58	47.6	22.17	14.21	40.58	17.2	30.54	8.5	22.02	-2.64	0.95	11.67
Rabat	Moracco	14.07	55.97	34.63	23.71	10.78	27.73	21.24	30.4	6.85	23.55	-2.77	0.17	11.62
Tehran	Iran	14.42	54.32	32.94	21.94	9.39	25.91	19.91	30.56	8.77	21.79	-2.62	1.07	11.67

FIGURE 9.

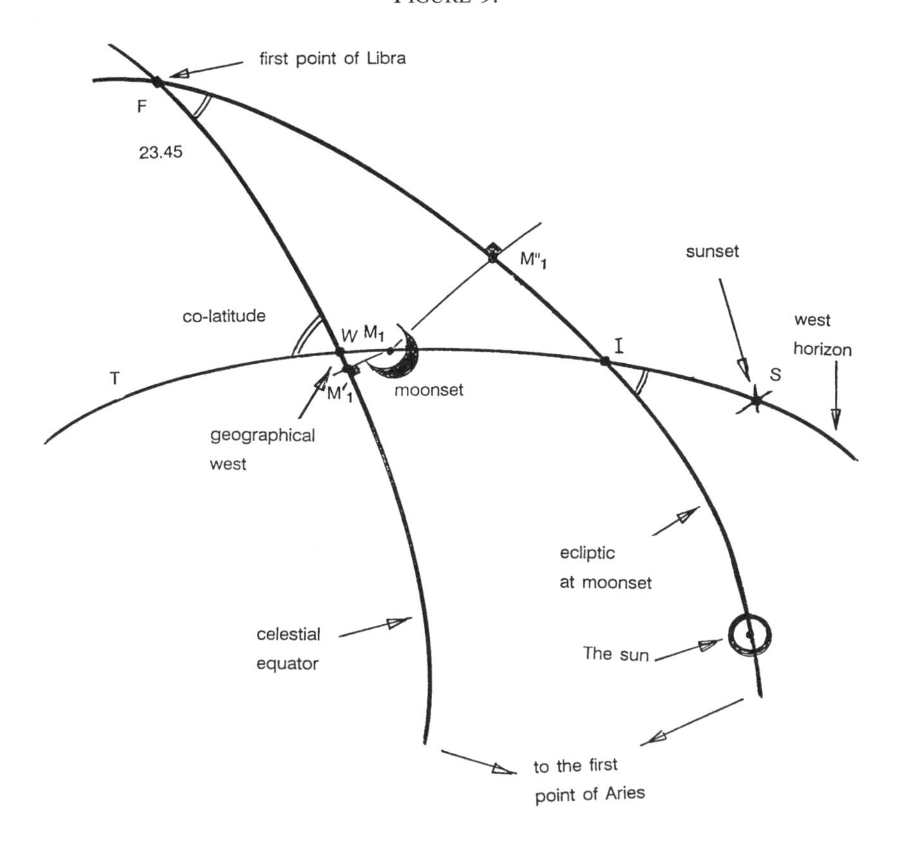

The position of moonset on Friday, Aug. 22, 1952 in Mecca and Tehran. See Table 10. Not drawn to scale.

FIGURE 10.

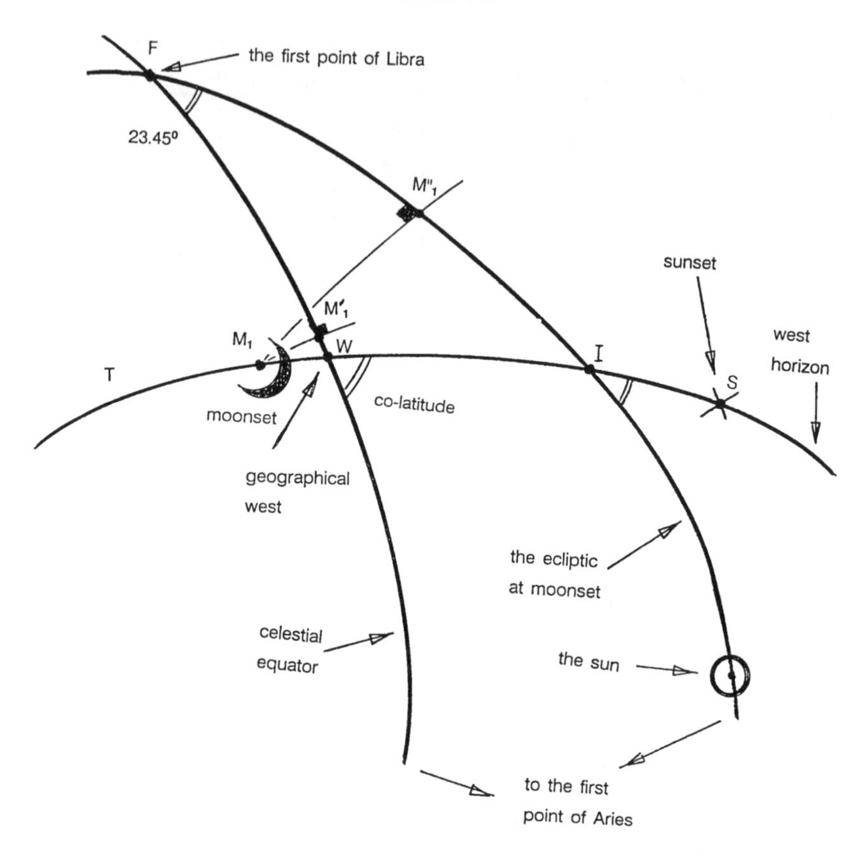

The position of moonset in Rabat on Friday, Aug. 22, 1952. See Table 10. Not drawn to scale.

TABLE 10.

Table (10)-Location of moonset on Friday, Aug. 22, 1952, see Fig. (9)&(10)

City	Country	angle FWT	angle FIT	arc M_1W	arc SM_1	arc M_1I	arc $M_1M'_1$	arc $M_1M''_1$	arc SW
		°	°	°	°	°	°	°	°
Mecca	Saudi Arabia	68.58	45.44	0.76	11.78	3.76	0.71	–2.68	12.54
Rabat	Morocco	55.97	32.79	0.04	14.1	5.18	-0.03	-2.8	14.07
Tehran	Iran	54.32	31.23	1.1	13.33	5.12	0.89	-2.65	14.42

FIGURE 11.

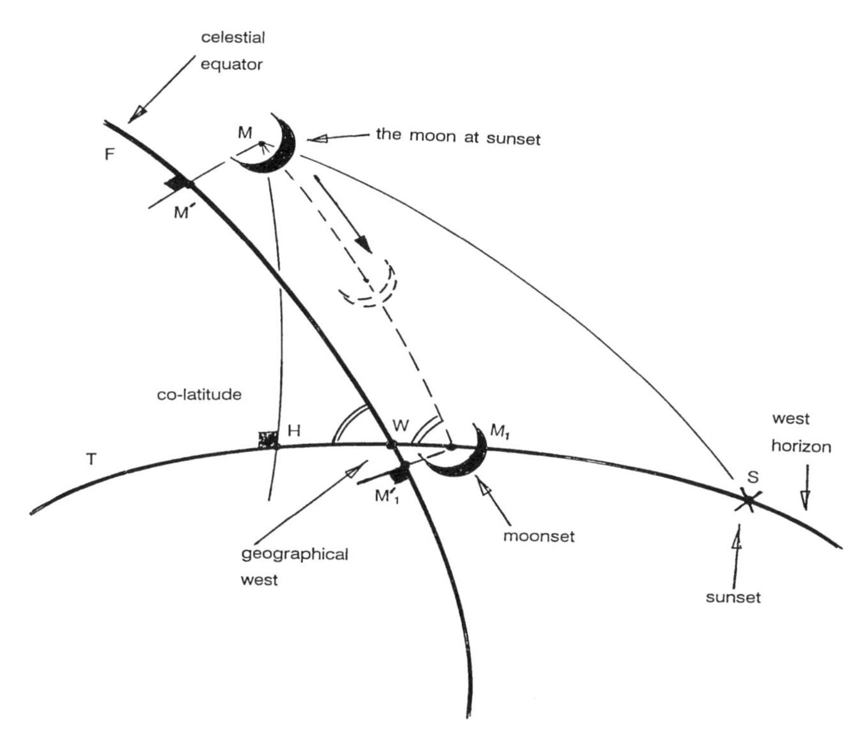

The location of the moonset in reference to celestial equator, the geographical west (W) and sunset (S), in Mecca and Tehran, on Friday, Aug. 22, 1952. See Table 11. Not drawn to scale.

TABLE 11.

Table (11)-Location of moonset in reference to celestial equator, geographical west (W) and sunset (S) on Friday, Aug. 22,1952 , see Fig. (11)&(12)

City	Country	angle FWT	angle MM_1W	arc SW	arc M_1S	arc MH	arc M_1W	arc SH	arc MM'	arc $M_1M'_1$	arc MS	arc MM_1	arc WH	Frac Ill.*
		°	°	°	°	°	°	°	°	°	°	°	°	°
Mecca	Saudi Arabia	68.58	69.56	12.54	11.78	14.21	0.76	17.2	0.95	0.71	22.17	15.19	4.65	3.7
Rabat	Moracco	55.97	56.86	14.07	14.1	10.78	0.04	21.44	0.17	-0.03	23.71	12.91	7.18	4 22
Tehran	Iran	54.32	55.27	14.42	13.33	9.39	1.1	19.91	1.07	0.89	21.94	11.46	5.49	3 62

* Frac. Ill. = Fraction Illuminated/Moon phase.

FIGURE 12.

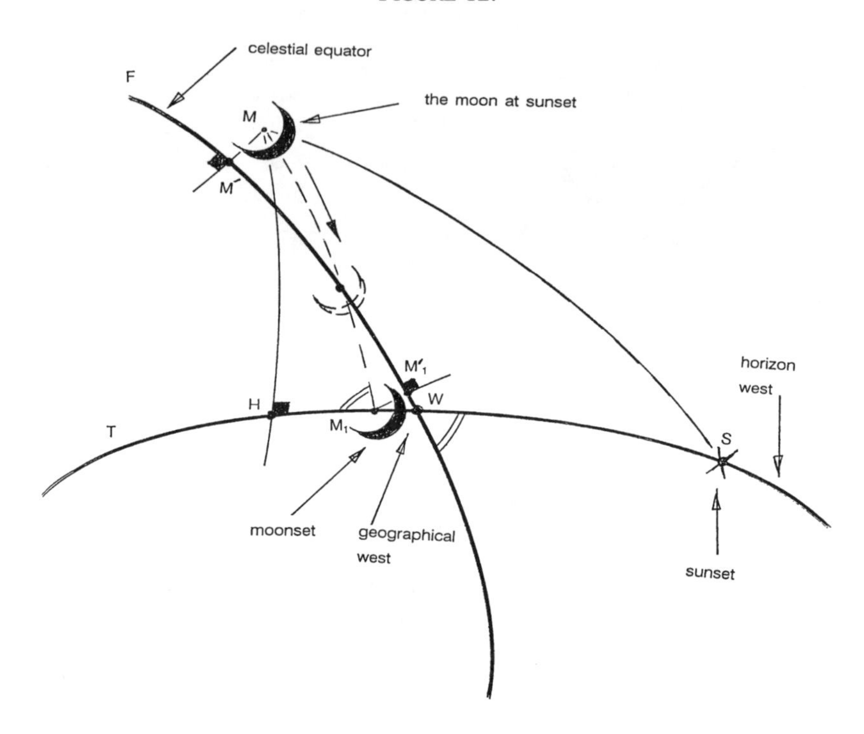

The location of the moonset in reference to celestial equator, geographical west (W) and sunset (S), in Rabat, on Friday, Aug. 22, 1952. See Table 11. Not drawn to scale.

Quelques aspects du problème de la transmission des connaissances scientifiques en mécanique

Mariam M. Rozhanskaya

Dans le cadre du thème du symposium je voudrais m'arrêter sur le problème de la transmission des connaissances scientifiques de l'Orient à l'Occident dans le domaine de la statique théorique et pratique (appliquée), c'est-à-dire principalement la science des balances et de la pesée au Moyen Age.

De toute l'abondance du matériel j'ai choisi trois problèmes, assez importants pour comprendre l'essence de cette transmission. L'histoire de chacun d'eux est le chaînon important dans la chaîne ininterrompue du développement des connaissances venant de la tradition antique à ce que nous appelons la préhistoire de la mécanique classique en Europe.

C'était l'étape " orientale ", si l'on peut la désigner comme ça, qui était nécessaire et la plus importante. Maintenant je vais exposer les problèmes eux-mêmes.

Notion de la gravité, de la force et du moment d'une force

Les sources principales pour ces problèmes sont le traité de Thābit b. Qurra (IX) *Kitāb fī-l-qarastūn* (Le livre al-qarastun)[1], les extraits des oeuvres mécaniques d'al-Qūhī (X) et d'Ibn al-Haitham (X-XI) et aussi d'Isfizārī (XI) qui nous sont parvenus exposés par al-Khāzīnī et le texte d'al-Khāzinī lui-même, c'est-à-dire *Kitāb mizān al-hikma* (Livre de la balance de la sagesse)[2].

1. Thābit b. Qurra, *Kitāb fī-l-qarastūn*, Texte arabe et trad. française par K. Jaouiche, Leiden, 1976 ; trad allem. " Die Schrift über den qarastun ", *Bibliotheca Mathematica*, 3, 12 (1912) ; trad. angl. de E.A. Moody et M. Clagett : *The Medieval Science od Weight,* version latine et trad. angl., Madison, 1952 ; trad. russe par B.A. Rosenfeld, " Matematicheskiye traktati ", *Nauchnoye nasledstvo*, vol. VIII, Moscou, 1984.

2. Al-Khazini, *Kitab mizan al-hikma*, Hyderabad, 1941 ; " Al-Khazini. Kniga vesov mudrosti ", Trad. russe par M.M. Rojanskaja, I.S. Levinova, *Nauchnoye nasledstvo,* VI, Moscou, 1983, 15-140.

Les notions de la force (*quwwa*) et de la gravité (*thiql*) sont examinées dans ces traités sous deux aspects :

1) dans le sens aristotélicien, c'est-à-dire ayant en vue la notion du mouvement naturel et de la tendance de l'inclination naturelle analogiquement à *ῥοπή* grec et celles du lieu naturel et du centre de l'Univers.

2) dans le sens archimédien, en partant du principe du levier, c'est-à-dire, en application aux poids suspendus aux extrémités du levier.

Adressons-nous maintenant aux deux notions que les auteurs mentionnés distinguent et pour lesquelles ils emploient les termes différents : *wazn* (poids) et *thiql* (gravité).

Schématiquement cela peut être présenté de la manière suivante :

a) Tous les corps ont un poids (*wazn*). C'est une catégorie constante qui se détermine par la pesée et se lie à la pression que le corps exerce sur le plateau de la balance pendant la pesée.

b) La gravité du corps, c'est une catégorie changeante, dépendant de la position du corps par rapport à un certain point.

Ainsi, en fonction du point de vue des auteurs, la notion de la gravité remonte aux aspects déjà mentionnés. Si ce point est le “ Centre de l'Univers ”, c'est la conception du changement de la gravité en fonction de la distance de ce point. Alors cette notion de la “ gravité ” remonte aux catégories aristotéliciennes du “ lieu naturel ” ainsi que le “ mouvement naturel ” et de “ l'inclination naturelle ”. C'est un certain analogue de la notion contemporaine de l'énergie potentielle.

Si ce point, c'est le point de la suspension ou de l'appui du levier, alors “ la gravité ” du corps change aussi en fonction de la position du poids sur le bras du levier, c'est-à-dire, de sa distance de ce point. Cette conception développée par Thābit b. Qurra et par al-Isfizārī, remonte à la mécanique hellénistique, aux *Problèmes mécaniques* de Pseudo-Aristote, à l'hypothèse que le même poids pèse différemment en fonction de sa position sur le bras du levier. Dans ce cas “ la gravité ” changeante d'un corps est en fait le moment de la force par rapport au point.

Thābit b. Qurra et al-Khāzīnī unissent dans leurs traités les deux aspects de la notion de la gravité changeante, c'est-à-dire, liées à la distance du “ Centre de l'Univers ”, ainsi qu'à la distance du poids sur le bras du levier du point de son appui ou de sa suspension.

Le premier concept n'a trouvé de développement ultérieur au Moyen Age ni en Orient, ni en Occident, mais il s'est développé beaucoup plus tard du fait du développement de la théorie de la gravitation.

Le second concept peut être considéré comme le prototype de la notion future de la “ gravité par rapport à la position ” (*gravitas secundum situm*), développée plus tard dans toute une série de traités *De ponderibus* connue en Europe médiévale, notamment dans les oeuvres de Iordanus de Nemore, ses élèves et successeurs qui ont fait la base de la statique européenne du Moyen Age[3]. Ce sont justement les traités *De ponderibus*, qui donnent comme postulat la différence entre le " poids " en tant que catégorie constante et la " gravité " en tant que catégorie changeante. C'est une tradition venant aussi de la statique arabe médiévale qui remonte à son tour aux " Problèmes mécaniques " et à d'autres traités hellénistiques plus tardifs sur la statique.

Il est à noter, que les termes latins *pondus* et *gravitas* sont la traduction littéraire des mots arabes *wazn* et *thiql*[4].

La balance et le problème de la solution des équations

Le deuxième problème, c'est la modification du levier-balance le plus simple à deux plateaux, en tant que moment de départ dans l'histoire du développement des méthodes de la solution des équations linéaires.

Il s'agit d'un procédé connu, qu'on appelle *al-jabr wa al-muqābala,* c'est-à-dire, il faut parfois le complément et l'opposition. La théorie des équations en général *al-jabr* consiste à transporter les expressions soustractives et *al-muqābala* — à réduire les termes semblables. Dans le cas le plus simple, quand il s'agit des équations linéaires, on doit laisser les membres inconnus d'une part de l'équation, transporter les membres connus de l'autre part et puis comparer les deux parts.

On peut supposer que ces deux termes, *al-jabr* et *al-muqābala* , sont liés à la pratique de la pesée, c'est-à-dire, à l'équilibre de la balance.

Le cas le plus simple, c'est quand l'inconnu (le fardeau) se trouve dans un plateau, suspendu à une extrémité de la balance et dans un autre plateau, suspendus à une autre extrémité se trouvent les poids. Cela se ramène seulement à " l'opposition ". Si la somme des poids à un plateau est égale au poids du fardeau dans un autre plateau, la balance est en équilibre (les deux parties sont " en opposition "), et cette somme, c'est la quantité inconnue trouvée.

Dans le cas plus compliqué, quand on pèse le fardeau sur la balance à deux plateaux, pour équilibrer la balance il faut parfois, mettre une partie du poids sur le plateau avec le fardeau (complètement). C'est justement ce qu'on appelle " complément " (le transfert des membres de l'équation à l'autre partie avec un signe contraire). Pour calculer le poids du fardeau dans ce cas il faut pour l'équilibre de la balance mettre quelques poids en arrière dans le plateau

3. Thābit b. Qurra, *Kitāb al-qarastūn*, Texte arabe et trad. française par K. Jaouiche, Leiden, 1976 ; trad allem. " Die Schrift über den qarastun ", *Bibliotheca Mathematica*, 3, 12 (1912) ; trad. angl. de E.A. Moody et M. Clagett : *The Medieval Science od Weight,* version latine et trad. angl., Madison, 1952, 33-35 ; trad. russe par B.A. Rosenfeld, " Matematicheskiye traktati ", *Nauchnoye nasledstvo*, vol. VIII, Moscou, 1984.

4. M.M. Rozhanskaya, " Statics ", in R. Rashed (éd.), *Encyclopedia of the History of Arabic Science*, vol. 2, London, New York, 1996, 614-642, esp. 619-623 ; trad française, Paris, 1997, 263-292, esp. 268-272.

avec les poids. On peut supposer, que la méthode d'*al-jabr wa al-muqābala* exposé dans le traité algébrique d'al-Khwārismī[5] et dont le nom *al-jabr* signifie maintenant le domaine des mathématiques appelé *algebra* est à l'origine liée à la pratique de la pesée à l'aide de la balance simple à bras égaux. Nous connaissons bien, qu'à son tour, le célèbre livre d'al-Khwārismī n'a cessé d'être source d'inspiration et l'objet de commentaire non seulement en arabe et en persan, mais aussi en latin et dans les langes de l'Europe de l'Ouest jusqu'au XVIIIe siècle[6].

Le problème de la pesée

C'est un problème célèbre dans l'histoire des mathématiques et de la mécanique. Le sens du problème, c'est le choix du nombre minimal des poids pour peser tous les fardeaux entiers plus petits ou égaux au poids donné. Voici comment cette histoire est exposée dans *Le Livre de la balance de la sagesse*[7].

Le lot des poids généralement admis dans la pratique de la pesée, c'est-à-dire, 1, 2, 5 ; 10, 20, 50 ; 100, 200, 500, en somme — 888 unités du poids, ne satisfait pas al-Khāzīnī. A l'aide de ce lot, si l'on met les poids sur un plateau, on peut peser tous les fardeaux de 1 jusqu'à 888, mais avec une certaine exclusion. Par exemple, on peut peser le fardeau donné le 3, parce que 3 = 1 + 2, mais on ne peut pas peser le 4 : 4 = 1 + 1 + 2 = 2 + 2 et le 9 : 9 = 1 + 1 + 2 + 5 = 2 + 2 + 5, etc. Pour cela il faut le double lot des poids généralement admis. Si l'on peut mettre les poids sur les deux plateaux, il ne faut pas avoir le double lot. Mais dans ce cas l'unique solution-pesée n'est pas respectée. Al-Khāzīnī met encore une condition, qui consiste en ce qu'on pouvait peser le fardeau donné à l'aide du lot des poids proposés par lui d'une seule manière, de sorte que chaque poids ne s'utilise qu'une fois. C'est pourquoi, pour satisfaire ces deux conditions, il propose deux autres lots de poids.

Si l'on ne met les poids que sur un plateau, il doit choisir ce lot de telle façon, qu'ils font la progression géométrique avec le premier membre 1 et le dénominateur 2.

Si l'on met les poids sur les deux plateaux, les poids doivent faire la progression avec le premier membre 1 et le dénominateur 3.

Le nombre des poids pour le premier cas est égal à 10, pour le deuxième cas - à 7.

5. Al-Khwarismi, *Kitab al-jabr wa al-muqabala*, 'Abd al-Hadi Abu Rida (éd.), 1939 (2 vols) ; R. Rashed, " L'algèbre ", *Histoire des sciences arabes*, vol. 2, Paris, 1997, 31-54, esp. 31-32.

6. A. Allard, " L'influence des mathématiques arabes dans l'Occident médiéval ", in R. Rashed (éd.), *Encyclopedia of the History of Arabic Science*, vol. 2, London, New York, 1996, 199-230 ; Al-Khwarismi, *Kitab al-jabr wa al-muqabala*, 'Abd al-Hadi Abu Rida (éd.), 1939, 31-32 (2 vols).

7. Al-Khazini, *Kitab mizan al-hikma*, Hyderabad, 1941, 109-110.

Si l'on désigne le fardeau donné pour les deux cas comme P, on peut le présenter comme cela :

1) $$P = a_0 + a_1 p_1 + a_2 p_2 + \ldots + a_9 p_9 = \sum_{i=0}^{9} a_i p_i \leq 1023 \text{, ou}$$

$$p_i = 1, 2, 2^2, \ldots, 2^9, a_i = 0 ou 1$$

2) $$P = a_0 + a_1 p_1 + a_2 p_2 + \ldots + a_6 p_6 = \sum_{i=0}^{6} a_i p_i \leq 1097 \text{, ou}$$

$$p_i = 1, 3, 3^2, \ldots, 3^6, a_i = -1, 0, 1$$

Dans le sens mathématique les deux cas d'al-Khāzīnī sont les cas particuliers du problème de la présentation du nombre naturel arbitraire sous la forme de la somme algébrique des puissances différentes d'un nombre naturel donné, c'est-à-dire, de la présentation d'un certain nombre naturel n sous la forme de la somme algébrique de m < n nombres naturels.

(Dans le premier cas m = 10, dans le deuxième cas m = 7).

Si nous allons nous adresser à la terminologie moderne, le problème est formulé de la façon suivante.

Il faut trouver une telle sous-multitude M* avec m éléments de la multitude M des nombres naturels (n éléments), pour qu'on puisse poser la multitude M* en somme de m < n éléments de la multitude M. Avec cela la sous-multitude M* doit contenir le nombre minimal des éléments et ne doit pas contenir deux éléments égaux.

Sans cette limitation le " problème de la pesée ", c'est un soi-disant " Problème de Bachet ", du nom de Bachet de Méziriac, qui a publié sa solution en 1612.

Al-Khāzīnī remarque, que si la condition de la seule possibilité n'est pas respectée, c'est déjà un autre problème. On peut supposer, qu'il avait en vue ce problème : par combien de procédés on peut présenter le nombre entier donné en somme des nombres entiers peu petits. A cette condition nous recevons le soi-disant " Problème de Leibnitz ".

(Dans l'aspect strict ce problème est posé justement par Leibnitz en 1666).

Il est évident que le " problème de la pesée " est d'origine orientale et remonte à la haute Antiquité. Ses traces mènent à la civilisation ancienne indienne (système de mesure et des poids formé sur la puissance de 2), à l'arithmétique égyptienne pratique (la multiplication des nombres entiers, quand on représente l'un des multiplicateurs en somme des nombres entiers à additionner du type 2^k), au modèle de Platon du cosmos avec ses racines pythagoriciennes[8].

Le problème de la pesée était probablement bien connu à l'Orient médiéval. Il y avait des traités, contenant ce problème, l'un d'entre eux était déjà connu au XIe siècle[9].

En Europe occidentale ce problème est rencontré dans les oeuvres de Leonardo Pisano, qui a étudié le deuxième cas d'al-Khāzīnī (la présentation du poids P = 40 suivant les puissances de 3 ; m = 4), mais répand ce problème sur n'importe quel n, jusqu'à l'infini (*ad infinitum*) : 1, 3, 3^2, ...[10].

Dans le traité de Jordan de Nemore *De algorismo* on trouve une proposition qu'on peut interpréter comme une version du " problème de la pesée " (présentation du nombre entier comme la somme des puissances de 10).

Presque sous cet aspect (comme chez Leonardo), ce problème on le rencontre dans les ouvrages de N. Chuquet (XV), de L. Paccoli (XV-XVI), de N. Tartaglia (XVI), de M. Stifel (XV-XVI), de R. Gemma Frisius (XVI), jusqu'à Bachet de Méziriac (la première publication imprimée) et plus loin jusqu'à l'étude de L. Euler (XVIII), qui l'a resolue et a trouvé une solution tout-à-fait rigoureuse[11].

8. M.M. Rozhanskaya, *La mécanique en Orient médiéval*, Moscou, 1976 (en russe) ; M.M. Rozhanskaya, " Statics ", in R. Rashed (éd.), *Encyclopedia of the History of Arabic Science*, vol. 2, London, New York, 1996, 614-642 ; trad française, Paris, 1997, 263-292 ; M.M. Rozhanskaya, " On a mathematical problem in al-Khazini's Book of the Balance of Wisdom ", in D.A. King, G. Saliba (éds), *From Deferent to Equant : a Volume of Studies in the History of Science in the Ancient and Medieval Near East in Honor of E.S. Kennedy*, New York, 1987, 427-436.

9. Tabarī Mohammed ibn Ayyūb, *Miftāh al-m'amalāt*, in M. Amin Riyahi (éd.) Teheran, 1970 ; J. Tropfke, *Geschichte des Elementarmathematik*. Bd. I : *Arithmetik und Algebra*, 4e éd. revue par K. Vogel, K. Reich, H. Gerike, Berlin, New York, 1980, 634.

10. [Leonardo Pisano], *Scritti di Leonardo Pisano matematico del secolo decimo terzo, pubblicati da baltassare Boncompagni*, I (Liber Abbaci), Rome, 1857.

11. M.M. Rozhanskaya, " On a mathematical problem in al-Khazini's Book of the Balance of Wisdom ", *op. cit.*, 427-436.

The Problems of physics in the works of ar-Razi and Ibn-Sina

Abdulhai Komilov

The Middle and the Near East, in other words, so-called Muslim East in the Middle Ages, have a special place in the history of the world culture. From the 9th to 15th century, the region was a place of the intensive development of science and culture. The literary, poetic and philosophical heritage of scientists who had lived in the region during this period has been studied in rather subtle detail. This can't be said, however, about the natural-scientific heritage of these scientism, first of all, about the physical and mathematical disciplines and, in particular, about physics.

Analyzing the period of Middle Ages in the Middle and Near East, the historians have paid a special attention to the period from the 9th to 11th centuries, that was termed in a different way by different authors : the Muslim Renaissance, the science and culture, the epoch of awakening and revival, the epoch of the rapid development of science and culture and so on. I will not focus on the problem of correct interpretation of these terms in the interests of time. I should mention, however, that this was a period of the explosive development of physic-mathematical sciences, how they were understood and defined in the Middle Ages.

I would also like to note that there exists a vast and extensive literature about the history of " science of nature " — this was the common understanding of physics in the antiquity and Middle Ages — in the Ancient World. There also are many studies on the history of physics during the European Renaissance. At the same time, the history of " science of nature " in the medieval East that had a deep impact on the development of natural science and philosophical thought during the scientific revolution in Europe is studied absolutely insufficiently.

Up to the present, there is no special study, analyzing the entire history of physics in the medieval East. Analyzed by the well-known historians, the history of physics and astronomy in the medieval Muslim East has been studied better, than the history of physics. Neither the *History of Physics* by Laue[1] and the *History of Physics* by Liozzi[2], nor the Russian books on the subject by Dorfman[3], Kudriavtsev[4] and Spasskii[5] contain any substantial information about the history of physics in the medieval Muslim East.

In my talk I will analyze briefly the physical views of the two important Iranian scientists — Abu Bakr Muhammad ibn Zakaria ar-Razi (865-925) and Abu Ali Ibn Sina (980-1037) who are mainly known as physicians and philosophers.

The core of the physical teaching of ar-Razi is formed by his ideas about the structure of matter, about space and time, and his studies in the fields of mechanics and geometrical optics. He stated that matter is eternal. And because matter is eternal, existing in space and time are also eternal. According to Aristotle, the time is the interval between the beginning and the end of a motion ; at the same time, according to ar-Razi, the time may be both absolute and limited. His limited time is, in its essence, the time of Aristotle, and the absolute time is the eternal category, like matter and absolute space.

Though in some issues of nature philosophy, and specifically, in the views on the structure of matter, ar-Razi follows Democritus, he doesn't agree with him on everything. This is true, in particular, with respect to the question of the existence of emptiness. What is the essence of this conception as it has been formulated by ar-Razi and what is the difference from the theory of D. ? According to D., the emptiness is a place where the atoms are staying and mixing ; for ar-Razi emptiness — is the absolute space that itself is a substance.

Ar-Razi regards two kinds of space : the absolute and partial, or incomplete. He links this conception to the eternity and the undestroyability of matter and time.

The followers of ar-Razi reported about his views on the absolute and limited time and space. According to Nazir Hosrov, speaking about the limited space, ar-Razi stated that the relationship between space and matter was the same as between the jug and its content, *i.e.* butter, water, etc.

The absolute space, in his view, is the emptiness which is unrestricted by anything and in which " things " — what means matter in the firm of various substances — are colliding. Things, in fact, emerge as a result of these collisions.

1. Laue, *History of Physics*, Göttingen, 1950.
2. Liozzi, *History of Physics*, Torino, 1965.
3. Moscow, 1974.
4. Moscow, 1974, 1982.
5. Moscow, 1963, 1977.

And the limited space is the totality of particles of atoms of air inside some large, but restricted volume in which " other things may be stored ". This is reflected in his example with jug and butter.

Similarly with his analysis of space, he makes a distinction between the two kinds of time. The time may be limited, literally the " partial time " and absolute.

The absolute time, in the physical and philosophical interpretation is one of the five eternal elements (God, soul, matter, time and space) that had existed always and will be forever. (And the partial time, in fact, is the time in the meaning of Aristotle, *i.e.* the interval between the beginning and the end of the motion).

A special treatise, entitled " About time and space ", also belongs to ar-Razi. This treatise had been only mentioned in various sources but hasn't yet been found.

With respect to geometrical optics, ar-Razi believed that the light was transmitted by rays emitted by the sources of light. This position contrasted with the views of ancient authors, in particular of Euclid and his followers, who thought that the process of visual perception was realized by means of special " visual rays " going out from the eyes. Ar-Razi's compatriot, Fakhr ad-Din ar-Razi, the well-known philosopher, stated : Ar-Razi had believed that the eye of the man couldn't be a source of rays and that the rays could be emitted only by fire and by stars. We also know about ar-Razi's treatment that hasn't come to our days : " The book of the character of vision, proving that the vision is not due to the rays emitted by eye and refuting the Euclid's book on optics ".

Ar-Razi's ideas about the mechanism of vision have been further developed in the works by (865-1039) and Ibn Sina (980-1037).

Along with Beruni, ar-Razi may be called with right one of the founders of the applied physics in the medieval Muslim East. The treatise, entitled " The physical balance ", that also came from his pen is devoted to the problem of determination of the specific gravity of metals scientists from next generations who dealt with the " science of nature ".

Many works have been devoted to Ibn Sina, the other great medieval scientist-Encyclopedist. But, despite the profusion of studies by Avicenna scientists, his physical legacy is, in my view, yet insufficiently studied. According to some evidence, his scientific heritage included about five hundred works, embracing practically all fields of his time's science.

The problems of physics have been analyzed in several special treatises : " The criterion of reason ", " The golden sawdust of nature ", " The discourse on the origins of thunder " and also — in his encyclopaedic works " The book of knowledge " and " The book of healing " and in his scientific correspondence with Aburaikhon Reruni.

The teaching of space, time and motion is mainly given in his encyclopaedic works : " The book of knowledge " and " The book of problem of essence and source of motion ", the rectilinear and circular " aspiration ", the incidence of heavy bodies, the theory of simple machines and their classification and also — the issues of acoustics and geometrical optics— have been analyzed in his encyclopaedic works " The book of knowledge ", " The book of healing " and " The canon of medicine ", as well as in his special treatises " The criterion of reason ", " The golden sawdust of nature " and " Discourse on the origins of thunder ".

Ibn Sina's dynamics took shape as result of assimilation of the dynamical conception of Aristotle, but Aristotle's teaching was, to a large extent, modernized, resulting in the own " Avicenna's conception ". Under the influence of Avicenna's teaching the so-called " Avicenna's school " had formed in the medieval East ; this " school " included such scientists, as (he died about 1164), (12th century) and many others.

The science of " simple " machines or to be more precise, the mechanics, was called or " science of crafty devices " in the most important work in this field.

In the treatise " The golden sawdust of nature " Ibn Sina dealt with some problems of acoustics, analyzing at the same time, the reason for the loudness of sound, the reason for its origination as a result of collision between two objects and the nature of the subsequent vibration of the air. Furthermore, he had, in principle reduced the acoustic phenomena on to the mechanical motion, as it was done by practically all scientists from next generations — up to the creators of modern theories. He also left a very important note about the strengthening and weakening of the sound — the phenomena linked by him to the acceleration and slowing down of the mechanical motion that gives rise to the sound and promotes its diffusion. His notes about the barrier placed on the way of sound and about the loudness of sound emitted by various metals and organic substances are also very interesting.

In same treatise Ibn Sina explains the reasons for the emergence of echo in mountains, makes an attempt for the explanation of why does the higher standing person hears better the voice of the lower standing, than vice versa ; he discusses the question of why the shout cannot be heard in desert and can be heard in mountains and dealt with other important issues of medieval acoustics.

In the treatise " The discourse on the origins of thunder " he compares — in my view, for the first time in history of physics — the velocities of light and sound. Though his examples are from most simple, they are interesting and understandable. Far example, he writes that the thunder and the lightning appear simultaneously but we first see the lightning and only then, hear the thunder.

In this way, Ibn Sina's and ar-Razi's works in science has left its mark in history of world culture. The scientists, physicians, poets and historians from later generations highly appreciated their encyclopaedic gifts that enabled them to address the most important problems of their time's science. The works of ar-Razi and Ibn Sina have fruitfully influenced the formation and development of the medieval " science of nature ", as well as of science and culture on the whole. This is only confirmed by the fact that from the period of Middle Ages, the scientific works of ar-Razi and Ibn Sina have been many times translated and published in many languages.

ARISTOTLE'S *METEOROLOGY* IN THE ARABIC WORLD

P. LETTINCK

Meteorological phenomena have been described and explained by various Arabic authors. We are concerned here with the treatises on meteorology written by the philosophers that were inspired by Aristotle's *Meteorology*[1]. These works include a number of letters by al-Kindī, the chapters on meteorological phenomena from the *Kitāb aš-Šifā'* by Ibn Sīnā and the *Short and Middle Commentary on the Meteorology* by Ibn Rušd[2]. Furthermore, there are the *Treatise on Meteorological Phenomena* by Ibn Suwār ibn al-K̲ammār († ±1030) and Ibn Bājja's *Commentary on the Meteorology.* These last two works will be published as part of a forthcoming book on Aristotle's *Meteorology* in the Arabic world[3]. In this book also the meteorological sections of the encyclopaedic works of Bahmanyār ibn al-Marzubān[4] (*Kitāb at-Taḥṣīl*), Abū l-Barakāt al-Baġdādī *(Kitāb al-Mu'tabar)* and Fak̲r ad-Dīn ar-Rāzī[5] *(Kitāb al-Mabāḥit̲ al-mašriqiyya)* will be studied. Also the Arabic paraphrase of Aristotle's *Meteorology* by Yaḥyā Ibn al-Biṭrīq, the compendium by Ḥunayn ibn Isḥāq and the work entitled *Olympiodorus' Commentary on Aristotle's Meteorology,* trans-

1. Aristotle, *Opera ex recensione Immanuelis Bekkeri*, Berlin, 1831 (2 vols) ; Aristotle, *The Arabic Version of Aristotle's Meteorology (Kitāb al-āt̲ār al-'ulwiyya li-Arisṭūṭālīs),* in C. Petraitis (ed.), Beyrouth, 1967 ; Aristotle, " Meteorologica ", in E.W. Webster (transl.), *The Works of Aristotle Translated into English,* vol. III, Oxford, 1923 ; Aristotle, *Meteorologica*, With an English Translation by H.D.P. Lee, Cambridge (Mass.), London, 1952 (" Loeb Classical Library " ; 397) ; Aristotle, " Meteorologie. Ueber die Welt ", in H. Strohm (transl.), *Aristoteles, Werke in deutscher Ubersetzung*, Bd. 12, I-II, Darmstadt, 1970.

2. Ibn Rušd, *Short Commentary* = " Kitāb al-al-āt̲ār al-'ulwiyya ", *Rasā'il Ibn Rušd,* vol. 4, Hyderabad, 1947. Another edition : *Kitāb al-al-āt̲ār al-'ulwiyya* (*Epitome Meteorologica*), in S.F. Abū Wāfia, S.A. 'Abd ar-Rāziq (eds), Cairo, 1994. Latin translation in : Aristotle, *Aristotelis omnia quae extant Opera,* vol. V.

3. P. Lettinck, *Aristotle's Meteorology and its Reception in the Arab World ; with an Introduction of Ibn Suwār's Treatise on Meteorological Phenomena and Ibn Bājja's Commentary on the Meteorology,* Leiden, 1999.

4. Bahmanyār ibn al-Marzubān, *at-Taḥṣīl* = *Kitāb at-Taḥṣīl,* in M. Muṭahharī (ed.), Teheran, 1970.

5. Fak̲r ad-Dīn ar-Rāzī, *al-Mabāḥit̲* = *Kitāb al-mabāḥit̲ al-mašriqiyya fī 'ilm al-ilāhiyya wa-ṭ-ṭabī'iyyāt,* Teheran, 1966 (2 vols).

lated by Ḥunayn ibn Isḥāq[6] and revised by Isḥāq ibn Ḥunayn will be discussed. The latter work is not an Arabic translation of Olympiodorus' Greek commentary[7], but appears to be a paraphrase and systematization of Olympiodorus' commentary, also containing features that are not in Olympiodorus. It is quite possible that what Ḥunayn translated was a Greek or Syrian work that was largely, but not exclusively, a paraphrase of Olympiodorus' commentary. In this paper we shall gives examples of how Aristotle's text of the *Meteorology* was turned into Ibn al-Biṭrīq's Arabic version, which is the version used by all Arabic authors on meteorology. We shall also mention some ideas with which the Arabic authors have contributed to the discussion of meteorological phenomena.

Aristotle's work became known in the Arab world by means of translations, of course. Sometimes this is a rather straightforward matter ; for instance in the case of the *Physics* the Arabic text (translated by Isḥāq ibn Ḥunayn) is a faithful rendering of the Greek. In the case of the *Meteorology* however, the Arabic version by Ibn al-Biṭrīq is rather different from the Greek : the order of chapters is different, there are chapters which do not correspond to any text by Aristotle, commentaries are added and on certain subjects different opinions are expressed. One has to conclude that it is not Aristotle's text which was rendered into Arabic, but a later, Hellenistic paraphrase, in which the text was adapted in several ways[8]. Furthermore, the Greek text of this paraphrase was probably first translated into Syrian, then from Syrian into Arabic. In several cases mistranslatings resulted in a text different from that of Aristotle. Also, the Greek author of the Hellenistic paraphrase may have misunderstood certain passages of Aristotle's text, with the same result.

Ibn al-Biṭrīq's " transformed " version of the *Meteorology* was used by all Arabic authors who wrote treatises on meteorology, either in the form of commentaries, or as more independent works. This means that they were under the impression that they read Aristotle's text, but in fact they read something different. Ibn Rušd, for instance, when he read the *Meteorology* in the version of Ibn al-Biṭrīq, thought that this represented Aristotle's text ; at the same time he also read the Arabic translation of Alexander's commentary on the *Meteorology* (extant in Greek, not in Arabic). Alexander gives a faithful rendering of Aristotle's text. Thus, Ibn Rušd would read two different statements on a certain subject, the one of Alexander and the other of what he thought was Aristotle, but in fact was the " transformed " Hellenistic-Arabic version. Then he

6. H. Daiber, *Ein Kompendium der aristotelischen Meteorologie in der Fassung des Ḥunayn ibn Isḥāq (Jawāmi' Abī Zayd Ḥunayn ibn Isḥāq al-'Ibādī li-kitāb Arisṭūṭālīs fī l-āṯār al-'ulwiyya)*, Amsterdam, Oxford, 1975 (Aristoteles Semitico-Latinus ; 1).

7. (Pseudo) Olympiodorus, " Tafsīr Alimfīdūrūs li-kitāb Arisṭāṭālīs fī l-āṯār al-'ulwiyya tarjamat Ḥunayn ibn Isḥāq ", in A. Badawī (ed.), *Commentaires en grec perdus sur Aristote et autres épîtres*, Beyrouth, 1971.

8. G. Endress, review of C. Petraitis, " The Arabic Version of Aristotle's Meteorology ", *Oriens*, 23-24 (1974), 497-509, esp. 506-509.

tries to harmonize both opinions, *i.e.* he does not choose between them, but says that there is some truth in both, or that they are different aspects of the same. This occurs, for instance in the discussion of the Milky Way, of certain properties of wind and of earthquakes). The same holds for Ibn Tibbon, who translated Ibn al-Biṭrīq's version into Hebrew, and also compared his text with Alexander's commentary.

An example of a mistranslating occurs in the chapter on winds. Ibn al-Biṭrīq's text reads : " There is calm weather (*i.e.* no wind) between the mountains (*fī wasaṭ al-jibāl*) "[9], where *fī wasaṭ al-jibāl* serves as translation of 361b28 : ἐν ταῖς ἀνὰ μέσου ὥραις (in the intervening periods, sc. between the coldness of winter and the heat of summer). Ibn Rušd in his *Middle Commentary* adopts Ibn al-Biṭrīq's reading and goes on to explain why no wind arises between mountains : because no suitable exhalation rises there, or because moistness is dominating[10]. Also Ibn Tibbon reads *fī wasaṭ al-jibāl* and notes that Alexander does not mention this issue[11].

An example of misunderstanding (possibly by the author of the Greek paraphrase) occurs in the explanation of the halo and the rainbow. Aristotle says that they are similar in certain respects and different in others. They are similar, as they both arise from reflection of sunrise against water drops or moisture in clouds ; they are different, because the halo is white, whereas the rainbow has various colours. This difference arises because in the case of the halo the reflection is from a light kind of mist, and then the colour of the reflected light is light, white. In the case of the rainbow the reflection is from a dark, dense cloud, and then the reflected light is coloured, for according to Aristotle, colours are a mixture of light and dark, or in other words, white light that has been weakened. (Aristotle gives another cause of the difference in colour between halo and rainbow, namely that the reflection in the former case is from mist which is nearby, in the latter case from a cloud which is further away, but this is not relevant for our discussion here).

This difference is explained in one phrase (373b35-374a4) : " Both (halo and rainbow) arise by reflection, but the one (rainbow) from reflection from a dark cloud..., the other (halo) from air (mist) that is lighter... ". This phrase was wrongly interpreted by the author of the Hellenistic paraphrase (or by the translator) ; apparently he thought that the phrase referred to the rainbow only,

9. Ibn al-Biṭrīq, *Meteor.* 69,6 = see C. Petraitis (ed.), *The Arabic Version of Aristotle's Meteorology (Kitāb al-āṯār al-'ulwiyya li-Arisṭūṭālīs),* Beyrouth, 1967 (Recherches publiées sous la direction de l'Institut de Lettres Orientales de Beyrouth ; serie. I, t. 39), see Aristotle, *The Arabic Version of Aristotle's Meteorology, op. cit.*

10. Ibn Rušd, *Middle Commentary = Talḵīṣ al-āṯār al-'ulwiyya,* in J.A. al-Alawī (ed.), Beirut, 1994. Partial Latin translation in : Aristotle, *Aristotelis omnia quae extant Opera,* vol. V, 107, 7-10.

11. Ibn Tibbon, *Otot ha-Shamayim,* II, 312, see R. Fontaine, *Otot ha-Shamayim ; Samuel Ibn Tibbon's Hebrew Version of Aristotle's Meteorology,* Leiden, New York, Köln, 1995 (Aristoteles Semitico-Latinus ; 8).

and concludes that the rainbow has various colours and that also a white colour appears in it which arises by reflection against a spray of mist formed from the cloud. This white colour, says Ibn al-Biṭrīq, is the yellow we may see in the rainbow[12]. The whole subject of colours in the rainbow is treated in a very confused way in Ibn al-Biṭrīq ; also Ibn Tibbon thought so, and therefore he also gave Alexander's commentary on this subject[13]. Also Ibn Rušd's treatment of the subject suffers from the confused version of Ibn al-Biṭrīq. He does not arrive at a completely consistent explanation, although he also adduced Alexander for a better understanding[14].

Similar confusions occur in Ibn al-Biṭrīq's text, e.g. in his discussion of the saltness of the sea, of winds, etc. Views which are different from Aristotle or additional explanations are presented by the Hellenistic paraphrase on subjects such as the number of exhalations rising from the earth, the Milky Way, the saltness of the sea, earthquakes, etc. All this again leads to confusion when Ibn Rušd and Ibn Tibbon read these passages and compared them with Alexander's commentary. Ibn Rušd tried to harmonize both explanations, as was said before.

Views presented in Arabic treatises on meteorology which could not be traced in previous works (Aristotle and his Greek commentators) are, for example, al-Kindī's explanation of precipitation and wind[15], Ibn Sīnā's idea that the tropics have the most moderate and temperate climate in the world[16] and Abū l-Barakāt's adoption of spiritual celestial forces that maintain various phenomena such as wind, comets and rainbow[17]. These three authors appear to have original and independent minds. Ibn Sīnā's ideas are basically Aristotelian, but he is quite critical. He makes his own observations and on the basis of these he rejects some of Aristotle's explanations, e.g. of the rainbow. He gives additional explanations for wind, earthquakes and thunder.

If one looks for progress in the direction of correct explanations one will not find them in the treatises that were inspired by Aristotle's *Meteorology* and its Greek commentators. One has to turn to the tradition based on Ptolemy's

12. Ibn al-Biṭrīq, *Meteor., op. cit.*, 92,10-93,12.

13. Ibn Tibbon, *Otot ha-Shamayim,* III, 189-229, see R. Fontaine, *Otot ha-Shamayim ; Samuel Ibn Tibbon's Hebrew Version of Aristotle's Meteorology,* Leiden, New York, Köln, 1995 (Aristoteles Semitico-Latinus ; 8).

14. Ibn Rušd, *Middle Commentary = Talḵīṣ al-āṯār al-'ulwiyya,* , I.A. al-Alawī (ed.), Beirut, 1994. Partial Latin translation in : Aristotle, *Aristotelis omnia quae extant Opera,* vol. V, 154,4-156,20.

15. al-Kindī, *Rasā'il = Rasā'il al-Kindī al-falsafiyya,* II, in M. Abū Rīda (ed.), Cairo, 1950, 1953, 70-75, 80-85 (2 vols).

16. Ibn Sīnā, *aš-šifā', Ṭab. 5 = Kitāb aš-šifā', aṭ-Ṭabī'iyyāt 5 : Al-ma'ādin wa-l-āṯār al-'ulwiyya,* in A. Muntaṣir, S. Zāyid, A. Ismā'īl, I. Madkūr (eds), Cairo, 1964, 27,16-30,15 ; Ibn Sīnā, *aš-šifā', Ṭab. 4 = Kitāb aš-šifā', aṭ-Ṭabī'iyyāt 4 : Fī l-af'āl wa-l-infi'ālāt,* in M. Qāsim, I. Madkūr (eds), Cairo, 1969.

17. Abū l-Barakāt al-Baġdādī, *al-Mu'tabar = Kitāb al-Mu'tabar fī l-ḥikma,* II, Hyderabad, 1939, 223,13-226,23 and 219,12-220,15 (3 vols).

Optics, namely the works of Ibn al-Haytham, with his experimental research on optical phenomena, and Kamāl al-Dīn al-Fārisī, who performed experiments studying the course of light rays in a transparent sphere (raindrop). In this way he found the correct explanation of the formation of the rainbow, the primary one as well as the secondary one. One had to wait for Descartes and Newton to get an explanation of the colours.

Medieval Arabic sources on the magnetic compass[1]

Petra G. Schmidl

Lorsque j'ai vu, dit l'auteur [Ibn Mâjid] dans le préambule, que les gens s'écartaient de la connaissance de la ķibla, que leurs mosquées s'écartaient de la direction de la ķibla et qu'il leur manquait la base pour connaître la ķibla, [...] j'ai déterminé par des preuves claires et faciles la direction de la ķibla par quatre moyens.

Le premier moyen est la longitude et la latitude de la Mekke illustre et, d'autre part, la longitude et la latitude du pays où se trouve l'orant.

Pour le deuxième moyen, on se sert de Al-Judayy (l'étoile polaire) [...].

Le troisième moyen suit la division de la boussole [...].

Le quatrième moyen, ce sont les côtés de la Ka'ba, c'est-à-dire les côtés de la Ka'ba (orientés d'après) les quatre vents[2].

Introduction

In this paper I present some medieval Arabic sources on the use of the magnetic compass to find the *qibla*, the sacred direction of Islam towards Mecca.

I present first with a short treatise by the Yemeni sultan al-Ashraf (*ca.* 1290), then a chapter by a Cairene astronomer called Ibn Simᶜûn (*ca.* 1300), and I conclude with some Islamic examples of astronomical instruments fitted with

1. This paper is a revised summary of my Master's thesis in History at Frankfurt University submitted in June 1995, of which an English translation is to appear.

2. G. Ferrand, *Instructions nautiques et routières arabes et portugaises du* XVe *et* XVIe *siècles*, Tome III : " Introduction à l'astronomie nautique arabe ", Paris, 1928 ; reprint Frankfurt, 1986, 209.

a magnetic compass[3]. I shall show that there was a continuous tradition of compass-making in the Islamic world from at least the 13th century to the 19th century.

AL-ASHRAF'S TREATISE ON THE MAGNETIC COMPASS

Al-Ashraf was the third Rasûlid sultan of the Yemen from 1295 until his death about a year later. Before he came to the throne he constructed several astrolabes, one of which is preserved in the Metropolitan Museum of Art in New York[4]. In addition, he wrote two substantial and sophisticated scientific

3. On the magnetic compass in the Arabic and Islamic world see E. Wiedemann, article " Maghnâṭîs, 2. The Compass ", *The Encyclopaedia of Islam*, 1st ed., Leiden, London, 1913-1936 ; reprint Leiden, 1987, and *The Encyclopaedia of Islam, new edition*, Leiden, 1960ff ; D.A. King, article " Ṭâsa ", *The Encyclopaedia of Islam, new edition*, Leiden, 1960ff ; P. Schmidl, " Two Early Arabic Sources on the Magnetic Compass " (to appear). In this summary I do not take into consideration the development in China and Europe or the use of the magnetic compass in Arabic and Islamic navigation.

On the use of the magnetic compass in China see for example M. Hashimoto, " The Origin of the Compass ", *Memoirs of the Research Department of Toyo Bunko*, 1 (1926), 69-92 ; J. Needham, " The Chinese Contribution to the Development of the Mariner's Compass ", *Scientia*, 55 (7) (1961), 225-233 (supp. 116-124) ; J. Needham, *Science and Civilisation in China*, vol. 4 (2), Cambridge, 1954-1986, 286ff (south-pointing chariot), vol. 3, 310ff (compass-sundials), vol. 4 (3), 562ff (use of the magnetic compass for navigation), vol. 4 (1), 229ff (magnetism, magnetic stones and their use) (6 vols).

For an overview of the European sources on the magnetic compass see for example A.C. Mitchell, " Chapters in the History of Terrestrial Magnetism ". Part 1 : " On the Directive Property of a Magnet in the Earth's Field and the Origin of the Nautical Compass ", *Terrestrial Magnetism and Atmospherical Electricity*, 37 (1932), 105-146 ; U. Schnall, article " Kompaß ", *Lexikon des Mittelalters*, Munich, Zurich, 1980ff ; M.J. Klaproth, *Lettre à M. le Baron A. de Humboldt, sur l'invention de la boussole*, Paris, 1854 ; reprint in : F. Sezgin *et al.* (eds), *Reprints of Studies on Nautical Instruments*, Islamic Geography, 15. Mathematical Geography and Cartography, 5, Frankfurt, 1992.

On Arabic and Islamic navigation see for example S. Soucek, article " Milâḥa, 2. In the Later Medieval and Early Modern Periods ", *The Encyclopaedia of Islam, new edition*, Leiden, 1960ff ; G.R. Tibbetts, article " Milâḥa, 3. In the Indian Ocean ", *The Encyclopaedia of Islam, new edition*, Leiden, 1960ff ; G.R. Tibbetts, *Arab Navigation in the Indian Ocean before the Coming of the Portuguese*, London, 1971 ; reprint 1981, especially on the magnetic compass 290ff.

4. On Sultan al-Ashraf see D.M. Varisco, *Medieval Agriculture and Islamic Science. The Almanac of a Yemeni Sultan*, Seattle, London, 1994, 12ff, and the literature cited there ; see further H. Suter, " Die Mathematiker und Astronomen der Araber und ihre Werke ", *Abhandlungen zur Geschichte der Mathematischen Wissenschaften mit Einschluß ihrer Anwendungen*, 10 (1900) (With additions in : *Abhandlungen zur Geschichte der Mathematischen Wissenschaften mit Einschluß ihrer Anwendungen*, 14 (1902), 157-185 (reprint in : *idem*, " Beiträge zur Geschichte der Mathematik und Astronomie im Islam. Nachdruck seiner Schriften aus den Jahren 1892-1922 ", in F. Sezgin (ed.), *Nachdrucke, Abteilung Mathematik*, Reihe B, vol. 1, n° 394, Frankfurt, 1986, 1-314, 160f (2 vols) ; D.A. King, *Mathematical Astronomy in Medieval Yemen. A Biobibliographical Survey*, Malibu, 1983, 27f ; D.A. King, " The Medieval Yemeni Astrolabe in the Metropolitan Museum of Art in New York City ", *Zeitschrift für Geschichte der arabisch-islamischen Wissenschaften*, 2 (1985) (suppl. in : *Zeitschrift für Geschichte der arabisch-islamischen Wissenschaften*, 4 (1987/88), 268-269, esp. 100 (reprint in : D.A. King, *Islamic Astronomical Instruments*, II, London, 1987 ; reprint Aldershot, 1995).

On al-Ashraf's astrolabe see L.A. Mayer, *Islamic Astrolabists and their Works*, Geneva, 1956, 83f ; a detailed description is given in D.A. King, " The Medieval Yemeni Astrolabe in the Metropolitan Museum of Art in New York City ", *op. cit.*, 99ff.

works, both extant[5]. One is an extensive astrological compendium *al-Tabṣira fî ᶜilm al-nujûm*[6], the other deals with the construction of astrolabes and sundials. It has the title *Muᶜîn* (or *Manhaj*) *al-ṭullâb fi'l-ᶜamal bi'l-asṭurlâb* and includes two short statements on the water-clock as well as on the magnetic compass and the determination of the *qibla*[7].

In this treatise on the magnetic compass al-Ashraf describes the making of " a bowl of silver or brass of medium size and with a broad rim like the rim of the astrolabe ". He marks on it the cardinal directions, south at the top, a scale for every five degrees subdivided into degrees, and the *qibla* of Taiz 20° *west* of north as well as the qibla of Aden 20° *east* of north (see fig. 1). Then al-Ashraf advocates filling the bowl with water and magnetizing a steel needle by rubbing it with a magnetic stone. The needle is inserted crosswise in a rush or reed and floats on the water in the bowl. He continues " Thereupon the needle rotates until it stops approximately on the meridian ". After determining the meridian with the aid of this floating compass one should take the direction of the *qibla* from a " twenty-by-twenty-table "[8]. After determining the meridian

5. A list of all treatises attributed to al-Ashraf is given in D.M. Varisco, *Medieval Agriculture and Islamic Science. The Almanac of a Yemeni Sultan*, *op. cit.*, 14ff ; see further D.A. King, *Mathematical Astronomy in Medieval Yemen. A Biobibliographical Survey, op. cit.*, 27ff.

6. On this treatise preserved in a single manuscript in Oxford, Bodleian Huntington 233 (Uri 905), and its contents see D.A. King, *Mathematical Astronomy in Medieval Yemen. A Biobibliographical Survey*, *op. cit.*, 28 ; D.M. Varisco, " Medieval Agricultural Texts from Rasulid Yemen ", *Manuscripts of the Middle East*, 4 (1989), 152 ; D.M. Varisco, *Medieval Agriculture and Islamic Science. The Almanac of a Yemeni Sultan*, *op. cit.*, esp. 16ff.

7. On this treatise preserved in two manuscripts in Cairo, TR (Taymûr riyâḍa) 105, and Tehran, Majlis 150 and its contents see D.A. King, *A Catalogue of the Scientific Manuscripts in the Egyptian National Library*, vol. 1, Cairo, 1981-1986, 581 and vol. 2, 362ff (2 vols) (in Arabic) ; D.A. King, *A Survey of the Scientific Manuscripts in the Egyptian National Library*, Winona Lake, Indiana, 1986, 132, n° E8 ; with more details — and with reference to the Tehran manuscript — D.A. King, *Mathematical Astronomy in Medieval Yemen. A Biobibliographical Survey*, *op. cit.*, 28f ; D.A. King, " The Medieval Yemeni Astrolabe in the Metropolitan Museum of Art in New York City ", *op. cit.*, 101f and 268f ; A. al-Azzawi, *History of Astronomy in Iraq and its Relations with Islamic and Arab Countries in the Post Abbasid Periods*, Baghdad, 1959, 234 (in Arabic) mentioning two manuscripts in Tehran, one in the library of the Majlis al-umma al-Îrânî and another in a private library.

Al-Ashraf's statement on the magnetic compass is preserved in three complete copies, and one fragment (see P. Schmidl, " Two Early Arabic Sources on the Magnetic Compass " (to appear), with an Arabic text and English translation ; E. Wiedemann, " Zur Geschichte des Kompasses bei den Arabern ", *Vehandlungen der Deutschen Physikalischen Gesellschaft*, 19/20 (1919), 665-667 (reprint in : *idem*, " Gesammelte Schriften zur arabisch-islamischen Wissenschaftsgeschichte ", in F. Sezgin *et al.* (eds), *Veröffentlichungen des Institutes für Geschichte der arabisch-islamischen Wissenschaften*, Reihe B, vol. 2, Nachdrucke 1 (1-3), Frankfurt, 1984, 883-885 (3 vols) ; reprint in : F. Sezgin *et al.* (eds), *Reprints of Studies on Nautical Instruments*, *op. cit.*, 271-273, with a partly German translation).

8. On such tables from the Yemen see D.A. King, *A Survey of the Scientific Manuscripts in the Egyptian National Library*, *op. cit.*, 192, n° Z25 ; D.A. King, " The Earliest Islamic Mathematical Methods and Tables for Finding the Direction of Mecca ", *Zeitschrift für Geschichte der arabisch-islamischen Wissenschaften*, 3 (1986), 82-149 (suppl. in : *Zeitschrift für Geschichte der arabisch-islamischen Wissenschaften*, 4 (1987-1988), 270) (reprint in : *idem*, *Astronomy in the Service of Islam, Aldershot*, 1993, XIV).

we finally count the angle on the rim of the bowl and determine the direction towards Mecca.

In one part of al-Ashraf's treatise there are some references which raise the question whether he knew the magnetic declination, that is the deviation of magnetic north from geographical north. It is not clear whether he is referring to the magnetic variation or to the result of the inaccuracy of the magnetic compass. In his diagram the cross formed by the needle and reed is aligned precisely in the cardinal directions (see fig. 1), and there are no hints of a corrected north-point. It seems more likely that al-Ashraf did not know the magnetic variation even if he had an idea that the magnetic needle does not point to true north or true south every time[9].

Further there are some difficulties with the *qibla* values in this treatise. In the diagram al-Ashraf gives the *qibla* of Taiz as 20° west of north and the *qibla* of Aden as 20° east of north (see fig. 1), but in the text the *qibla* for the middle of the Yemen (Aden, Taiz, and Zabid) is given as 27° east of north (see Table 1).

TABLE 1.

The *qibla*-values given in al-Ashraf's treatise on the magnetic compass.

	Aden	Taiz	Zabid
text	27 NE	27 NE	27 NE
figure[10]	20 NE	20 NW	-

Earlier in this manuscript, in his treatise on sundials, al-Ashraf gives also a *qibla* for the Yemen of 27° but without mentioning the quadrant, and he also gives the longitudes for Mecca and for the Yemen. This *qibla*-value can be recomposed approximately and explains the value in the text. The values in the diagram could be explained perhaps by the lack of the quadrant in the manuscript : the copyist was not aware which quadrant was be the right one.

A unique example of the very kind of bowl described by al-Ashraf is a ceramic bowl for a floating compass, made in Damascus about 1520 (see fig. 2). It is marked with a degree scale on the inner side of the rim of the bowl as well as with qibla-values for 40 cities. The geographical data on this piece shows that it belongs to a Persian or Central Asian tradition that predates this particular instrument by at least two centuries[11].

9. For a detailed discussion see P. Schmidl, " Two Early Arabic Sources on the Magnetic Compass " (to appear) and the literature cited there.

10. Labelled as *qiblat Ta^c^īzz* and *qiblat ^c^Adan*.

11. See D.A. King, " Two Iranian World Maps for Finding the Direction and Distance to Mecca ", *Imago Mundi*, 49 (1997), 62-82 ; D.A. King, *World-Maps for Finding the Direction and Distance of Mecca. Examples of Innovation and Tradition in Islamic Sciences*, Leiden (to appear).

The existence of the Damascene compass bowl attesting to a much older tradition and the existence of a treatise on the magnetic compass from the Yemen, which was an important trading center about 1300[12], point to the origin of the floating compass somewhere in Persia or Central Asia. Actually, one of the leading astronomers in the Yemen at the end of the 13th century, Muḥammad ibn Abî Bakr al-Fârisî, was a Persian[13].

But more significant than the problematic question about the origin of the idea of the magnetic compasses the indisputable fact that al-Ashraf's treatise at the end of the 13th century contains instructions on the construction and use of the magnetic needle to find the qibla.

THE CHAPTER ON THE MAGNETIC *QIBLA*-INDICATOR BY IBN SIMᶜÛN

A treatise on time-keeping written by an Egyptian astronomer called Ibn Simcûn about 1300 contains the chapter on the magnetic *qibla*-indicator[14]. It is partly based on the better-known compendium of Abû ᶜAlî al-Marrâkushî (Cairo, about 1280) which deals with spherical astronomy and astronomical instruments, but al-Marrâkushî did not mention the magnetic compass[15].

12. On the Yemen in Rasulid times see G.R. Smith, article " Rasûlids ", *The Encyclopaedia of Islam, new edition*, Leiden, 1960ff. and the literature cited there.

13. On al-Fârisî see in general D.A. King, *Mathematical Astronomy in Medieval Yemen. A Bio-bibliographical Survey*, *op. cit.*, 23 ; and also C. Brockelmann, *Geschichte der arabischen Literatur*, 2nd edition, Leiden, 1943, 3 vols and 2 suppls, vol. 2, 474 and suppl. 1, 866f ; H. Suter, " Die Mathematiker und Astronomen der Araber und ihre Werke ", *op. cit.*, 218, note 72 and p. 175, n° 349.

14. On Ibn Simᶜûn see D.A. King, article " Ṭâsa ", *The Encyclopaedia of Islam, new edition*, Leiden, 1960ff ; H. Suter, " Die Mathematiker und Astronomen der Araber und ihre Werke ", *op. cit.*, 162, n° 398 ; D.A. King, *A Survey of the Scientific Manuscripts in the Egyptian National Library*, *op. cit.*, 60, n° C24 ; D.A. King, " On the Role of the Muezzin and the Muwaqqit in Medieval Islamic Society ", in F.J. Ragep, S.P. Ragep (eds), *Tradition, Transmission, Transformation, Proceedings of Two Conferences on Pre-modern Science Held at the University of Oklahoma*, Leiden, 1996, 298f. (Collection des travaux de l'Académie Internationale d'Histoire des Sciences ; 37).

The " Chapter on the use of the qibla instrument for every locality " is only known in a single copy preserved in Leiden, Or. 468 (see P. Voorhoeve, *Handlist of Arabic Manuscripts in the Library of the University of Leiden*, Leiden, 1957, 153 ; D.A. King, " The Earliest Islamic Mathematical Methods and Tables for Finding the Direction of Mecca ", *op. cit.*, 131).

15. On the author see D.A. King, article " al-Marrâkushî ", *The Encyclopaedia of Islam, new edition*, Leiden, 1960ff. Al-Marrâkushî's compendium of spherical astronomy and astronomical instruments *Kitâb Jâmiᶜ al-mabâdi' wa 'l-ghâyât fî cilm al-mîqât* was investigated first by father and son Sédillot (J.-J. Sédillot, *Traité des instruments astronomiques des Arabes composé au XIIIe siècle par Aboul Hhassan Ali de Maroc intitulé Jâmic al-mabâdi' wa-l-ghâyât*, Paris, 1834-1835 ; reprint Frankfurt, 1984) contains the first part on spherical astronomy and sundials ; L.-A. Sédillot, " Mémoire sur les instruments astronomiques des Arabes ", *Mémoires de l'Académie Royale des inscriptions et belles-lettres de l'Institut de France*, 1 (1844), 1-229 (reprint Frankfurt, 1989) summarizes the second part on astronomical instruments).

As a first part of constructing the *qibla*-indicator Ibn Simᶜûn describes the making of a circular plate out of a sort of paper (*waraq mutamâsik*), probably *papier mâché* or a substance, probably wood (*qarᶜ*). The round plate is marked with a scale for every degree and some qibla-directions. Ibn Simᶜûn remarks that " sometimes a horizontal sundial (*basîṭa*) for a specific latitude " is fitted on the instrument. Further he says " sometimes a diagram of the exalted Kaaba is made " in the middle of the circular plate (see fig. 3).

Then he fixes two needles, probably magnetized before and probably on the back of the circular plate to balance it, which he puts " on a cone of glass or brass ". Afterwards the circular plate with the needles and the cone is put on a short vertical axis " implanted in a box of ebony or brass " and the whole is fitted with a glass cover (see fig. 4-6).

On the whole, the text is intelligible, but there are some problems in the details, among other things the precise nature of the materials involved and the interpretation of some of the technical terms used.

Finally it is not clear whether he is referring to the magnetic variation when he speaks of " the deviation of the magnetite from the meridian (*inḥirâf dhalik al-maghnâṭîs ᶜan al-khaṭṭ)* ". As with al-Ashraf, the fact that the needle does not indicate the true north-south-line is not necessarily the result of an observation of the magnetic variation[16]. But more important than specific aspects of the construction and the materials is the fact that Ibn Simᶜûn's treatise about 1300 provides the earliest written evidence from the Islamic world of a dry compass and its detailed construction.

The instrument described by Ibn Simᶜûn is very similar to the " qibliyyas " mentioned by Lane in his *Account of the Manners and Customs of the Modern Egyptians about 1830* : " The astrolabe and the quadrant are almost the only astronomical instruments used in Egypt. Telescopes are rarely seen here ; and the magnetic needle is seldom employed, except to discover the direction of Mekkeh ; for which purpose, convenient little compasses (called " ḳibleeyehs "), showing the direction of the *ḳibleh* at various large towns in different countries, are constructed, mostly at Dimyáṭ (Damietta) : many of these have a dial, which shows the time of noon, and also that of the *'aṣr* (prayer), at different places and different seasons "[17].

Already in Lane's time these instruments had a history of over 500 years in Egypt. Only very few Egyptian pieces survive, all of them late. One of these is a round Egyptian *qibla*-indicator made about 1600, now in London : all the parts which Ibn Simᶜûn mentions are represented on it, namely, *qibla*-directions, a diagram of the Kaaba, a sundial and a magnetic compass (see fig. 7).

16. See P. Schmidl, " Two Early Arabic Sources on the Magnetic Compass " (to appear).

17. E.W. Lane, *An Account of the Manners and Customs of the Modern Egyptians [...]*, vol. 1, 5th edition, London, 1871, 275 (2 vols).

Some Islamic Examples of Astronomical Instruments Fitted with a Magnetic Compass

To show that there existed an Islamic tradition of compass making is not easy. Sometimes on instruments the magnetic compass is lost, sometimes it is doubtful whether the magnetic compass is really Islamic or influenced by European traditions. Often enough the magnetic compass is definitely Islamic, but sometimes it is really European in design though the instrument may be Islamic.

The *ṣandûq al-yawâqît*, a compendia or multifunctional device by Ibn al-Shâṭir of Damascus, 1366, and now preserved in Aleppo, seems to be the earliest Islamic instrument originally fitted with a dry magnetic compass, but the magnetic compass is lost (see fig. 8)[18].

Some instruments are typically Islamic. Consequently the compass is probably Islamic too. The 'equatorial' (semi-circle) (*dâ'irat al-mu^c addal*), an astronomical instrument invented by al-Wafâ'î in Cairo in the middle of the 15th century, was originally fitted with a magnetic compass, so the magnetic compass is probably Islamic (see fig. 9)[19].

Other magnetic compasses are in the Islamic tradition. Clues to such a provenance of the magnetic compass are a south-pointing needle, a bird-shaped magnetic needle (see fig. 10), or a fish-shaped magnetic needle.

Other magnetic compasses are influenced by European instrument-making. The markings of the cardinal directions with 'spikes', one side natural, the other darkened and probably the shape of the needle indicates an European influence (see fig. 11).

Yet more obviously influenced by European trends are Islamic instruments with a magnetic compass inscribed in Latin letters (see fig. 12).

18. On the author see D.A. King, article " Ibn al-Shâṭir ", *Dictionary of Scientific Biography*, New York, 1970-1980. For a discussion of the instrument, and two medieval texts on the use of the compendium, likewise incomplete, see L. Janin, D.A. King, " Ibn al-Shâṭir's Ṣandûq al-Yawâqît : An Astronomical 'Compendium' ", *Journal for the History of Arabic Science*, 1 (2) (1977), 187-256 (reprint in : D.A. King, *Astronomy in the Service of Islam, Aldershot, op. cit.*, XII). On the magnetic compass see especially *ibid.*, 192, 195f, 204, 209, 215ff.

19. See on this instrument M. Dizer, " The Dâ'irat al-Mucaddal in the Kandilli Observatory, and Some Remarks on the Earliest Recorded Islamic Values of the Magnetic Declination ", *Journal for the History of Arabic Science*, 1 (1977), 257-262. Al-Wafâ'î (on whom see H. Suter, " Die Mathematiker und Astronomen der Araber und ihre Werke ", *op. cit.*, 177, n° 437 ; D.A. King, *A Survey of the Scientific Manuscripts in the Egyptian National Library*, *op. cit.*, 70ff, n° C61) determines a value for the magnetic variation for the first time in the Islamic world in the 15th century (M. Dizer, " The Dâ'irat al-Mucaddal in the Kandilli Observatory, and Some Remarks on the Earliest Recorded Islamic Values of the Magnetic Declination ", *op. cit.*, 260, gives al-Wafâ'î's value as 7 degrees east of north ; see further L. Janin, D.A. King, " Ibn al-Shâṭir's Ṣandûq al-Yawâqît : An Astronomical 'Compendium' ", *op. cit.*, 204).

SUMMARY

Both the treatises of al-Ashraf and Ibn Sim^{c}ûn describe the magnetic compass " in the service of Islam ". Al-Ashraf's treatise is the earliest evidence in the Islamic world of a floating compass used for astronomical and religious not for navigational purposes, and Ibn Sim^{c}ûn gives the earliest description of a dry compass in the Islamic world, for exactly the same purpose. Both descriptions are integral parts of astronomical treatises. The surviving magnetic compasses likewise attest to an Islamic tradition of such devices as integral parts of the astronomical instruments from at least the 13th century to the 19th century.

FIGURES

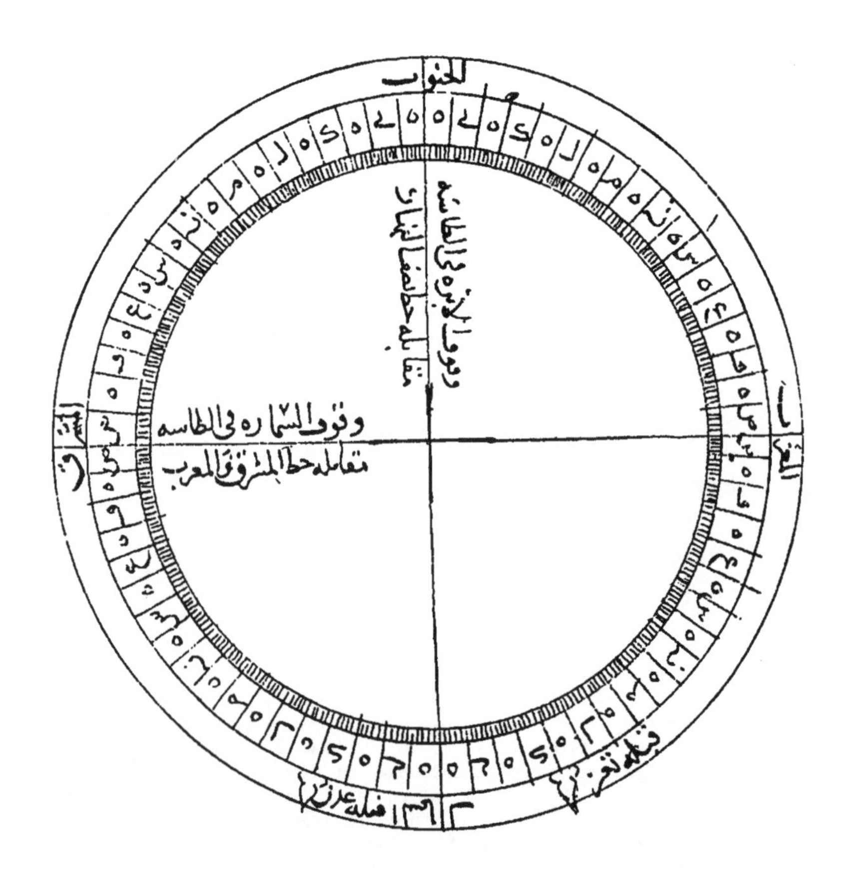

1. Al-Ashraf's diagram of the compass bowl (taken from MS Cairo TR 105, fol. 145v ; courtesy of the Egyptian National Library, Cairo).

2. Damascene compass bowl about 1520 (courtesy of the Institut du Monde Arabe, Paris).

3. The one diagram of Ibn Sim^{c}ûn's quibla-indicator (taken from MS Leiden Or. 468, fol. 190r ; courtesy of the Universiteitsbibliotheek, Leiden).

4. Part of the diagram of Ibn Sim^{c}ûn's quibla-indicator, look at the side (taken from MS Leiden Or. 468, fol. 190r ; courtesy of the Universiteitsbibliotheek, Leiden).

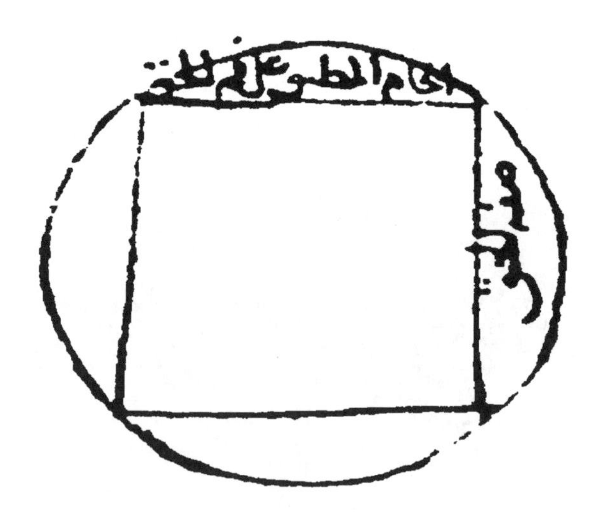

5. Part of the diagram of Ibn Sim^{c}ûn's quibla-indicator, look from above (taken from MS Leiden Or. 468, fol. 190r ; courtesy of the Universiteitsbibliotheek, Leiden).

6. The other diagram of Ibn Simᶜûn's quibla-indicator as in the manuscript (taken from MS Leiden Or. 468, fol. 190r ; courtesy of the Universiteitsbibliotheek, Leiden).

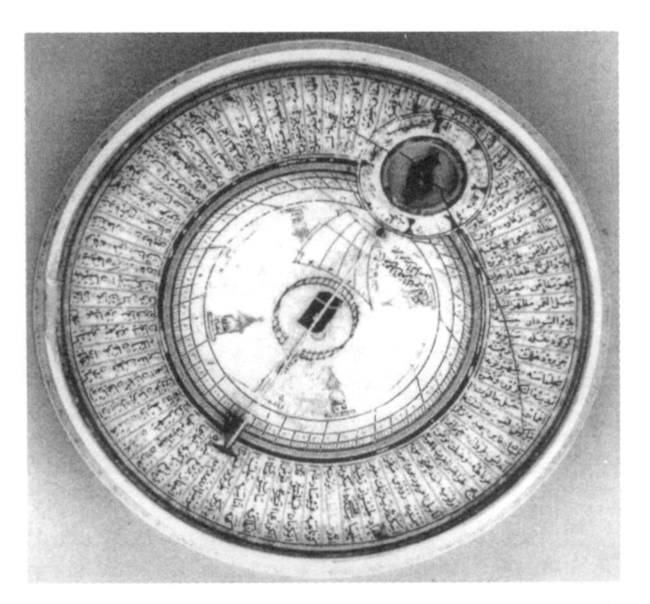

7. Circular Egyptian quibla-indicator with Kaaba, sundial, and magnetic compass (courtesy of the British Museum, London).

8. Ibn al-Shâṭir's *Ṣandûq al-yawâqît.*

9. 'Equatorial' (semi-circle) (*dâ'irat al-muᶜaddal*)
courtesy of History of Science Museum in Kandilli Observatory, Istanbul).

10. Throne of the astrolabe made by Muḥammad Mahdî Yazdî (Isfahan, about 1650) with bird-shaped magnetic needle (courtesy of the National Maritime Museum, Greenwich).

11. One of the two Safavid world-maps from the late 17th century with a complete magnetic compass (Private collection, photos courtesy of Christie's of London).

12. Instrument of unknown use with an European or rather French, magnetic compass (courtesy of the Museum of the History of science, Oxford).

Pionnier dans la recherche du degré de qualité des médicaments composés. Le médecin et philosophe arabe al-Kindī (IXe siècle)

Nouha Stéphan[1]

Dans son épître intitulée *Fī 'illat al-baḥḥārīn li-l-amrāḍ al-ḥādda*[2] (Sur la cause des jours de crise dans les maladies aiguës), al-Kindī exige du médecin d'être philosophe et de savoir appliquer en médecine aussi bien les sciences mathématiques que les sciences naturelles. Dans l'introduction à cette oeuvre nous lisons : " ... Il se peut que la plupart de ceux qui veulent être médecin ne connaissent ni les sciences mathématiques ni les sciences naturelles, et ils n'ont besoin dans l'art médical que d'apprendre les livres et de les mémoriser. Ceci empêche ces gens de savoir les causes de la médecine, ses origines. Mais ceci est acquis par le philosophe qui possède ces sciences et ce qui paraît être démontré au philosophe ne présente pas de doute, puisqu'il possède la connaissance des savants des générations précédentes "[3]. (m.t.)

Nous retrouvons l'application des mathématiques en médecine dans une autre oeuvre d'al-Kindī connue par les érudits modernes sous le titre *Fī ma'rifat quwa al-adwiya al-murakkaba* (Sur la connaissance du degré de qualité des médicaments composés). Ce sont les résultats de nos recherches sur cette oeuvre que nous voulons présenter. La critique externe du traité portera sur son attribution à al-Kindī, en se basant sur les traditions manuscrites arabe et latine ainsi que sur les publications au XVIe siècle de la traduction latine du texte ; en remarquant l'absence de titre dans l'unique manuscrit arabe, l'autonomie de notre texte sera mise pour la première fois en doute. Quant à la critique interne, elle consistera à dégager l'apport d'al-Kindī par rapport à Galien

1. L'étude des oeuvres pharmaceutiques d'al-Kindī forme le sujet d'une thèse de doctorat d'université préparée à l'Université Paris 7 sous la direction de Prof. Roshdi Rashed.

2. Le texte en question fut édité, traduit en allemand et commenté par F. Klein-Franke, " Die Ursachen der Krisen bei akuten Krankheiten. Eine wiederentdeckte Schrift al-Kindis ", *Israël Oriental Studies*, 4 (1975), 161-188.

3. Voir F. Klein-Franke, " Die Ursachen der Krisen bei akuten Krankheiten. Eine wiederentdeckte Schrift al-Kindis ", *op. cit.*, 171.

et à élucider son emploi, dans son raisonnement spéculatif, des lois de proportions selon le pythagoricien grec du IIe siècle Nicomaque de Gérase. En fait, ces lois lui serviront d'outil pour élaborer sa théorie concernant le degré de qualité des médicaments composés.

CRITIQUE EXTERNE

Le texte et son attribution à al-Kindī

Le manuscrit unique qui nous est parvenu est parvenus fut copié en 1759 et notre texte couvre les folios 25 à 35 du volume manuscrit n° 838 appartenant au fonds arabe de la Staats- und Universitätsbibliothek von München. Il est attribué explicitement à al-Kindī. En effet, tout de suite après la *basmalla* et sans aucune mention de titre, nous lisons *Qāla al-Kindī* (al-Kindī a dit).

Vient à l'appui de cette attribution du texte à al-Kindī, la mention de son nom dans le titre de trois des cinq manuscrits comprenant la traduction latine du texte[4]. La comparaison de cette traduction latine et du texte arabe manuscrit qui nous est parvenu, a révélé que la traduction latine suit assez fidèlement le texte arabe à l'exception d'un long passage qu'on ne retrouve pas dans le texte arabe et sur lequel nous reviendrons plus loin. Quant aux nombreuses éditions imprimées de la traduction latine durant le seizième siècle à Strasbourg, Lyon, Padoue et Venise, elles viennent dissiper tout doute d'attribution du texte à al-Kindī[5], et attestent, par ailleurs, la grande diffusion de cette oeuvre dans le monde médiéval latin.

Suite à l'examen attentif du passage divergent contenu dans la traduction latine, nous avons abouti à la conclusion, qu'il n'est ni l'addition du copiste latin[6], ni celle du traducteur potentiel de ce texte, Gérard de Crémone[7], mais

4. Nous avons pu localiser un manuscrit de la traduction latine de notre texte dans la collection du Couvent des Dominicains de Vienne. Ceci porte la liste des manuscrits de la traduction latine à cinq, les quatre autres ont été signalés et localisés par Mc Vaugh (éd.) dans *Arnuldi de Villanova Opera medica omnia, II, Aphorismi de gradibus*, Granada, Barcelone, 1975, 263-295. Ces pages comprennent aussi l'édition du texte latin.

5. Pour les références des publications de la traduction latine, voir K. Lantzsch, *Abū Jūsuf Al-Kindī und seine Schrift De medicinarum compositarum gradibus*, Aus der Inaugural-Dissertation der Medizinischen Fakultät der Universität Leipzig, 1921 ? Le livre de N. Rescher, *Al-Kindī. An Annotated Bibliography*, Pittsburgh, 1964 complétera la liste des publications donnée par K. Lantsch.

6. Un copiste ne peut être l'auteur de ce développement arithmétique, qui reflète des connaissances mathématiques très approfondies.

7. Nous sommes d'avis que le traducteur potentiel de cette traduction est Gérard de Crémone. D'une part, ce titre de la traduction latine est signalé dans la liste des oeuvres de Gérard dressée par ses élèves et d'autre part, un des manuscrits est conservé dans un volume manuscrit comprenant des textes traduits par Gérard. Or Gérard de Crémone est connu pour sa traduction littérale des textes aussi bien scientifiques que philosophiques. Pour sa méthode de traduction voir les articles de H. Hugonnard-Roche, " Les oeuvres de logique traduites par Gérard de Crémone ", *Gerardo da Cremona, Annali della biblioteca statale e libreria civica di Cremona*, XLI (1990), 45-56, et de D. Jacquart, " Les traductions médicales de Gérard de Crémone ", *Gerardo da Cremona, Annali della biblioteca statale e libreria civica di Cremona*, *op. cit.*, 57-70.

qu'il s'agit d'une omission survenue dans la transmission du texte arabe. L'omission vient-elle du médecin copiste Miḫā'il Ibn Šukrallāh Zanad ou bien existait-elle déjà dans la version qu'il a copié en 1759 ? (voir fig. I). Dans l'état actuel de la recherche, l'auteur de cette omission reste anonyme.

Absence de titre

L'absence du titre, que nous avons constaté dans le seul manuscrit arabe conservé, mériterait d'être soulignée. Nous avons été amené à mettre en question que notre texte soit une oeuvre autonome.

FIGURE I.

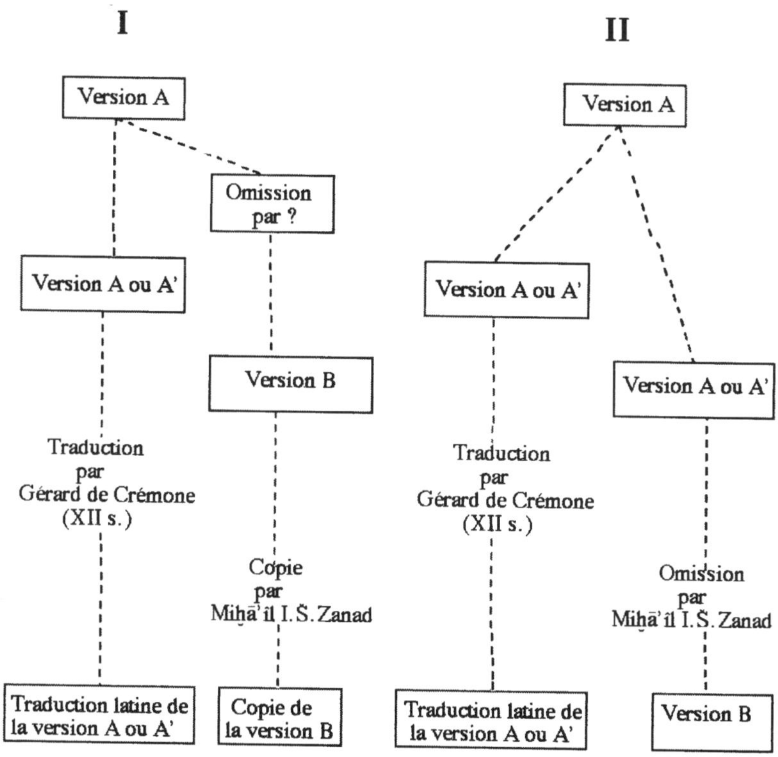

Schéma montrant les deux explications proposées pour la transmission du texte arabe.

Un élément favorable à la supposition que le texte ne forme qu'un chapitre est que le texte arabe nous est parvenu sans titre et présente une fin brusque qui rappellerait plutôt la fin d'un chapitre. Un autre élément pourrait être avancé : le médecin et philosophe du XIIe siècle Ibn Rušd (Averroès), dans sa réfutation de la théorie du degré de qualité des médicaments composés d'al-Kindī, dans son livre *al-Kulliyāt*[8] *(Colliget),* parle d'une *maqāla* (chapitre), puis, plus loin dans le même texte, d'un *kitāb* (livre), où al-Kindī aurait exposé cette théorie, sans pour cela citer un titre. Cette double référence de *maqāla* et de *kitāb* de la part d'Ibn Rušd laisserait penser à ce que cette théorie est signalée dans un chapitre d'un traité. Si ce texte devait être un chapitre d'un traité on pourrait supposer qu'il constituerait " très probablement " un chapitre de l'*aqrābāḏ īn*[9] (ou pharmacopée) d'al-Kindī, dont une sélection nous est parvenue[10].

Si certains éléments nous laissent penser que le texte est un chapitre d'une oeuvre d'al-Kindī, probablement l'*aqrābāḏīn*, un autre élément laisse penser à une oeuvre complète : le texte arabe ne nous est parvenu qu'isolément, il en est de même de la traduction latine et des diverses éditions imprimées de cette traduction latine.

CRITIQUE INTERNE

Théorie pharmacologique d'al-Kindī

Dès l'introduction de son traité, al-Kindī dévoile qu'il se propose de parachever la théorie pharmacologique des anciens (c'est effectivement celle du médecin grec du IIe siècle Galien qu'il citera plus loin dans le texte) relative au degré de qualité des médicaments simples et des médicaments composés et dit : " Quand j'ai constaté que les anciens ont pris le soin de parler de chacune des qualités des médicaments simples isolément, quant à la chaleur, à la froideur, à l'humidité ou à la sécheresse, et y ont distingué des limites au nombre de quatre, et les ont appelées, en chacune de ces qualités, premier degré, deuxième, troisième et quatrième, mais ont renoncé à en faire autant pour les médicaments composés...

j'ai donc pensé que parvenir à la connaissance du degré de qualité des médicaments composés offrirait de grands avantages, car il est nécessaire que les degrés de qualité du médicament composé varient en plus ou en moins selon que varient les degrés de qualité de ses composants, et il est impossible que ses degrés de qualité se réduisent aux degrés de qualité de quelques uns de ses composants à l'exclusion des autres "[11]. (m.t.)

8. Voir Ibn Rušd, *Kitāb al-Kulliyāt*, Édition critique par J.M. Forneas et C. Alvarez Morales, Madrid, 1987, 391.

9. Un contemporain d'al-Kindī, Sabūr Ibn Sahl, a consacré les quatre premiers chapitres de son *aqrābāḏīn* aux règles à respecter pour préparer des médicaments composés. Il y signale les degrés de qualité des médicaments simples selon Galien. Voir O. Kahl (éd.), *Sābūr ibn Sahl's (d. 255/ 869) Dispensatorium parvum (al-Aqrābā ḏ īn al-saghīr)*, Thèse de doctorat, Manchester, 1992.

10. al-Kindī, *The Medical formulary or Aqrābāḏ īn of al-Kindī. Translated with a study of its Materia Medica*, in Martin Levey (éd.), Madison, Milwaukee, London, 1966.

11. Voir Manuscrit arabe n° 838 de la Staats- und Universitätsbibliothek von München, fol. 25a.

L'originalité qu'al-Kindī revendique est, d'une part, une confirmation de sa méthode d'aborder ses sujets, qui consiste à apporter tout ce qu'ont dit les anciens à ce sujet et de parachever là où ils n'ont pas tout dit[12]. Elle est d'autre part, un exemple qui reflète l'existence au IXe siècle, dans le domaine de la médecine et de la pharmacie arabes, d'une dialectique recherche traduction : le médecin al-Kindī, essaie de combler une lacune remarquée dans les oeuvres médicales traduites et élabore une théorie concernant le degré de qualité des médicaments composés.

Cet exemple vient s'ajouter à ceux présentés par Roshdi Rashed dans son étude sur l'existence au IXe siècle de cette dialectique recherche traduction, exemples qui relèvent eux des domaines des mathématiques et de l'optique[13].

Mais retournons à notre texte et voyons en résumé les différentes étapes de l'élaboration de la théorie pharmacologique d'al-Kindī. Après avoir dit que les anciens n'ont pas parlé du degré de qualité des médicaments composés, al-Kindī propose de combler cette lacune en étudiant tout d'abord, le tempéré absolu où les deux qualités primaires actives — chaleur et froideur — ainsi que les deux qualités primaires passives — sécheresse et humidité — sont à égales quantités. Il recherchera, en une seconde étape, la relation qui lie entre eux les différents degrés d'une qualité dominante et c'est là qu'il a recours à la loi des proportions du pythagoricien du IIe siècle Nicomaque comme nous le montrerons plus loin. Se basant sur sa spéculation mathématique, il avancera que la progression géométrique de raison 2 est la relation qui lie les différents degrés d'une même qualité dominante entre eux.

Il reprend le même raisonnement pour le médicament tempéré pris dans son action sur le corps humain tempéré et arrive à des résultats que nous avons réuni sous forme de tableau (Fig. II) pour faciliter la compréhension.

Pour al-Kindī :

FIGURE II.

Rapport des parties de qualités contraires et du degré de qualité d'un médicament simple selon al-Kindī.

Médicament	Qualité dominante (par ex. chaleur)	Qualité contraire (par ex. froideur)
tempéré	1/2 partie	1/2 partie
1er degré	1 partie	1/2 partie
2e degré	2 parties	1/2 partie
3e degré	4 parties	1/2 partie
4e degré	8 parties	1/2 partie

12. Pour plus de détails sur cette méthode d'al-Kindī, on consultera son épître " Philosophie Première ", dans M. Abū Rīdā (éd.), *Rasā'il al-Kindī al-falsafiyya*, vol. I, Le Caire, 1950, 123.

13. Voir R. Rashed, " Problems of the transmission of Greek scientific thought into Arabic : examples from mathematics and optics ", *History of Science*, 27 (1989) 199-209.

A partir de cette relation qui lie les différents degrés d'une qualité dominante dans les médicaments simples, le calcul du degré de qualité d'un médicament composé, pour al-Kindī, sera un simple calcul : on additionne séparément les parties chaudes, humides et sèches des différents médicaments simples entrant dans la composition d'un médicament composé. En dégageant leur relation, on pourra trouver le ou les degré(s) de qualité d'un médicament composé.

Cette théorie élaborée permettait trois applications principales pour lesquelles al-Kindī avance des exemples représentés sous forme de tableaux. La première application est le calcul du degré d'un médicament composé connaissant le ou les degré(s) de qualité des médicaments simples qui le composent ainsi que leur quantité. Cette même théorie permettait aussi de préparer des médicaments tempérés en utilisant les médicaments simples de qualités contraires dans des proportions bien définies. Quand à la troisième application — qui est à notre avis la plus importante — elle consiste à pouvoir donner avec précision le sous-degré de qualité d'un médicament composé. Comparée à la théorie pharmacologique de Galien qui donne pour une qualité dominante quatre degrés et trois sous-degrés les séparant ce qui ramène à douze possibilités, la théorie d'al-Kindī présente une multitude de possibilités de sous-degrés qui peuvent être exprimées avec une plus grande précision.

Influence de Nicomaque

La tendance pythagoricienne d'al-Kindī est attestée dans un grand nombre de ses oeuvres et l'application des mathématiques en médecine n'en est qu'un exemple. En effet, les oeuvres philosophiques d'al-Kindī dévoilent aussi cette tendance comme par exemple son traité intitulé : *fī kammiyat kutub Aristūtālī* (Sur le nombre de livres d'Aristote) où il rapporte que les sciences mathématiques sont les sciences propédeutiques nécessaires pour l'acquisition de la science philosophique. Dans ce même livre, al-Kindī insiste aussi que l'arithmétique est la science qu'il faut apprendre en premier parmi les sciences mathématiques[14].

La comparaison du texte latin qui porte un développement supplémentaire signalé dans l'introduction de notre étude, avec l'*Introduction Arithmétique* de Nicomaque a montré qu'al-Kindī possédait bien les connaissances en ce qui concerne les lois de proportions, leur ecthèse et leur résolution à l'égalité.

Nous avons comparé la méthode de Nicomaque selon laquelle à partir de trois termes égaux on peut avoir toutes les progressions, avec le développement donné par al-Kindī. Le résultat de la comparaison a montré qu'al-Kindī présente la même méthode en se contentant de ne présenter que la première progression pour chaque genre.

14. Voir Kindī, *Rasā'il al-Kindī al-falsafiyya*, vol. II, in M. Abū Rīdā (éd.), Le Caire, 1950, 365-384.

Au sujet de l'ecthèse des progressions, le développement de Kindī, dont nous donnons une citation, trahit l'influence de Nicomaque :

" Composons donc, après la progression du double, les espèces restantes, en alignant dans le tableau la [progression du] double et inscrivant le plus grand nombre le premier et le plus petit le dernier ; puis, sous ce premier [terme], écrivons son semblable ; sous le moyen, un nombre égal à sa [propre] valeur plus celle du premier ; sous le dernier, sa [propre] valeur plus celle du premier plus le double du moyen : on obtiendra ainsi une progression qui ajoute une partie. ...

On forme aussi [quoique, c.à.d. ensuite] la [progression] qui ajoute des [c.à.d. deux] parties : [pour cela] nous disposons la [progression] qui ajoute une partie comme nous l'avons fait pour la [progression du] double, et nous opérons jusqu'au bout comme nous avons fait d'abord.

On forme ensuite la [progression du] double qui ajoute une partie ; [pour cela,] nous disposons la [progression] qui ajoute une partie, telle quelle, sans la renverser, et nous faisons comme nous avons fait d'abord.

On forme ensuite la [progression du] double qui ajoute des parties : [pour cela,] nous disposons la [progression] qui ajoute des [c.à.d. deux] parties, [telle quelle] sans la renverser, et nous opérons jusqu'au bout comme nous l'avons fait jusqu'ici... "[15]. (Trad. L. Gautier)

Nous rapportons dans un but comparatif un passage de l'Introduction Arithmétique de Nicomaque où il est dit :

" De ces multiples bien ordonnés eux-mêmes, une fois inversés naissent immédiatement..., les épimores... du premier multiple inversé, le double, naît le premier épimore, l'hémiole ; du second, le triple , naît le second dans les épimores...

En partant d'un autre début, les épimores eux-mêmes étant disposés comme ils ont poussé, avec cependant une inversion voici que naissent les épimères qui viennent naturellement après eux... Si les épimores rangés en bon ordre ne sont pas inversés, mais disposés à l'endroit, voici que sont engendrés au moyen des mêmes règles les multiplépimores... De ceux qui sont engendrés à partir des épimores inversés, c'est-à-dire des épimères, et de ceux qui le sont à partir des épimores sans inversion, c'est-à-dire des multiplépimores, voici de nouveau que de la même façon, au moyen des mêmes règles, sont engendrés directement et par inversion les nombres qui mettent en évidence les relations restantes "[16]. (Trad. J. Bertier)

Al-Kindī reprend aussi la méthode nicomaquienne pour ramener les progressions proportionnelles à trois termes égaux et dit : " Quand maintenant

15. Voir L. Gautier, *Antécédents gréco-arabes de la psychophysique*, Beyrouth, 1938, 49 nbp3.

16. Consulter Nicomaque de Gérase, *Introduction arithmétique*, Introd., trad. et notes par J. Bertier, Paris, 1978, 90-95.

nous voudrons en réduisant ces espèces les ramener à l'égalité, nous retrancherons du moyen la valeur du petit, puis du grand le double de ce qui est resté du moyen plus la valeur du petit ; et s'il reste trois nombres égaux, nous avons trouvé ce que nous cherchions. Sinon, les [trois termes] sont sortis de l'espèce dans laquelle ils furent en rapport [réductible à légalité] et sont passés dans une espèce qui est plus éloignée de la nature que l'espèce de laquelle a été faite cette réduction. Ensuite, nous opérerons encore de la même manière que d'abord. Si les trois nombres restants sont égaux exactement, nous avons obtenu ce que nous voulions. Sinon, ils sont passés dans une autre espèce qui est plus éloignée de la nature que l'[espèce cherchée] "[17]. (Trad. L. Gautier)

En fait de réduction à l'égalité de trois termes proportionnels selon une relation donnée Nicomaque dit : " Que te soient donnés trois termes dans une relation et proportion quelconque ... à la seule condition que le moyen terme apparaisse relativement au petit dans le même rapport que le grand relativement au moyen ; retire toujours le petit terme du moyen, qu'il se trouve premier ou dernier, pose le petit comme premier terme, place comme deuxième terme le reste du second après la soustraction, et une fois le premier terme et deux fois le second terme ainsi obtenus étant soustraits du reste ... fais du reste le troisième terme, et les nombres qui naissent seront dans une autre relation antérieure par nature. Si de nouveau, de ces trois mêmes termes et de la même façon, tu enlèves le reste du terme, les trois termes que tu réexamines seront réduits en trois autres plus pythméniques, et tu trouveras toujours cette consécution, jusqu'à ce qu'ils aient été ramenés à l'égalité ; en conséquence de quoi, il y a toujours nécessité, c'est clair, de soutenir que l'égalité est de toute façon l'élément de la quotité relative "[18]. (Trad. J. Bertier)

Les nombreux points de vue identiques entre al-Kindī et le pythagoricien Nicomaque mériteraient d'être le point de départ d'une recherche exhaustive sur l'influence de la tradition nicomaquienne sur l'oeuvre d'al-Kindī.

En montrant par quelques citations qu'al-Kindī a utilisé la théorie des proportions de Nicomaque, nous espérons avoir contribué à mieux connaître les deux traditions al-kindienne et nicomaquienne.

17. Voir L. Gautier, *Antécédents gréco-arabes de la psychophysique, op. cit.*, 49 nbp 3.
18. Voir Nicomaque de Gérase, *Introduction arithmétique, op. cit.*, 95-97.

SÉPARATION DE LA MÉDECINE ET LA PHARMACIE : PLAIDOYER D'AL-RĀZĪ

Mehrnaz KATOUZIAN-SAFADI

Le texte *Kitāb ṣaydalat al-ṭibb*[1] (le livre de la pharmacie en médecine) inaugure le chapitre des médicaments composés de l'encyclopédie médicale *Kitāb al-ḥāwi fī al-ṭibb* qui a été rassemblée peu après la mort du savant et du médecin, al-Rāzī ou Rhazès du monde latin (*ca.* 865-925). Ce livre a été édité[2] et l'introduction du chapitre XXII sur les médicaments composés nous a servi de texte de référence. Après l'analyse de ce texte, nous donnerons une première traduction en langue française.

Le biobibliographe Ibn Abi Usaybi'a (XIIIe s.) rapporte que ce même texte intitulé *al-ṣaydala fī ṭibb* forme un chapitre de *Kitāb al-fāḫir* d'al-Rāzī. Ce texte nous est également parvenu comme fragment isolé et A.Z. Iskandar (1959) a édité ce fragment et en a donné la première traduction anglaise[3]. Une traduction en langue allemande a également été donnée par M.M. Kanawati (1975)[4]. Nous ferons appel à ces deux éditions et à leur traduction.

1. Ce titre est cité par les biobibliographes anciens. Dans *al-fihrist* Ibn Nadīm écrit que le livre *al-ḥāwi* est divisé en douze parties et que la septième partie est nommée *fī ṣydana al-ṭibb*. Ibn Abi 'Uṣayba rapporte que la cinquième partie du livre *kitāb al-ğame'* s'intitule *fī ṣydana al-ṭibb*. La traduction littérale de *ṣaydala al-ṭibb* serait " le livre de la pharmacie de la médecine ". Ce titre ne nous semblant pas correct nous traduisons le titre par " le livre de la pharmacie en médecine ". A son livre concernant les médicaments simples, al-Birūnī donne un titre similaire : " le livre de la pharmacie en médecine " *Kitāb al-ṣaydala fī al-ṭibb*.

2. Rāzī (Abū Bakr Muḥhammad b. Zakariyya b. Yaḥya al-), *Kitāb al-Hāwī fī al-ṭibb (Rhazes' Liber Continens)*, in M. 'Abdu'l Mu'id Khan (éd.), Hydarabad, Dā'irat al-ma'ārif al-osmania, 1371/1952-1394/1974 (Osmania Oriental Publications Bureau ; vol. XXII) (1971). Le nombre des chapitres de *Kitāb al-ḥāwi fī al-ṭibb* varie selon les manuscrits et selon les bibliographes.

3. A.Z. Iskandar, *A study of ar-Rāzī's medical writings with selected texts and english translations*, A Thesis submitted for the degree of Doctor of Philosophy in the University of Oxford, Oxford, 1959.

4. M.M. Kanawati, *Ar-Rāzī Drogenkunde und Toxicologie im " Kitāb al ḥāwi " (Liber Continens) unter Berücksichtigung der Verfälschungs- und Qualitätskontrolle*, Inaugural Dissertation, Marburg, Philipps-Universität Marburg/Lahn, 1975.

L'introduction du chapitre des médicaments composés est un véritable plaidoyer pour la séparation de la pharmacie et de la médecine. Et, dès les premières lignes l'auteur qualifie d'*ignorant* celui qui n'a pas saisi la nécessité d'une telle séparation. Dans ces pages, l'auteur s'efforce de déterminer les frontières entre les deux professions *al-ṣaydala* (la pharmacie) et *al-ṭibb* (la médecine), les deux mots composant l'intitulé même du texte. Le champ d'activité de chacun de ces deux métiers ou arts est délimité à plusieurs reprises. Le médecin est celui qui " connaît l'effet des médicaments sur le corps humain et la connaissance des causes et des symptômes des maladies ". De même le domaine de la pharmacie est défini par opposition à celui du médecin : " Et il n'est pas non plus permis, d'appeler médecin celui qui a des bonnes connaissances des diverses sortes de médicaments, leurs formes, leurs couleurs, leur pureté et leur falsification ".

L'auteur souligne que de multiples arts sont au service de la médecine mais le médecin ne doit se consacrer qu'à son art spécifique. La variété des outils nécessaires au médecin est énumérée (les divers remèdes, et les divers instruments tels que les lancettes, les ciseaux, les scarificateurs, etc...). Les médicaments, tout comme les instruments médicaux, ne sont que des outils au service du médecin, et la pharmacie n'est qu'un des arts au service de la médecine. Or, aucun de ces outils et aucun de ces arts ne mérite un plaidoyer sauf la pharmacie.

Al-Rāzī rappelle au médecin comment, en approfondissant ses propres connaissances et ses pratiques rationnelles, il peut palier à ses méconnaissances en pharmacie et se protéger des erreurs volontaires ou involontaires du pharmacien. Les médicaments et divers composés sont considérés sous deux angles : " leurs propriétés internes " ou les *ḫawāṣ*, dont la connaissance n'est pas du domaine de l'activité du médecin et " leurs effets apparents ", auxquelles le médecin peut accéder grâce à son expérience[5].

Selon Al-Rāzī, seulement certains médecins ont la capacité de se consacrer en même temps à la médecine et à la pharmacie. Al-Rāzī qui appartient à cette catégorie capable d'acquérir ces deux spécialités, veut défendre le médecin contre sa propre tentation et contre l'ignorance d'un public plus large se doutant de la prescription médicamenteuse d'un médecin non connaisseur de la pharmacie. Il semble que la tentation et le danger d'une confusion des champs de spécialisation proviennent essentiellement de la pharmacie dont la délimitation ne semble pas encore être affirmée.

5. Selon la médecine ancienne, les cinq sens permettaient de découvrir certaines actions des médicaments sur le corps humain. *Cf.* Al-Rāzī (Abū Bakr Moḥammad b. Zakariyya b.Yaḥya al-), *al-Mansūrī fī al-ṭibb*, Edité et commenté par H.B. al-Siddīqī, al-Kuwait, Mansurāt ma'had al-maḫṭūṭāt al-'arabiyya, 1979. Au chapitre III de ce livre l'auteur donne une description de ces qualités apparentes.

En effet, dans l'introduction de sa pharmacopée (*aqrābāḏīn*)[6], Šābūr ibn Sahl, médecin du IXe siècle s'adresse à " ceux qui veulent préparer des médicaments ", sans mentionner spécifiquement le titre du pharmacien. Si nous comparons ce texte à l'introduction d'al-Bīrūnī (mort en 1048) dans *Kitāb al-ṣaydala fi al-ṭibb*[7], nous nous apercevons que presque un siècle après al-Rāzī, cette séparation de la pharmacie et de la médecine semble être bien établie et al-Bīrūnī y développe les diverses activités au sein de la pharmacie et donne déjà les prémices d'une histoire de la pharmacie.

Nous avons souligné ci-dessus que dans notre texte l'auteur s'acharne à plaider avec vigueur pour une séparation de deux champs d'activité, la pharmacie (*ṣaydala*) et la médecine (*ṭibb*). Il emploie explicitement les termes *al-ṣaydala* (la pharmacie) et *al-ṣaydalānī* (le pharmacien). Or dans les traductions anglaise et allemande mentionnées plus haut, le mot *ṣaydala* a été traduit par *materia medica*. L'origine de cette erreur semble provenir d'un article de M. Meyerhof (1933)[8] qui pour la première fois a défini *ṣaydala* par *materia medica*, et lors de leur traduction, A.Z. Iskandar et M.M. Kanawati se réfèrent tous deux à cet article de M. Meyerhof. Une telle traduction qui limite le vaste champ de " la pharmacie " à " la matière médicale " trahit la vision d'al-Rāzī. En effet, la matière médicale (*al-adwiya al-mufrada*) ne couvre qu'une fraction du champ des connaissances et des activités rattachées à la pharmacie, et al-Rāzī évite une telle confusion, en énumérant la diversité des connaissances et des pratiques nécessaires au pharmacien.

Dans un certain nombre de ses oeuvres, l'auteur souligne que la connaissance en général et la médecine en particulier sont le produit d'une accumulation des acquisitions théoriques et expérimentales dans le temps. Ainsi au chapitre quatre du livre *al-manṣūrī* sous le titre de *miḥnat -al-ṭabib* (l'examen du médecin), l'auteur précise qu'une vie entière ne suffit pas à l'acquisition des connaissances de son époque sans le recours aux *anciens,* car il a fallu que des milliers d'hommes travaillent aux cours des milliers d'années pour constituer ce savoir. Dans le texte examiné ici, l'auteur insiste sur le temps nécessaire à investir par le médecin dans son propre champ et dit que " le médecin ne doit pas délaisser son art qui lui est spécifique et se pencher sur d'autres arts " et précise que " le médecin doit s'occuper de son art et le maîtriser comme d'autres spécialistes le font ". Ce plaidoyer est une mise en garde du médecin

6. Šābūr Ibn Sahl, *Sābūr ibn Sahl's (d. 255/869) Dispensatorium parvum (al-Aqrābāḏīn al-saghīr)*, Edited and commented by O. Kahl, Leiden, 1992 ; Šābūr Ibn Sahl, *al-Aqrābāḏīn,* MSS Téhéran Malik 4234.

7. Al-Bīrūnī (Abū al-Rayḥān Muḥammad b. Aḥmad), *Kitāb al-Ṣaydana fī al-ṭibb*, Edition avec commentaire de 'Abbās Zaryāb, Téhéran, 1981 (Publications de l'Université de Téhéran ; n° 572, Section Histoire des Sciences ; n° 3) ; Al-Bīrūnī (Abū al-Rayḥān Muḥammad b. Aḥmad), *Kitāb al-Ṣaydana fī al-ṭibb. Al-Bīrūnī's Book on Pharmacy and Materia medica*, Edited with english translation by H.M. Sa'id, Karachi, 1973.

8. M. Meyerhof, " Das Vorwort zur Drogenkunde des Bīrūnī ", *Quellen und Studien zur Geschichte der Naturwissenschaften und der Medizin*, 3 (1933b), 157-208 Texte arabe 1-18.

contre toute dispersion et une incitation à la spécialisation dans le domaine de la médecine. Une connaissance approfondie de la pratique pharmaceutique de al-Rāzī est indispensable pour comprendre les diverses motivations de l'auteur pour écrire ce plaidoyer. Nos recherches doivent se poursuivre afin d'établir l'influence et la portée de ce texte d'al-Rāzī sur ses contemporains.

TRADUCTION

Livre *Kitāb al-Hāwi fī al-ṭibb* d'al-Rāzī, volume XXII

Au nom de Dieu clément et miséricordieux.
Livre de la pharmacie en médecine

Abū Bakr Muḥammad Zakariyyā al-Rāzī a dit :

“ la connaissance des médicaments, et la distinction de leur bonne ou mauvaise qualité, de leur pureté ou de leur falsification ne sont pas nécessaires au médecin, comme le croient les gens ignorants mais, le médecin est plus digne et méritant en les dominant. Pour cela, j'ai pensé qu'il faut rassembler “ ce qui se rapporte à ” cet art, bien qu'il ne soit pas une branche nécessaire de la médecine, dans un livre qui lui est consacré, pour qu'il soit connu. Ainsi, nous avons collecté chaque art, et nous lui avons consacré un livre ”[9].

Si quelqu'un considère que “ la pharmacie ” est une des branches propres de la médecine, au même titre que la connaissance de l'action des médicaments sur le corps humain et la connaissance des causes et des symptômes des maladies, et qu'il ne considère pas la “ pharmacie ” comme un art au service de la médecine, tels que le sont tous les métiers les uns pour les autres, celui-ci sera alors obligé de considérer de nombreux “ autres ” métiers comme branches de la médecine, le médecin ayant besoin de ces arts de la même façon qu'il a besoin de s'intéresser à la pharmacie. En effet, le médecin a besoin de nombreux remèdes dont des condiments aromatiques, des fruits, des gommes et des huiles essentielles. De même, en thérapie, en cas de besoin, il pourra prendre des outils et des instruments, comme des lancettes, des forceps, des sondes, des aiguilles, des tenacules, des ciseaux, des clystères, des scarificateurs de cou (ou des ventouses)[10] et d'autres “ outils ” de ce genre, trop nombreux pour être énumérés.

Et il ne faut pas considérer ceux-ci comme une branche spécifique de la médecine mais comme un art au service de la médecine comme le sont les métiers les uns par rapport aux autres. Et il n'est pas non plus permis d'appeler médecin celui qui a de bonnes connaissances des diverses sortes de médica-

9. Pour cette phrase, l'édition de A.Z. Iskandar (AZI) nous paraissant plus cohérente, nous en donnons la traduction.

10. Pour les instruments médicaux voir : M.S. Spink, G.L. Lewis, *On surgery and instruments, Abulcasis abul -Qasim Khalaf ibn 'Abbas al-Zahrawi, a definitive ed. of the Arabic text with English translation and commentary*, London, 1973.

ments, de leurs formes, de leurs couleurs, de leurs puretés et de leur falsification. En revanche, est appelé médecin celui qui connaît les effets de ces médicaments sur le corps humain. Il est aussi celui qui, lorsqu'on lui présente un médicament qu'il n'a jamais vu et qu'il ne connaît ni de nom ni de citation, sera capable de le reconnaître en examinant tous ses effets apparents. Et si nous soulignons les effets apparents c'est parce que les médicaments ont aussi des effets internes " non apparents " et celles-ci se nomment les propriétés, *al-ḫawāṣ*, et le médecin ne peut les déduire.

Par exemple, tel est le cas de l'attraction de l'aimant pour le fer, et de la cessation de cet effet lorsque l'aimant est traité à l'ail et le rétablissement de l'effet premier si l'aimant est lavé au vinaigre ; ou le cas de la répulsion de la roche dite " anti-vinaigre " pour le vinaigre ; ou encore l'édulcoration du vinaigre si l'on y plonge la litharge (*mordāsinğ*), ou le noircissement des corps que la litharge provoque lorsqu'elle est " accidentellement " associée à la pâte épilatoire (*nūra*), et ainsi de suite. Et s'il ne peut pas déduire ces effets par la voie médicale, il est plus à même de les déduire par la voie de l' expérience, s'il est compétent dans cet art, et s'il est excellent[11].

Et le médecin ne doit pas délaisser son art qui lui est spécifique et se pencher[12] vers d'autres arts car, s'il le fait, comme les arts sont liés les uns aux autres, il sera contraint d'aborder beaucoup d'autres arts. Cependant, " il peut le faire " modérément et de sorte que cela ne l'occupe pas excessivement, en discernant la nécessité de ce qui lui est plus convenable et plus spécifique[13].

Le médecin doit s'occuper de son art et le maîtriser comme d'autres spécialistes le font. Ces spécialistes portent " principalement " intérêt à leur propre art et ne s'intéressent à d'autres arts, qui s'en rapprochent, que dans les moments de loisir et de repos. Après avoir dominé intégralement son art, le médecin peut faire de même, en y puisant juste ce dont il a besoin. Cependant, s'il trouve en lui la capacité et le plaisir d'y consacrer plus de temps, " il peut le faire ". Car s'il est possible à un homme de s'entraîner dans plusieurs arts cela est meilleur[14].

Certaines personnes pensent que le médecin qui connaît l'effet d'un médicament mais qui ignore son espèce ne peut pas l'utiliser avec assurance et confiance. Car prétendent-ils, il n'est pas à l'abri que le vendeur lui fournisse autre chose que ce qu'il désirait. Cela constitue le meilleur " argument " pour ceux qui voudraient que l'art de la pharmacie soit une branche de la médecine.

11. Les quelques mots qui se succèdent sont incompréhensibles dans les deux éditions *bāle ġan li jami'nās*. C'est nous qui donnons cette lecture.

12. Pour ce mot *yağida* nous avons opté pour l'édition de A.Z.I., *yaḫdima*.

13. Ce passage est incompréhensible dans les deux éditions et nous donnons ici une interprétation possible.

14. *Ibidem*.

Et nous pouvons leur répondre ; que le médecin pourrait éviter cela de deux manières : en premier, par sa pratique et par son art[15]. Car, s'il a demandé un médicament *muḥallilan*[16] et qu'on lui donne un astringent ou " un médicament " d'un goût âpre ou un médicament obstruant fort, il saurait assurément que le médicament n'est pas celui qu'il a prescrit. Et si le médecin demande un médicament *mumsikan et muġaiyran*[17], et " le médecin " le trouve " au goût " piquant ou amer[18] comme précédemment, le médecin juge ainsi que c'est le pharmacien qui s'est trompé.

Et la seconde est la voie de l'expérience rationnelle ; en effet s'il veut acheter un médicament composé de plusieurs (ingrédients) que " les vendeurs " nomment par leurs " propres " noms, il ne peut y avoir erreur ni négligence, s'il les achète dans les lieux de ventes " spécifiques ". Il en fera de même pour connaître leur bonne qualité et leur pureté. Et cela n'est pas du tout difficile ou contraignant sauf dans les villages et les hameaux. Et il peut aussi exister (d'autres situations) semblables au second cas, que nous ne voulons pas exposer ici, par souci de brièveté.

Toutefois, nous ne nions pas que la connaissance de cet art, c'est à dire la pharmacie, si elle s'ajoute " à l'art médical " rend le médecin plus accompli, parce que cet art est très proche de l'art médical. Cependant, nous disons que la pharmacie n'occupe pas une place parmi les grandes branches de la médecine que nous avons citées.

Galien a dit dans la première partie du cinquième traité sur les natures de drogues[19] : " A considérer le métier des pharmaciens, des herboristes, de ceux qui préparent les décoctions et les onguents, des lotions, de ceux qui pratiquent les lavements et les saignées, et qui ouvrent les plaies, si le médecin pratique de temps en temps certaines de ces activités, il le fait tout comme le menuisier qui manipule parfois la rame et le mât, qui sont des objets d'activité des marins ".

15. *Kitāb al-adwiya al-mufrada,* BNF : Mss Ar. 2857 fol. 1-31 et 61-139. Dans ce livre au chapitre II, folio 69, Galien écrit que le goût des drogues révèle leurs effets sur le corps. Cette théorie pharmacologique a été adoptée et approfondie par les médecins médiévaux arabes.

16. Al-Rāzī, *Kitāb al-Qūlanǧ*, (*livre de la colique*), Edition critique et traduction de S.B. Hammami, publié par les soins d'Aleppo University et ALECSO, 1403H./1983. Dans ce livre *muḥallilan* qualifie un produit qui peut dissoudre, décomposer, résoudre. Al-Rāzī dans son livre *al-manṣūrī* (au chapitre III) donne la liste des médicaments qui sont *muḥallilan.* Certains de ces médicaments sont diurétiques. A.Z. Iskandar traduit ce terme par *discutient* ou résolutif.

17. Les définitions précises des termes médicaux *mumsekan* et *muġriyyan* ne nous sont accessibles dans l'état actuel de nos recherches. A.Z. Iskandar les traduit par le terme " constipant ". Il nous semble que le terme " constipant " ne couvre pas les notions sous-tendues par les termes arabes.

18. Selon al-Rāzī dans son livre *al-manṣūrī* (au chapitre III), les médicaments et les aliments amers et piquants sont chauds et asséchants.

19. Galien, *Kitāb al-adwiya al-mufrada,* BNF : Mss Ar. 2857 fol. 1-31 et 61-139. Une lecture de ce texte de Galien, ne nous a pas permis de repérer un passage similaire à celui évoqué par al-Rāzī. Nos recherches se poursuivent pour savoir s'il s'agit d'un autre texte de Galien.

ARABIC *MATERIA MEDICA* IN BYZANTIUM DURING THE 11th CENTURY A.D. AND THE PROBLEMS OF TRANSFER OF KNOWLEDGE IN MEDIEVAL SCIENCE

Alain TOUWAIDE

INTRODUCTION

If Greek knowledge, and especially Science, was transmitted to the Arabic World from the 9th century A.D. onwards, and contributed in this way to the rise of Arabic Science, this one, on its turn, appeared in the Byzantine World from the 11th century, after it had re-elaborated and freshly developed the material from which it was born[1].

1. This paper has been prepared in the context of the research programme devoted to the study of cultural identities among the groups of the Balkanic area during the Middle Ages and the Early Modern Times, which is currently in progress at the Consejo Superior de Investigaciones Científicas, Instituto de Filologia (Madrid), and which is granted by the Spanish Ministry of Scientific Research (ref. PB 95-013).

I wish to express here my thanks to both these institutions, and especially to the Director of the programme, Prof. Pedro Bádenas de la Peña (CSIC), who associated me to this research and entrusted me with the sub-programme dealing with the History of Medical Sciences.

For the clarity, we have adopted here some conventions :

- the Greek and Arabic names are quoted in their current English form ;

- the Greek and Arabic terms quoted (plant names) are transcribed into Latin alphabet, according to their orthographical (and not phonetical) form (being a transliteration) ;

- the works of any type (monographs, articles, or others) quoted in the notes are identified so as to constitute an overview of the current status of research in the field, as well as a tool for further research on the topic;

- the dates of Arabic authors, works, political figures and facts are given in their equivalent in the chronology A.D. and not A.H. ;

- when several works are quoted in the notes on the same point, they appear in chronological order of publication.

While the first transfer of knowledge has already been, and still is, studied[2]

2. It must be stressed that the body of knowledge studied here has to be referred to as Arabic Science and not *Islamic Science*, as often mistakingly written (see, e.g. : A. Tihon, " Tables islamiques à Byzance ", *Byzantion*, LX (1990), 401-425 (reprint in : A. Tihon, *Etudes d'astronomie byzantine,* Collected Studies Series, CS 454, n° VI (1994)) ; it is, indeed, the Science elaborated in Arabic by scientists of any origin, but Arabic-speaking within the context of the Empire ruled by the dynasties of Arabian origin (above all that of the ᶜAbbasids), especially during the Golden Era *(i.e.* the 9th and 10th centuries). *Islamic Science* exists, however, and especially Islamic Medicine, but it is constituted by the body of knowledge derived, directly or not, from the Qur'ân and from its prescriptions in matter of health. The bibliography on the development of Arabic Science from Greek Science is immense ; see, among others and recently : G. Saliba, " Translations and translators, Islamic ", in J.R. Strayer (ed.), *Dictionary of the Middle Ages*, vol. 12, New York, 1989, 127-133 ; L.E. Goodman, *The Translation of Greek materials into Arabic*, in M.J.L. Young, J.D. Latham, R.B. Serjeant (eds), 1990, 477-497 ; M.J.L. Young, J.D. Latham, R.B. Serjeant, " Religion, Learning, and Science in the ᶜAbbasid Period ", *The Cambridge History of Arabic Literature*, Cambridge, 1990 ; L.I. Conrad, " Arab-Islamic Medicine ", in W.F. Bynum, R. Porter (eds), *Companion Encyclopedia of the History of Medicine*, vol. 1, London, New York, 1993, 676-727, esp. 686-708 ; G. Strohmaier, " La ricezione e la tradizione : la medicina nel mondo bizantino e arabo ", *Storia del pensiero medico occidentale*, A cura di M.D. Grmek, vol. 1 : " Antichità e Medioevo ", Roma, Bari, 1993, 167-215 ; G. Saliba, *Arabic Science and the Greek Legacy, From Baghdad to Barcelona. Studies in the Islamic Exact Sciences in Honour of Prof. Juan Vernet* (Anuari de Filologia [Universitat de Barcelona], XIX [1996] B-2, Instituto Millás Vallicrosa de Historia de la Ciencia Arabe), vol. 1, Barcelona, 1996, 19-37 ; and G. Troupeau, " Les débuts de la médecine arabe : les grandes familles de médecin, Hunayn, la formation du vocabulaire ", *La médecine* (1996), 41-45, all with further bibliography.

On this topic, we have to mention that the role traditionally attributed to the so-called *Academy of Gondishapûr* in Persia (on this point, see, e.g. : H.H. Schöfler, " Die Akademie von Gondishapur. Aristoteles auf dem Wege in dem Orient ", *Logoi*, Band 5, Stuttgart, 1980 ; and, recently, P. Gignoux, " Introduction socio-culturelle ", *Splendeur des Sassanides. L'Empire perse entre Rome et la Chine [224-642]. 12 février au 25 avril, Bruxelles, 1993*, 1993, 31-43, esp. 42) and to the *Bayt al-Hikma* (or : *House of Wisdom*) in Baghdâd (on this point, see, e.g. : M. Ullmann, " Islamic Medicine ", *Islamic Surveys*, 11, Edinburgh, 1978, 9) has been recently re-evaluated ; see, for Gondishapûr : V. Nutton, " Jundîshâbûr ", *La médecine* (1996), 22 ; and V. Nutton, L.I. Conrad, *Jundîshâbûr, from Myth to History*, Princeton (forthcoming) ; and, for Baghdâd : M.-G. Balty-Guesdon, " Le " Bayt al-Hikma " de Baghdad ", *Arabica*, XXXIX (1992), 131-150 ; and M.-G. Balty-Guesdon, *" Bayt al-Hikma " et politique culturelle du Calife Al-Ma'mûn*, vol. 1, in L.R. Angeletti, A. Touwaide (eds), 1994, 275-291.

For the sake of completeness, we mention here the most significant references among the many at disposal on the *History of Arabic Pharmacology* and *Materia Medica*, in the Middle-East, as well as in the West (*i.e.* in Al-Andalus) : M. Meyerhof, " Esquisse d'histoire de la pharmacologie et botanique chez les Musulmans d'Espagne ", *Al-Andalus*, III (1935), 1-41 ; M. Meyerhof, " Die literarischen Grundlagen der arabischen Heimittellehre ", *Ciba Zeitschrift,* 85 (1942), 1321-1356 (English translation : " Arabian Pharmacology ", *Ciba Symposia*, 6 [1944], 1847-1876) ; S.K. Hamarneh, " Origins of Arabic Drug and Diet Therapy ", *Physis*, XI (1969), 267-286 (reprint in : S.K. Hamarneh, *Health Sciences in Early Islam,* vol. 2, Collected papers by S.K. Hamarneh, in M.A. Anees (ed.), Blanco (Texas), 1984, 205-215, 2 vols) ; F. Sezgin, *Geschichte des arabischen Schrifttums*, Band 3 : " Medizin, Pharmazie, Zoologie ", *Thierheilkunde bis ca. 430 H.*, Leiden, Köln, 1970, passim ; M. Ullmann, " Die Medizin im Islam ", *Handbuch der Orientalistik*, Erste Abteilung : " Der Nahe und der Mittlere Osten ", Ergänzungsband VI, Erster Abschnitt, Leiden, Köln, 1970, passim ; S.K. Hamarneh, " A History of Arabic Pharmacy ", *Physis*, XIV (1972), 5-54 (reprint in : S.K. Hamarneh, *Health Sciences in Early Islam, op. cit.*, 73-97) ; J. Samsó, " Las ciencias de los antiguos en Al-Andalus ", *Colecciones MAPFRE 1492, XVIII : Colección Al-Andalus*, 7, Madrid, 1992, 110-123, and 267-277 ; and, recently : T. Fahd, " Botany and Agriculture ", in R. Rashed, R. Morelon (eds), *Enclyclopedia of the History of Arabic Science*, vol. 3, London, New York, 1996, 813-852, esp. 817-819, and 821-839 ; J. Moulierac, " Les thériaques et la polypharmacie ", *La médecine* (1996), 101-103 ; N. Stephan, " La pharmacie médiévale d'expression arabe ", *La médecine* (1996), 83-91 ; and A. Touwaide, " La matière médicale : Dioscoride, une autorité incontestée ", *La médecine* (1996), 97-99.

the second, instead, remains largely unknown, although it was revealed to the attention of the Historians of Sciences as earlier as 1939[3].

In this paper we would like to analyse this topic in the field of the History of Medicine and, particularly, of *Materia Medica*. We shall focus on a manuscript which evidences, as we would like to show, the transfer of Arabic Medical Science into the Byzantine World during the 11th century, hoping to bring the phenomenon to light ; in doing so, we would also like to contribute to the study of the transfer of knowledge between cultures, from a theoretical point of view.

In this perspective, the paper will be divided into four main parts :

i) current status of knowledge of the topic ;

ii) description of the manuscript used as a source for our study ;

iii) elements of the codex which evidence a transfer of knowledge from the Arabic World ;

iv) a temptative theoretical conceptualisation of the general phenomenon of transfer of knowledge between Medieval cultures.

We can anticipate the conclusion of our study from now : while it is generally assumed that Oriental drugs did not appear in Byzantium before the work by Symeon Seth (second half of the 11th century A.D.), they seem to have been present earlier on the Byzantine market and to have been commonly used by physicians in their practice ; they appeared in written treatises only after that, in what could be considered a further phase of a phenomenon of assimilation which appears, thus, to have been by far larger than previously considered.

Analysis

Current status of knowledge

In the field of Medicine, it is generally assumed that the first Byzantine work in which appear Arabic drugs is the *Syntagma de alimentorum facultatibus*, written by Symeon Seth (11th-12th centuries) probably during the second

3. See : A. Kouzes, " Quelques considérations sur les traductions en grec des oeuvres médicales orientales et principalement sur les deux manuscrits de la traduction d'un traité persan par Constantin Meliteniotis ", *Praktika tês Akadêmias Athênôn*, 14 (1939), 205-220, who focuses mainly on the 14th century, however.

For the 11th century, but in the field of Astronomy, the fact has not been stressed before 1962 (see : J. Mogenet, " Une scolie inédite sur les rapports entre l'astronomie arabe et Byzance ", *Osiris*, XIV (1962), 198-221).

For a recent overview on this *reverse influence* (*i.e.* from Baghdâd to Constantinople), see : M.-H. Congourdeau, " Le monde byzantin ", *La médecine* (1996), 271-273.

half of the 11th century[4]. The paucity of these informations can be complemented by that of another scientific field, Astronomy. We learn that the first trace of the Arabic influence in Byzantium is a *scholion* in the margins of the manuscript *Vaticanus graecus* 1594 (Ptolemy, *Almagest*, 9th century) : although this note dates back to the 13th century (?), it is a copy of an earlier one of *ca.* 1032 which reported an astronomical observation done in Damascus in 829, and evidences the use, by the author, of a Greek adaptation of the astronomic tables by a certain Alêm[5], identified with ibn al-Aᶜlam (dead in 985 A.D.)[6].

4. On this point, see : K. Krumbacher (1897), " Geschichte der byzantinischen Litteratur von Justinian bis zum Ende des Oströmischen Reiches (527-1453) ", *Handbuch der klassischen Altertums-Wissenschaft*, Neunter Band, 1. Abteilung, München, 1890, 615, and G. Harig (1969), " Von den arabischen Quellen des Siemon Seth ", *Medizinhistorisches Journal*, 2 (1967), 248-268. On Symeon Seth, see principally : K. Krumbacher (1897), " Geschichte der byzantinischen Litteratur von Justinian bis zum Ende des Oströmischen Reiches (527-1453) ", *Handbuch der klassischen Altertums-Wissenschaft*, Neunter Band, 1. Abteilung, München, 1890, 615, 896 ; I. Bloch, " Byzantinische Medizin ", in M. Neuburger, J. Pagel (eds), *Handbuch der Geschichte der Medizin*, Begründet von Th. Puschmann, Jena, 1902, (reprint : Hildesheim, New York, 1971), 492-588, 563-564 ; M. Brunet, *Simeon Seth, médecin de l'empereur Michel Doukas. Sa vie, son oeuvre*, Bordeaux, 1939 ; H. Hunger, *Die hochsprachliche profane Literatur der Byzantiner*, Zweiter Band : " Philologie, Profandichtung, Musik, Mathematik und Astronomie, Naturwissenschaften, Medizin, Kriegswissenschaften, Rechtsliteratur " (= *Handbuch der Altertumswissenschaft*, Zwölfte Abteilung : *Byzantinisches Handbuch*, vol. 2, Füfnter Teil, Zweiter Band, München, 1978, 308-309 ; and, recently : A. Kazhdan, " Seth, Simeon ", *The Oxford Dictionary of Byzantium*, vol. 3, in A.P. Kazhdan (ed.), New York, Oxford, 1991, 1882-1883, with further bibliography. Seth is known for his multidisciplinary activity : Literature (see, among others : H.-G. Beck, " Geschichte der byzantinischen Volksliteratur ", *Byzantinisches Handbuch*, Zweiter Teil, Dritter Band, München, 1971, 41-44 ; P. Bádenas de la Peña, *Barlaam y Josafat. Redacción bizantina anónima, Edición a cargo de Selección de lecturas medievales*, 40, Madrid, 1993, XXI-XXXIV), Astronomy and Astrology (see recently : H. Hunger, *Die hochsprachliche profane Literatur der Byzantiner,* vol. 2, *op. cit.,* 241 ; and A. Tihon, " Tables islamiques à Byzance ", *Byzantion*, LX (1990), 401-425, esp. 404-405 (reprint in : A. Tihon, *Etudes d'astronomie byzantine,* n° VI, *op. cit.*) and Medicine (see recently : K.-H. Leven, " Seth, Symeon ", in W.U. Eckart, C. Gradmann (eds), *Ärztelexikon. Von der Antike bis zum 20. Jahrhundert*, München, 1995, 330-331 (Beck'sche Reihe, 1095)), not only with the *Syntagma de alimentorum facultatibus* (edition : B. Langkavel, " Simeoni Sethii ", in B. Langkavel (ed.), *De alimentorum facultatibus*, Leipzig, 1868, but also with a botanical lexicon (edition : A. Delatte, *Anecdota Atheniensia et allia*, II : Textes relatifs à l'histoire des sciences (= *Bibliothèque de la Faculté de Philosophie et Lettres de l'Université de Liège*, fascicule LXXXVIII), Liège, 1939, 339-361 ; see : H. Hunger, *Die hochsprachliche profane Literatur der Byzantiner,* vol. 2, *op. cit.,* 275). A monograph about him and his activity would be necessary, which could take into account the data provided by a renewed analysis of his work and, among others, of his sources ; " La Médecine ", *A l'ombre d'Avicenne. La médecine au temps des Califes. Exposition présentée du 18 novembre 1996 au 2 mars 1997, Paris, Institut du Monde Arabe*, Paris, Gand, 1996.

5. On this point, see : J. Mogenet, " Une scolie inédite sur les rapports entre l'astronomie arabe et Byzance ", *Osiris*, XIV (1962) ; J. Mogenet, " Sur quelques scolies de l'Almageste ", *Le Monde Grec. Pensée, littérature, histoire, documents. Hommages à Claire Préaux*, Bruxelles, 1975, 302-311 ; J. Mogenet, " L'influence de l'astronomie arabe à Byzance du IXe au XIVe siècle ", *Colloques d'Histoire des Sciences*, I (1972) & II (1973) organisés par le Centre d'Histoire des Sciences et des Techniques de l'Université Catholique de Louvain (Université de Louvain ; Recueil de travaux d'histoire et de philologie ; 6e Série, Fascicule 9), Louvain, 1976, 44-55, esp. 48-49, with a summary in : A. Tihon, " L'astronomie byzantine (du Ve au XVe siècle) ", *Byzantion*, LI (1981), 603-630, esp. 611 (reprint in : A. Tihon, *Etudes d'astronomie byzantine,* n° I, *op. cit.*) ; A. Tihon, " Tables islamiques à Byzance ", *Byzantion*, LX (1990), 401-425 (reprint in : A. Tihon, *Etudes d'astronomie byzantine,* n° VI, *op. cit.*), 402-403 ; A. Tihon, *Etudes d'astronomie byzantine*, Collected Studies Series, CS 454, Aldershot, 1994.

6. See A. Tihon, " Sur l'identité de l'astronome Alim ", *Archives internationales d'histoire des sciences*, 39 (1989), 3-21 (reprint in : A. Tihon, *Etudes d'astronomie byzantine,* n° IV, *op. cit.*).

Another Greek manuscript bears a text on Astronomy probably written around 1060-1072. Not only it used Arabic sources, but also, if not above all, it demonstrates a very good level of Hellenization of Arabic technical terms. From this, it has been argued that Arabic astronomical texts could have largely circulated in Byzantium[7].

Around the same period, Symeon Seth wrote astronomical tables, from which it can be stated that he knew Arabic Astronomy. Now, he is known for having travelled in Egypt in 1058[8].

Finally, the manuscript *Vaticanus graecus* 1056 (14th century) presents astrological material probably gathered during the 12th century. It has allowed to consider that Arabic Astronomy was fully acclimatized in Byzantium during the 12th century, probably as a result of a transfer during the previous century, *i.e.* the 11th[9].

The manuscript Athos, Megistês Lauras, Ω *75*

A closer examination of Byzantine medical documents in manuscripts[10] leads to identify a codex which is of prime importance for our topic : that currently conserved in the library of the Monastery Megistês Lauras, on Mount Athos[11] (Greece), under the signature Ω 75.

7. See : A. Tihon, " Tables islamiques à Byzance ", *Byzantion*, LX (1990), 401-425 (reprint in : A. Tihon, *Etudes d'astronomie byzantine*, n° VI, *op. cit.*), 403-404.

8. *Idem,* 404-405.

9. *Idem,* 405-413.

10. For the Greek medical manuscripts, see the specifical invento ries : H. Diels, *Die Handschriften der antiken Ärzte*, II. Teil : " Die übrigen griechische Ärzte ausser Hippokrates und Galenos ", *Abhandlungen der Königlichen Preussischen Akademie der Wissenschaften*, Philosophisch-historische Klasse, Jahre 1905, Abhandlung IIII, Berlin, 1905 (with a publication under separate title : *Die Handschriften der antiken Ärzte. Griechische Abteilung*, Berlin, 1906) ; H. Diels, *Die Handschriften der antiken Ärzte*, II. Teil : " Die übrigen griechische Ärzte ausserr Hippokrates und Galenos ", *Abhandlungen der Königlichen Preussischen Akademie der Wissenschaften*, Philosophisch-historische Klasse, Jahre 1906, Abhandlung I, Berlin, 1906 (with a publication under separate title : *Die Handschriften der antiken Ärzte. Griechische Abteilung*, Berlin, 1906), and H. Diels, *Die Handschriften der antiken Ärzte*, I. und II. Teil : " Bericht über den Stand des interakademischen Corpus medicorum antiquorum und erster Nachtrag zu den in den Abhandlungen 1905 und 1906 veröffentlichten Katalogen ", *Abhandlungen der Königlichen Preussischen Akademie der Wissenschaften*, Philosophisch-historische Klasse, Jahre 1907, Abhandlung II, Berlin, 1908, which aim to cover all the field. For the manuscripts supposed to have been copied in Southern Italy, see : A.M. Ieraci Bio, " La trasmissione della letteratura medica greca nell'Italia meridionale fra X e XV secolo ", *Contributi alla cultura greca nell'Italia meridionale*, I, A cura di A. Garzya (= *Hellenica et Byzantina Neapolitana*, XIII), Napoli, 1989, 133-257. Finally, for the manuscripts produced after the fall of Constantinople (1453), in the West as well as in the Ottoman Greece, see : G. Karas, " Oi Epistemes Sten Tourkokratia - Cheirografa Kai Entupa ", vol. 3 : " Oi Epistemes Tes Zoes ", *Kentpo Neoelle Nikon Ereunon E.I.E.*, 48, Athêna, 1994. See also : R.E. Sinkewicz, W.M. Hayes, " Manuscript listings for the authored works of the Palaeologan Period ", *Greek Index Project Series*, 2, Toronto, 1989 ; R.E. Sinkewicz, " Manuscript Listings for the Authors of Classical and Late Antiquity ", *Greek Index Project Series*, 3, Toronto, 1990 ; and R.E. Sinkewicz, " Manuscript listings for the authors of the Patristic and Byzantine Periods ", *Greek Index Project Series*, 4, Toronto, 1992, which are not particularly devoted to medical texts, but cover all the field of ancient texts.

11. Brockhaus, *Die Kunst in den Athos-Klöstern*, Leipzig, 1891.

Known in the bibliography since 1897[12], it has been studied only in recent time, however[13]. It is a thick book (293 ff.) of small format (192 x 121 mm.), on parchment[14]. Although it is not dated, it can be considered to have been

12. See : M. Wellmann, " Kratevas ", *Abhandlungen der Königlichen Gesellschaft der Wissenschaften zu Göttingen, philologisch-historische Klasse*, Neue Folge, Band 2, n° 1, Berlin, 1897, 23, note 30, for the first mention of the codex (in our knowledge). It must be stressed, however, that the German scholar did not know it personally ; he had at disposal (as he honestly confessed) the data provided by Dr C. Friedrich, who made an autoptic analysis of the codex.

13. On this manuscript and its text, see : S. Spyridon-Eustratiades, " Catalogue of the Greek Manuscripts in the Library of the Lavra of Mount Athos, with notices from other Libraries ", *Harvard Theological Studies*, XII, Cambridge, 1925, 343 ; S. M. Pelekanides, P.K. Chrestos, Ch. Mauropoudos-Tsioume, S.N. Kadas, A. Katsaros, *Oi thêsauroi tou Agiou Orous*, Seira 1 : " Eikonografêmena cheirografa ", tomos 3 : M. Megistês Lauras, M. Pantokratoros, M. Docheiariou, M. Karakalou, M. Filotheou, M. Agiou Paulou, Athênai, 1979, 258-259 ; A. Touwaide, *Les deux traités de toxicologie attribués à Dioscoride. La tradition manuscrite grecque*, Edition critique du texte grec et traduction, Louvain-la-Neuve, 1981 (5 vols), (unpublished PhD Thesis) : vol. 1, 7-16, vol. 2, 263-278, vol. 3, 7-16 & 263-278 ; and G. Christodoulou, " O Athônikos kôd. Meg. Lauras Ω 75 tou Dioskoridê. Palaiografikê episkopêsê ", in G. Christodoulou (ed.), *Summikta kritika*, Athêna, 1986, 131-199 (with the complements by A. Touwaide, " Un manuscrit athonite du Peri Ules Iatrikes de Dioscoride : l'Athous Megistis Lavras Ω 75 ", *Scriptorium*, XLV (1991), 122-127). There is also the earlier work of E. Kourilas, *Dioskorideioi meletai kai o Lauriôtikos Dioskoridês*, Athênai, 1935, passim, which has to be taken into consideration with some caution, because of the unreliability of the data it provides.

On its illustrations, see E. Kourilas, *Dioskorideioi meletai kai o Lauriôtikos Dioskoridês*, Athênai, 1935, 84-91 ; K. Weitzmann, *Illustrations in Roll and Codex*, Princeton, 1947, 86, 166-167 ; K. Weitzmann, " The Greek Sources of Islamic Scientific Illustrations ", in G.C. Miles, *Archaeologica Orientalia in Memoriam Ernst Herzfeld*, New York, 1952, 244-266 (reprinted in : K. Weitzmann, *Studies in Classical and Byzantine Manuscript Illuminations,* in H. Kessler (ed.), Chicago, London, 1971, 20-44, which is quoted here), 30-31 ; K. Weitzmann, " Das klassische Erbe in der Kunst Konstantinopels ", *Alte und Neue Kunst*, 3 (1954), 41-59 (English translation [quoted here] : *The Classical Heritage in the Art of Constantinople*, in : K. Weitzmann, *Studies in Classical and Byzantine Manuscript Illuminations, op. cit.*, 126-150), 146 ; K. Weitzmann, " Ancient Book Illumination ", *Martin Classical Lectures*, 15, Cambridge (Mass.), 1959, 13-14 ; K. Weitzmann, *Studies in Classical and Byzantine Manuscript Illuminations, op. cit.* ; G. Galavaris, " The Illustrations of the Liturgical Homilies of Gregory Nazianzenus ", *Studies in Manuscript Illumination*, 6, Princeton, 1969, 166-167 ; Z. Kádár, *Survivals of Greek Zoological Illuminations in Byzantine Manuscripts*, Budapest, 1978, 25, 53-55 ; and, recently, the tables 147-165 in S. M. Pelekanides, P.K. Chrestos, Ch. Mauropoudos-Tsioume, S.N. Kadas, A. Katsaros, *Oi thêsauroi tou Agiou Orous,* Seira 1 : " Eikonografêmena cheirografa ", tomos 3 : M. Megistês Lauras, M. Pantokratoros, M. Docheiariou, M. Karakalou, M. Filotheou, M. Agiou Paulou, Athênai, 1979, 104-111.

This manuscript has still not received the general study it deserves, neither from the point of view of its text or from that one of its illustrations, although it is a source of high interest not only for the history of the tradition of Dioscorides' *De materia medica*, but also for the history of cultural life in Byzantium during the 11th century.

14. Our analysis results from an autoptic examination of the manuscript, during a ten days journey on Mount Athos in 1978.

We take the opportunity to thank the many authorities who allowed us to stay in the Monastery of the Megistês Lauras and to see personally this important codex, from the Patriarchate of Constanti nople to the Police of Karyes, on Mount Athos.

Since then, we have had the opportunity to study the manuscript on a microfilm put at our disposal by Prof. S. Philianos, Laboratory of Pharmacognosy, University of Athens. We would like to thank him warmly for this facility, which has allowed us to deepen considerably our knowledge of the manuscript.

copied during the first half of the 11th century, according to a palaeographical analysis[15].

Its source (a manuscript used also for the copy of the codex New York Pierpont Morgan Library, M 652 [10th century], surely of Constantinopolitan origin)[16], and its later history (mainly the source of the restoration of its damaged parts [see below], and its copies, all Constantinopolitan) allows to state that it was made in Constantinople[17].

Its text is partially illustrated (ff. 6-135, and 167-184) with polychrome representations of plants disposed side by side on the full width of the page (up to five on the same width, with an average of three). Its illumination is not fully achieved : from f. 185 onwards spaces were left for illustrations, which were not filled.

15. See : A. Touwaide, *Les deux traités de toxicologie attribués à Dioscoride. La tradition manuscrite grecque. Edition critique du texte grec et traduction*, Louvain-la-Neuve, 1981 (5 vols), (unpublished PhD Thesis) : vol. 1, 7-12, vol. 3, 7-12. Similarly, see the data of the analysis by J. Leroy, in : G. Christodoulou, " O Athônikos kôd. Meg. Lauras Ω 75 tou Dioskoridê. Palaiografikê episkopêsê ", *op. cit.*, 136-137, note 25 (end 10th/beginning 11th centuries). The datation proposed by G. Galavaris (end 10th/beginning 11th centuries) (in : G. Christodoulou," O Athônikos kôd. Meg. Lauras Ω 75 tou Dioskoridê. Palaiografikê episkopêsê ", *op. cit.*, 138, note 25 cont.) does not seem to be probable, among others because it relies on the examination of few photographic reproductions of illustrations of the manuscript.

16. On this point, see : A. Touwaide, *Les deux traités de toxicologie attribués à Dioscoride. La tradition manuscrite grecque. Edition critique du texte grec et traduction*, vol. 2, Louvain-la-Neuve, 1981, 246-251 (5 vols) (unpublished PhD Thesis), with also a summary in : A. Touwaide, " L'authenticité et l'origine des deux traités de toxicologie attribués à Dioscoride. I. Historique de la question. II. Apport de l'histoire du texte ", *Janus*, 38 (1983), 1-53, and A. Touwaide, " Les deux traités de toxicologie attribués à Dioscoride. Tradition manuscrite, établissement et critique d'authenticité ", *Tradizione e ecdotica dei testi medici tardo-antichi e bizantini. Atti del Convegno internazionale, Anacapri, 29-31 ottobre 1990*, A cura di A. Garzya, Napoli, 1992, 291-339. On the New York manuscript, see, among the abundant bibliography, the following recent works : G. Vikan, *Illuminated Greek Manuscripts from American Collections*, Princeton, 1973, 66-67 ; J. Irigoin, " Une écriture du Xe siècle : la minuscule bouletée ", *La paléographie grecque et byzantine. Paris, 21-25 octobre 1974, Colloques internationaux du C.N.R.S.*, n° 559, Paris, 1977, 45-54, 195 ; A. Touwaide, *Les deux traités de toxicologie attribués à Dioscoride..., op. cit.*, vol. 1, 57-68, vol. 3, 57-68 ; H.C. Evans, W.D. Wixom (eds), *The Glory of Byzantium. Art and Culture in the Middle Byzantine Era A.D. 843-1261, Exhibition held at the Metropolitan Museum of Art, New York, from March 11th through July 6, 1997*, New York, 1997, 237-238, all with further bibliography ; full fac-simile edition (very rare sepia reproduction without name of author [*sic*]): Dioscurides, " Pedanii Dioscuridis Anazarbi ", *De Materia Medica Libri VII. Accedunt Nicandri et Eutecnii, Opuscula Medica. Codex Constantinopolitanus saeculo X exaratus et picturis illustratus, olim Manuelis Eugenici, Caroli Rinuccine Florentini, Thomae Phillipps Angli, nunc inter Thesauros Pierpont Morgan Bibliothecae asservatus*, Lutetiae Parisiorum, 1935 (2 vols). This manuscript should be submitted to a deep study and, hopefully, be reproduced in a new colour fac-simile edition.

17. While we previously thought that the manuscript was produced in Constantinople or at Mount-Athos (A. Touwaide, *Les deux traités de toxicologie attribués à Dioscoride..., op. cit.*, vol. 1, 15-16, vol. 3, 15-16), further work (still unpublished) has shown that the manuscript is Constantinopolitan ; it was, indeed, the source of 14th century codices, the Constantinopolitan origin of which can be identified with certainty. G. Galavaris (in : G. Christodoulou, " O Athônikos kôd. Meg. Lauras Ω 75 tou Dioskoridê. Palaiografikê episkopêsê ", n° 24, *op. cit.*, 138) affirms that the manuscript is not Constantinopolitan; if he eliminates also Egypt, he does not propose any other location. Besides the fact that he does not present any argument supporting his thesis, this cannot be taken into consideration at least for the reason we saw above (note 13) ; moreover, the evidence of the copies of the manuscript is so strong that it allows us to conclude with very few doubts that it is of Constantinopolitan origin.

Later on (possibly during the Latin occupation of Constantinople [1204-1261]), the manuscript was partially damaged : the inferior corner of the external border of the page on ff. 2-261 bears marks (of water [?]) and has been partially cut ; it was restored with paper and the lost text was rewritten seemingly at the end of the 13th century.

The texts originally contained in the manuscript[18] are mainly the *De materia medica*[19] of Dioscorides (1st century A.D.)[20], followed by the two toxicological treatises ascribed to the same author[21] ; and the paraphrases of the poems of Nicander, Oppian and Dionysius, by Eutecnius (3rd [?] 4th [?] century A.D.)[22].

18. Detailed identification of the texts contained in the manuscript in : A. Touwaide, *Les deux traités de toxicologie attribués à Dioscoride..., op. cit.*, vol. 1, 14 ; vol. 3, 14 ; and G. Christodoulou, " O Athônikos kôd. Meg. Lauras Ω 75 tou Dioskoridê. Palaiografikê episkopêsê ", *op. cit.*, 141-143.

19. Edition of the Greek text : M. Wellmann, *Pedanii Dioscuridis Anazarbei, De materia medica libri quinque*, Berlin, 1906-1914 (3 vols). (reprint : Berlin, 1958) ; ancient (17th century) English translation of the full text : R.T. Gunther, *The Greek Herbal of Dioscorides. Illustrated by a Byzantine, A.D. 512 ; Englished by John Goodyer, A.D. 1655, Edited and first printed, A.D. 1933...*, Oxford, 1934 (reprints : London, New York, 1959, 1968) ; modern English translation of the Preface, with commentary : J. Scarborough, V. Nutton, " The Preface of Dioscorides' Materia Medica. Introduction, Translation, and Commentary ", *Transactions and Studies of the College of Physicians of Philadelphia*, 5th Series, 4 (1982), 187-227.

20. For a full study on Dioscorides, see the classical, but now obsolete work of M. Wellmann, " Dioskurides 12 ", *RE*, V, 1 (1903), col. 1131-1142 ; recently : J.M. Riddle, *Dioscorides on Pharmacy and Medicine*, Austin, 1985 (History of Science Series ; n° 5). For shorter presentations : J.M. Riddle, " Dioscorides ", in C.G. Gillispie (ed.), *Dictionary of Scientific Biography*, vol. 4, New York, 1978, 119-123 ; J.M. Riddle, *Quid pro quo : Studies in the history of drugs*, Collected Studies Series, CS 367, Aldershot, 1992 ; A. Cutler, J. Scarbrough, " Dioskorides ", in A.P. Kazhdan (ed.), *The Oxford Dictionary of Byzantium*, vol. 1, New York, Oxford, 1991, 632 (both with further bibliography). For the biographical elements alluded to in the *Preface* of *De materia medica*, see : J. Scarborough, V. Nutton, " The Preface of Dioscorides' *Materia Medica*. Introduction, Translation, and Commentary ", *Transactions and Studies of the College of Physicians of Philadelphia*, 5th Series, 4 (1982). The data traditionally presented are rightly questioned in : M.-H. Marganne, " Les références à l'Egypte dans la Matière médicale de Dioscoride ", *Serta Leodiensia Secunda. Mélanges publiés par les Classiques de Liège à l'occasion du 175e anniversaire de l'Université de Liège,* Liège, 1992, 309-322, 309-313, 322. A renewed study of Dioscorides' biography would be welcomed, relying on a fresh analysis of his treatise.

21. Edition of the Greek text : A. Touwaide, *Les deux traités de toxicologie attribués à Dioscoride...*, vol. 4, *op. cit.* Study of their authenticity: A. Touwaide, " L'authenticité et l'origine des deux traités de toxicologie attribués à Dioscoride. I. Historique de la question. II. Apport de l'histoire du texte ", *Janus*, 38 (1983) ; A. Touwaide, " Les deux traités de toxicologie attribués à Dioscoride. Tradition manuscrite, établissement et critique d'authenticité ", *Tradizione e ecdotica dei testi medici tardo-antichi e bizantini. Atti del Convegno internazionale, Anacapri, 29-31 ottobre 1990*, A cura di A. Garzya, Napoli, 1992.

22. Critical editions of the paraphrases of Nicander by M. Papathomopoulos, " Eutekniou parafraseis eis ta nikandrou Thêriaka kai Alexifarmaka ", *Panepistêmion Iôanninôn, Filosofikê Scholê*, Seira " Peleia ", 2, Iôannina, 1976 ; for that one of Oppian : I. Gualandri, " Incerti auctoris in Oppiani Halieutica Paraphrasis ", *Testi e documenti per lo studio dell'Antichità*, XVIII, Milano, Varese, 1968 ; for that of Dionysius : M. Papathomopoulos, " Anônumou parafrasis eis ta Dionysiou Ixeutika ", *Panepistêmion Iôanninôn, Filosofikê Scholê*, Seira " Peleia ", 3, Iôannina, 1976.

On Eutecnius, the only recent notice at disposal is the very brief one by Gärtner (1967). Although the evidence provided by the sources is very limited, it would be worthwhile to try to deepen the study and perceive better the biography of this author, remained till now quite mysterious.

The version of Dioscorides' text presented in the manuscript is not the one considered to be original (*i.e.* that in five *books* into which the matter is analytically divided)[23], but the so-called *alphabetical in five books*[24]. From a detailed analysis, it appears that it was deeply revised in the manuscript, at its macro- and microscopic levels.

As for the macroscopic level, additional material was introduced into the text of the source : not only notes[25], but also entire chapters. These are explicitly identified as such in the table of contents of f. 6[recto], where their inventory is preceded by the following affirmation[26] : " There are also these chapters, found in another book and not present in those reproduced "[27]. These chapters were copied separately at the end of *De materia medica, i.e.* between book 5 and the pseudo-book 6 (*i.e.* the toxicological treatise on poisons), on ff. 185[recto]-190[verso].

At a microscopic level, the character itself of the text is often different from that of the other manuscript (the New York one) copied from the same source[28], so that we may consider that the text of the Athos manuscript is not a faithful reproduction of its source, but an erudite revision[29].

23. For the versions of Dioscorides' text, see, principally : M. Wellmann, *Pedanii Dioscuridis Anazarbei, De materia medica libri quinque*, vol. 2, Berlin, 1906-1914, V-XXIV (3 vols). (reprint : Berlin, 1958), the *Introduction* to the edition (although this history of the text is not complete and is quite approximative, it is still valid, at least for the general features of the history of the text ; given that new material has emerged since then, a history of Dioscorides' text which would take into consideration if not the totality of the data, at least a great deal, would be more than welcome) ; and, recently : A. Touwaide, " L'authenticité et l'origine des deux traités de toxicologie attribués à Dioscoride. I. Historique de la question. II. Apport de l'histoire du texte ", *Janus*, 38 (1983), 16-19 (for a new attempt).

On the concept of *books* (which is probably not original, but reflects a division of the work into papyrus rolls), see : A. Touwaide, " La thérapeutique médicamenteuse de Dioscoride à Galien : du " pharmaco-centrisme " au " médico-centrisme ", in A. Debru (ed.), *Galen on Pharmacology. Philosophy, History and Medicine, Proceedings of the Vth International Galen Colloquium, Lille, 16-18 March 1995. Studies in Ancient Medicine*, 16, Leiden, New York, Köln, 1997, 255-282, esp. 261-263.

24. On this version, see : A. Touwaide, " L'authenticité et l'origine des deux traités de toxicologie attribués à Dioscoride. I. Historique de la question. II. Apport de l'histoire du texte ", *Janus*, 38 (1983), 17-18.

25. See : G. Christodoulou, " O Athônikos kôd. Meg. Lauras Ω 75 tou Dioskoridê. Palaiografikê episkopêsê ", *op. cit.*, 146-151, for the inventory of these additions.

26. Reproduction of the inventory in : G. Christodoulou, " O Athônikos kôd. Meg. Lauras Ω 75 tou Dioskoridê. Palaiografikê episkopêsê ", *op. cit.*, 184-185.

27. In Greek : *Eisi kai ta eurethenta en eterô bibliô, ta mê egkeimena en tois metagrafeisi bibliois tauta.*

28. For a reconstruction of this archetype (Pseudo-Dioscorides, treatises of toxicology), see : A. Touwaide, *Les deux traités de toxicologie attribués à Dioscoride..., op. cit.*, vol. 2, 246-251 ; vol. 3, 256-251.

29. On this point, see, for Dioscorides' *De materia medica*: G. Christodoulou (ed.), *Summikta kritika*, Athêna, 1986, 148-158 (inventory of the interventions) ; and, for the two treatises of toxicology ascribed to Dioscorides : A. Touwaide, *Les deux traités de toxicologie attribués à Dioscoride...*, vol. 2, *op. cit.*, 264-271 (analysis of the interventions).

The problem we have to cope with is to know whether this work of revision (at macro- and microscopic levels) has to be attributed to the copyist of our manuscript or to that of another one, anterior to the Athos codex. In the latter case, the copyist of the *Athous* would have literally reproduced his source.

According to the principles of ecdotic, it seems that the revision cannot be attributed to the copyist of our manuscript : some of the peculiarities of the *Athous* are wrong lessons, *i.e.* they constituted *lessons* which are not the suitable ones (even if they can be right from other points of view, grammatical, lexical or syntactical) and, in certain cases, they are not right, but contains errors which do not appear in the other copy of the same archetype. Normally, the interventions with no suitable lessons and the errors were improved by subsequent copyists, on one hand, and, on the other, right and wrong lessons were not created by the same copyist. Consequently, it seems that the revision of the text cannot be attributed to the copyist of the *Athous* himself, but to an anterior one. In this case, it has to be located, with some probability, between the epoch of copy of the New York Dioscorides (end 10th century) and that of the Athos one (mid 11th century), *i.e.* during the first half of the 11th century. The place was most probably Constantinople, because of the location of the *Athous* itself and its copies, as well as of the New York manuscript. For the convenience, this reconstructed Constantinopolitan manuscript will be called, from now onwards, Athos text.

Elements of the manuscript which evidence a transfer of knowledge from the Arabic World

Once we admit that the text of the Athos text was deeply revised, we have to focus on the interventions of its copyist, in order to determine whether or not they have been made from other source than the main one, and, if this is the case, to identify this (or these) possible source(s).

As for the microscopic interventions, they may be divided into two groups, determined by the fact that they can be found or not in other manuscripts of Dioscorides' and Pseudo-Dioscorides' text.

A certain number of these microscopic interventions can be identified in the text of other manuscripts[30]. They demonstrate the presence, in the Athos text, of what is called a contamination, *i.e.* the association of lessons from different manuscripts ; consequently, they imply that several sources have been consulted, among which one was the main — the archetype of the New York manuscript and the Athos text — and others the secondary ones.

The fact is not without importance, because it allows to locate the constitution of the Athos text in a centre of a certain level : not only several copies of

30. For the text of Dioscorides, see : G. Christodoulou (ed.), *Summikta kritik, op. cit.,* 148-151 ; for the text of the two toxicological treatises ascribed to Dioscorides : A. Touwaide, *Les deux traités de toxicologie attribués à Dioscoride...,* vol. 2, *op. cit.,* 268-269.

the same text (*i.e.* the corpus of Dioscorides and Pseudo-Dioscorides treatises) were available, but also they were used to improve the text they contain ; moreover, it demonstrates a will and an interest in updating the actual versions of the classical texts in circulation.

Other peculiarities of the Athos text cannot be identified in no other manuscript[31], and seem, thus, to be original, *i.e.* to result from personal interventions of the copyist of our manuscript himself. They demonstrate at least that the copyist had a high level of education, and, possibly too, that he worked in a scriptorium linked with a medical centre (a hospital or another type of institution for health care)[32].

Concerning the macroscopic level of the text, the additional notes (no more than nine in total)[33] appear only in the *De materia medica* and provide the following data : synonyms of a phytonym (one case), therapeutical indications (two cases), preparation of medicinal products (four cases), properties according to Galen (two cases).

Their sources (as far as they have been identified) are interesting : in one case, the note is introduced with the expression : *in other* (*i.e.* : in other manuscript)[34] ; in another one, the data of the note are similar to the text of Aetius ; in another, some terms (non Classical, but Byzantine and not erudite) reveal a possible link with personal experience, if not with popular tradition ; in the two notes dealing with Galen's text, his name is explicitly stated.

As for the additional chapters, almost half of them deal with the following plants[35] :

31. For these lessons, see : G. Christodoulou (ed.), *Summikta kritika, op. cit.*, 151-158 for Dioscorides' text, and A. Touwaide, *Les deux traités de toxicologie attribués à Dioscoride, op. cit.*, vol. 2, 263-267 ; vol. 3, 263-267, for the Pseudo-Dioscorides.

32. On this point, see below. On the fact that the hospitals of Byzantium were often linked with a scriptorium and a library, especially from the Middle-Byzantine Period, see : T.S. Miller, " The Birth of the Hospital in the Byzantine Empire ", *Bulletin of the History of Medicine*, Supplements, New Series, n° 10 (Baltimore, 1985), passim. On Byzantine health institutions, see, besides : D. Constantelos, *Byzantine Philanthropy and Social Welfare*, New Brunswick (New Jersey), 1968.

33. Inventory and reproduction of their text in : G. Christodoulou (ed.), *Summikta kritika, op. cit.*, 146-148, and 158-159.

34. Greek text : *en allô.*

35. On the oriental origin attributed to these plants, see below. In the list, we quote the plant or drug names (in alphabetical order of their transcription into Latin alphabet), with the following elements : Greek name (in transliteration) as it appears in the manuscript (for an inventory, see : G. Christodoulou (ed.), *Summikta kritika, op. cit.*, 184-185) ; between brackets, location in the manuscript (see : *Ibidem*) ; Linnean identification according to current bibliography; we have used the following works, abbreviated as follows (quoted here in alphabetical order of the abbreviations) : A = Andre (1985) ; H = Herzhoff (1993) ; M = J.I. Miller, *The Spice Trade of the Roman Empire. 29 B.C. to A.D. 641*, Oxford, 1969 ; S = Stirling (1995-1997) ; English current name (according to A.K. Bedevian, *Illustrated Polyglottic Dictionary of Plant Names in Latin, Arabic, Armenian, English, French, German, Italian and Turkish Languages...*, Cairo, 1936, and A. Issa, *Dictionnaire des noms des plantes en latin, français, anglais et arabe*, s.l., s.d.).

BALSAMON (f. 187^{recto}) : *Commiphora opobalsamum* Engl. (A : p. 33, *s.v. balsamum* 1 ; M : p. 103-104 ; S : vol. 1, p. 111, *s.v. balsamum*), *C. gileadensis* (H : p. 82, s.v.) ;

KAGKAMON (f. 188^{recto}) : *Commiphora kataf* Forsk. (?), *Styrax benzoin* Dryander (?) (A : p. 47, *s.v. cancamum*) ; *Sturax benzoin* Dryander (M : p. 44-45) ; no identification (S : vol. 2, p. 17, *s.v. cancamon*) ;

KALAMOS ARÔMATIKOS (f. 187^{recto}) : *Acorus calamus* L. (A : p. 45, *s.v. calamus* 3 ; H : p. 83-84, s.v. ; M : p. 91-93 ; S : vol. 2, p. 6-7, *s.v. calamus* I 1) ;

KASSIA (f. 186^{verso}) : *Cinnamum cassia* Blume (A : p. 52, *s.v. casia* ; S : vol. 2, p. 44, *s.v. cas<s>ia* 3) ; *Cinnamomon cassia* Bl., *C. zeylanicum Nees* (H : p. 84-85, *s.v. kasia* and *kinamômon*) ; *Cinnamum cassia* Blume, *C. macrophyllum*, and other species (M : p. 46-50) ;

KOSTOS (f. 186^{verso}) : *Saussurea lappa* G. B. Clarke (A : p. 76, *s.v. costum* ; H ; p. 85-86, *s.v. koston* ; M : p. 84-86 ; S : vol. 2, p. 145, *s.v. costum* 1) ;

KUFI (f. 188^{recto}) : compound drug ;

MALABATHRON (f. 186^{recto}) : *Pogostemon patchouli* Pell. (A : p. 151-152, *s.v. malobat(h)rum* 1 ; M : p. 74-77) ;

NARDOS (f. 185^{recto}) : *Nardostachys jatamansi* D.C. (A : p. 170, *s.v. nardum* 1 ; H : p. 89-90 ; M : p. 88-91) ;

NARDOS OREINÊ Ê THULAKITIS (f. 186^{recto}) : *Valeriana tuberosa* Jacq. (A : p. 170, *s. v. nardum* 5 ; M : p. 118, nr 65).

From a closer examination, we reach the conclusion that these additions do not come from heterogeneous sources, but they reproduce the text of Dioscorides' original chapters dealing with these drugs in *De materia medica*[36]. Moreover, we discover that the sequence of these chapters in the Athos text corresponds to that of the version of Dioscorides' work considered to be original ; if we quote the chapters as they are ordered in our manuscript and, consequently, in the Athos text, instead of listing them in alphabetical order as we have done till now, it appears that the references of the related passages of Dioscorides' work are listed in a sequential order :

nardos	f. 185^{recto}	I, 7
nardos oreinê	f. 186^{recto}	I, 9
malabathron	f. 186^{recto}	I, 12
kassia	f. 186^{verso}	I, 13
kostos	f. 186^{verso}	I, 16
kalamos arômatikos	f. 187^{recto}	I, 18

36. References of these chapters in Diocorides' text (according to the edition of M. Wellmann, *Pedanii Dioscuridis Anazarbei, De materia medica libri quinque*, Berlin, 1906-1914 (3 vols). (reprint : Berlin, 1958) : balsamon I, 19 ; fou I, 11 ; kagkamon I, 24 ; kalamos arômatikos I, 18 ; kassia I, 13 ; kostos I, 16 ; krokomagma I, 27 ; kufi I, 25 ; malabathron I, 12 ; mion (falso pro mêon) ê athamantikon I, 3 ; nardos I, 7 ; nardos oreinê ê thulakitis I, 9.

balsamon	f. 187^{recto}	I, 19
kagakamon	f. 188^{recto}	I, 24
kufi	f. 188^{recto}	I, 25

This statement is significant, all the more because the version of Dioscorides' work contained in the Athos text is not the original version of the treatise, but an alphabetic one ; consequently, if the sequence of the chapters of the Athos text dealing with Oriental drugs reflects the original one, it indicates that a manuscript of the original version was consulted and reproduced for these chapters in the constitution of the Athos text.

Besides being a supplementary proof that the copyist of our manuscript used other sources than his main one (the archetype reproduced also in the New York manuscript), this changes the perspective in which we analysed the addition of these chapters : they were not added to the text of Dioscorides, but re-introduced in it, being omitted previously. This new formulation of the problem rises the question of knowing why and when these chapters were cancelled.

On this point, we have to remind that the alphabetic recension in five books of the archetype reproduced in the New York Dioscorides and in the Athos text results from a double, if not a triple work of revision : its first book reproduces, indeed, the so-called herbal of the *Vindobonensis medicus grecus* 1, *i.e.* the selection of chapters dealing with vegetals presented in alphabetical order in the Vienna manuscript[37] ; this, on its turn, is generally considered as a re-elaboration of an alphabetical version of the full text of Dioscorides.

As for the other books, they were constituted by recuperating the other chapters from a full version of the text, divided into four main sections, according to the nature of the drugs : animals, liquids of all types, tree, wines and minerals ; and, within these main sections, the chapters were ordered alphabetically.

Once added these sections to the herbal of the *Vindobonensis*, they were presented as four books of a set of five, the first of which being the herbal of the *Vindobonensis*, so that this division corresponded exactly to that of the original version of Dioscorides' treatise, with its five books.

On the occasion of these deep restructuring, chapters of the original De materia medica were omitted, like those we are dealing with. According to Ancient Classical culture and Dioscorides' *De materia medica*, the plants and drugs analysed in these chapters did not come from Greece (modern continen-

37. On this version of the text, as well as on the manuscript itself, see, among the many works dealing with these topics : H. Gerstinger, " Dioscurides. Codex Vindobonensis med. gr. 1 der Österreichischer Nationalbibliothek. Kommentarband zu der Faksimileausgabe ", *Codices Selecti Phototypice Impressi*, vol. XII*, Graz, 1970.

tal and insular Greece, as well as Mediterranean and Central Turkey), but from other regions[38] :

plant or drug	origin according to Dioscorides	origin according to modern Botany
balsamon	Judaea[39]	Arabia, Somalia, Palestina[40]
kagkamon	Arabia[41]	South-Eastern Asia[42]
kalamos arômatikos	India[43]	from the Black Sea to Japan[44]
kassia	Arabia[45]	China, Ceylon[46]
kostos	Arabia, India, Syria[47]	Cachemire[48]
kufi	Egypt[49]	-

38. In the following list, we give the origin of the plants and drugs as stated by Dioscorides and in modern literature. In the notes, we indicate : the reference of the location in Dioscorides' text (according to the edition of M. Wellmann, *Pedanii Dioscuridis Anazarbei, De materia medica libri quinque*, Berlin, 1906-1914 (3 vols). (reprint : Berlin, 1958) ; the references of modern botanical literature dealing with the distribution of the plants considered to correspond to the ancient ones from J.I. Miller, *The Spice Trade of the Roman Empire. 29 B.C. to A.D. 641*, Oxford, 1969 (abbreviated : M), Ferrara Pignatelli (1991) (abbreviated : FP), historical bibliography on plants and drugs.

39. Dioscorides, I, 19 (= vol. 1, 24, l. 5).

40. M, 103-104 ; FP, 184-186. See : Wagler, " Balsambaum ", *RE*, II (1896), col. 2836-2839 ; H. Gams, " Balsam ", *DKP*, 1 (1964), col. 818 ; M.G. Raschke, " New Studies in Roman Commerce with the East ", in H. Temporini Herausgegeben, *Aufstieg und Niedergang der römischer Welt*, vol. II, 9, Berlin, New York, 1978, 604-1361, esp. 926-927, note 1116 ; J. Fortes, " Los fitónimos griegos ", *Estudios de lingüística y paleoetnobotánica*, Barcelona, 1980 (2 vols) (unpublished PhD. Thesis), 192-193 ; C. Hünemörder, " Balsam ", *NP*, vol. 2 (1997), col.

41. Dioscorides, I, 24 (= vol. 1, 28, l. 8).

42. M, 44-45 ; FP, 193-196.

43. Dioscorides, I, 18 (= vol. 1, 23, l. 13).

44. M, 91-92 ; FP, 203-205. See : Stadler, " Kalamos 3 ", *RE*, X, 2 (1919), col. 1542-1543 ; K. Ziegler, " Kalamos 2 ", *DKP*, 3 (1969), col. 53 ; M.G. Raschke, " New Studies in Roman Commerce with the East ", in H. Temporini (ed.), *Aufstieg und Niedergang der römischer Welt*, vol. II, 9, Berlin, New York, 1978, 928, note 1121 ; J. Fortes, " Los fitónimos griegos ", *Estudios de lingüística y paleoetnobotánica*, Barcelona, 1980 (2 vols) (unpublished PhD. Thesis), 311, n° 8.

45. Dioscorides, I, 13 (= vol. 1, 17, l. 8).

46. M, 46-50 ; FP, 214-219. See : Olck, " Casia ", *RE*, III, 2 (1899), col. 1637-1650 ; H. Gams, " Casia ", *DKP*, 1 (1964), col. 1065 ; J. Fortes, " Los fitónimos griegos ", Estudios de lingüística y paleoetnobotánica (Barcelona, 1980) (2 vols) (unpublished PhD. Thesis), 325, 341-342 ; C. Hünemörder, " Casia ", *NP*, vol. 2 (1997) col. 1101-1102.

47. Dioscorides, I, 16 (= vol. 1, 21, l. 21 [Arabia] ; 22, l. 1 [India], l. 3 [Syria]).

48. M, 84-86 ; FP, 243-245. See : J. Berendes, *Des Pedanios Dioskurides aus Anazarbos Arzneimittellehre in fünf Büchern*, Stuttgart, 1902, 41-42 (reprints : Vaduz, 1970 ; Graz, 1988) ; M.G. Raschke, " New Studies in Roman Commerce with the East ", in H. Temporini (ed.), *Aufstieg und Niedergang der römischer Welt*, vol. II, 9, Berlin, New York, 1978, 1012-1023, note 1489 ; J. Fortes, " Los fitónimos griegos ", *Estudios de lingüística y paleoetnobotánica*, Barcelona, 1980 (2 vols) (unpublished PhD. Thesis), 366.

49. Dioscorides, I, 25 (= vol. 1, 28, l. 19).

plant or drug	origin according to Dioscorides	origin according to modern Botany
malabathron	India[50]	China, Ceylon[51]
nardos	Eastern and Western side of a mountain in Asia[52]	Himalaya and Western side of the Pamir and the Hindu Kush[53]
nardos oreinê	Cilicy, Syria[54]	Cilicia, Syria[55]

To identify the reason why these chapters were cancelled, we need to bear in mind that the re-elaborations of Dioscorides' text we have mentioned seem all resulting from a same intention : a practical one, aimed to transform the *De materia medica* into a handbook for daily use, perhaps of common physicians. If the chapters we are dealing with were cancelled, it was probably because the plants and drugs they described were no more available or, at least, not easily available ; consequently, the text dealing with these plants seemed to be useless and it was omitted.

The re-introduction of these drugs into the Athos text could seem to indicate their presence in the Byzantine World, *i.e.* their importation, possibly renewed, from the East. But it is not necessarily the case : it could result, only, from a textual borrowing from a more complete manuscript, in this case a copy of the original version. This interpretation might seem very probable, all the more because, during the period of production of the Athos text, Byzantium is known to have lived a cultural Renaissance (in fact, an internal revival of previous culture).

Another component of the Athos manuscript is significant in our perspective and provides the clue of the problem we are coping with : the illustrations of some plants. On f. 6^{recto} appear the representations of *malabathron, cinnamômon*, and *nardos* (plate 1). Besides the fact that these drugs are among those described by Dioscorides as oriental and appear as such in the list of the chap-

50. Dioscorides, I, 12 (= vol. 1, 16, l. 16).

51. M, 46-50 ; FP, 214-219. See : Steier, " Malabathron ", *RE*, XIV, 1 (1928), col. 818-823 ; K. Ziegler, " Malabathron ", *DKP*, 3 (1969), col. 923 ; M.G. Raschke, " New Studies in Roman Commerce with the East ", in H. Temporini (ed.), *Aufstieg und Niedergang der römischer Welt*, vol. II, 9, Berlin, New York, 1978, 910, note 1045 ; 1012-1023, note 1489 ; J. Fortes, " Los fitónimos griegos ", *Estudios de lingüística y paleoetnobotánica*, Barcelona, 1980 (2 vols) (unpublished PhD. Thesis), 420.

52. Dioscorides, I, 7 (= vol. 1, 11, l. 10-11).

53. M, 88-91 ; FP, 322-325. See : Steier, " Nardus ", *RE*, XVI, 2 (1935), col. 1705-1715 ; K. Ziegler, " Nardos ", *DKP*, 3 (1969), col. 1572 ; M.G. Raschke, " New Studies in Roman Commerce with the East ", in H. Temporini (ed.), *Aufstieg und Niedergang der römischer Welt*, vol. II, 9, Berlin, New York, 1978, 926-927, note 1116 ; J. Fortes, " Los fitónimos griegos ", *Estudios de lingüística y paleoetnobotánica*, Barcelona, 1980 (2 vols) (unpublished PhD. Thesis), 458-460.

54. Dioscorides, I, 9 (= vol. 1, 13, l. 25).

55. M, 118, n° 65.

ters added in the Athos text to that one of its source, their representations result to be particularly reach of information.

From a comparison between our manuscript and all the extant illustrated copies of Dioscorides' Greek text currently known[56], it appears that these plants are not represented in no other manuscript than the *Athous* one[57].

From a further comparison (with all the extant codices noticed of the Arabic versions of Dioscorides' treatise)[58], it results that the representation of *malabathron* (upper, left) reminds that of the same drug in the manuscript of Leiden, Universiteitsbibliotheek, signature or. 289[59], f. 8recto (plate 2). In the Leiden manuscript, the *malabathron* is represented, in conformity with Dioscorides' description[60], as a green leaf floating on water, with a lanceolate shape and a marked central nerve. In our manuscript, the general feature is the same, although there are two main differences : a first is the fact that the leaf does not float on water ; water is represented, however, by means of a sort of dark-blue/green ribbon, manifestly suggesting a river. As for the second difference, the leaves are represented vertically, like trees along a river.

Similarities and differences converge on suggesting that the copyist (and/or illustrator) of the *Athous*[61] depended upon an image like that one of the Leiden manuscript, without knowing the true appearance of the whole vegetal : he reproduced, indeed, the general shape it has in the *Leidensis* (similarity), but transformed it mistakenly into a tree (difference), obviously because he did not know the vegetal in nature, but depended on written and illustrated records ; at

56. There is not, at present, an inventory of these manuscripts. We have prepared a list, to be published in : A. Touwaide, " Les représentations de plantes du Traité de matière médicale de Dioscoride et la botanique du Ier siècle de notre ère ", Herausgegeben von W. Haase, *Aufstieg und Niedergang der römischer Welt*, vol. II, 37, 4, Berlin, New York (forthcoming).

57. There are no lists of the plant illustrations by manuscripts, except for the *Vindobonensis* (among the different published, see the most recent in H. Gerstinger, " Dioscurides. Codex Vindobonensis med. gr. 1 der Österreichischer Nationalbibliothek. Kommentarband zu der Faksimileausgabe ", *Codices Selecti Phototypice Impressi*, vol. XII*, Graz, 1970, 10-28). We have prepared a synoptic table of plant representations in all the Greek extant manuscripts sources of the subsequent traditions ; it will be published as an appendix of A. Touwaide, " Les représentations de plantes du Traité de matière médicale de Dioscoride et la botanique du Ier siècle de notre ère ", vol. II, *op. cit.*, 37, 4.

58. For the inventory of the illustrated Arabic manuscripts of Dioscorides' treatise, see : E.J. Grube, " Materialen zum Dioskurides Arabicus ", in R. Ettinghausen (ed.), *Aus der Welt der islamischen Kunst. Festschrift für Ernst Kühnel*, Berlin, 1959, 163-194, and, recently : M.M. Sadek, *The Arabic Materia Medica of Dioscorides*, Saint-Jean-Chrysostome (Québec), 1983, which does not bring any new element, however.

59. On this manuscript, see mainly : M.M. Sadek, *The Arabic Materia Medica of Dioscorides,* Saint-Jean-Chrysostome (Québec), 1983. Although it is more problematic than helpful (see : G. Strohmaier, review of Sadek (1983), *Gnomon*, 57 (1985), 743-745), this study has to be used, being the only one at disposal.

60. See : I, 16 (= vol. 1, 16, l. 15-17).

61. Or : of its source. Contrarily to what happens for the text, it does not seem possible, in the current status of the research, to determine if the illustration has been executed or copied in the *Athous* manuscript; in other words : if it is original or a reproduction of an earlier source.

the most, he saw the drug (*i.e.* the dry leaf), but surely not the plant as a whole[62].

The image of the nardos in the Laura manuscript (plate 1 : under, left and right) allows to reach similar conclusions : it may be compared to that of the same plant in the *Leidensis*, f. 6[recto] (plate 3). Here too, the similarities and differences testify the dependence of our manuscript on an image like that one : the same general form of the plants, but multiplication of the species by means of a change of colours (green and brown), so as to represent the two varieties described in Dioscorides' text[63].

It is probably not casual that these illustrations appear in our manuscript on the leaf which bears the index of the additional chapters : contrarily to what happens for other drugs, the figures do not appear before or next to the text related to the illustrations, and no blank space is left on the side of this text, as it happened in some cases ; instead, these figures were copied here, where a blank space were left in the book, *i.e.* on the lower part of the index in which they are listed. Moreover, these figures probably did not appear in the secondary source by means of which the text of the primary source has been improved.

However, these illustrations do not signify necessarily that the drugs represented were available on the Byzantine market ; they indicate only that Arabic manuscripts were probably at disposal in Constantinople or, at least, were known in the scriptorium in which our manuscript was copied.

The third figure of the same page (plate 1 : upper part, right), representing the *kassia* (*i.e.* the cinnamon), is significant : contrarily to the previous ones, it is not a reproduction of an Arabic illustration, since no representation of cinnamon appears in Dioscorides' Arabic manuscripts[64] ; on the contrary, it is clearly a drawing of a stick of cinnamon put into the earth as though it were a tree, with a green leaves-like top. This seems to demonstrate that the illustration was made from a true cinnamon stick, transformed into a tree as we have told and exactly as it was the case for the malabathron leaf.

It must be stressed that the representation of kassia created in this way is in full accordance with the system of illustration in Arabic manuscripts : while in Greek manuscripts, plants are represented entirely and alone, in an analytical

62. The vertical tree-like representation of the leaf could also suggest that the illustrator of the manuscript is not the copist of the text, because of the contradiction between textual (leaf on water) and iconic (tree-like leaf) data (at least, it would signify that the artist who represented the plant did not read the text he had to illustrate).

63. I, 7 (= vol. 1, 11, l. 8 - 12, l. 1), where Dioscorides mentioned a brownish colour (11, l. 12).

64. Once again, there is not at disposal a list of the illustrations of the extant Arabic illustrated Dioscorides (which are the sources of the subsequent copies). We have prepared a synoptic table, which hoping to publish it in a near future.

way, in thc Arabic codices there is a tendency to represent the plants inserted into their natural context, in a realistic way[65].

A further possible sign of an Arabic influence on our manuscript could be the presence of figures of black people. The question is problematic : human figures in manuscripts of materia medica have been interpreted as creations of the Mid-Byzantine Period[66] and are still considered to be explanatory additions[67]. It can be argued that this type of figure is ancient, because it appears in Arabic manuscripts, as well as in our Athos codex, in representations of plants linked with godnesses of Classical Mythology. Although this is not the place to discuss this topic[68], we have to say that this indicates that illustrations like those are most probably ancient (godnesses of Classical Mythology) and were furtherly transformed (depaganized), so as to be compatible with the new cultural context born with the affirmation of Christianism.

This kind of figures (which appears also in some other cases where human figure is indispensable, like, for example, to represent human urine) was almost abandoned in Greek manuscripts[69], while it has been particularly developed in Arabic codices[70].

It happens in our Athos codex (or, rather : it returns in the text of Dioscorides), with no less than thirteen illustrations, among which five with black people[71]. Even if the second element is not necessarily significant (better : not

65. For a comparison of these two types of plant illustration, see A. Touwaide, *Farmacopea Araba Medievale. Codice Ayasofia 3703*, Milano, 1992-1993 (4 vols), where a page (at least) of all the extant Greek and Arabic illustrated manuscript sources of the subsequent tradition are reproduced in full colour.

66. See : K. Weitzmann, *Illustrations in Roll and Codex*, Princeton, 1947, 52 ; K. Weitzmann, " The Greek Sources of Islamic Scientific Illustrations ", in G.C. Miles, *Archaeologica Orientalia in Memoriam Ernst Herzfeld,* New York, 1952, 23 ; K. Weitzmann, " Ancient Book Illumination ", *Martin Classical Lectures*, 15, Cambridge (Mass.), 1959, 15.

67. See : G. Orofino, " Gli erbari di età sveva ", in M. Oldoni, G. Orofino, A. De Martino, M. Pasca, E. Alfinito, M.A. D'Aronco (eds), *Gli erbari medievali tra scienza, simbolo, magia. Scrinium. Quaderni ed estratti di " Schede Medievali "*, 10, Palermo, 1994, 325-346, 342-343 ; G. Orofino, " Vedere la natura. Dal ritratto strumentale al ritratto d'ambiente ", *Vedere i Classici. L'illustrazione libraria dei testi antichi dall'età romana al tardo medioevo*, A cura di M. Buoncore, Roma, 1996, 69-76, 72.

68. We prepare a study on this point.

69. Few human figures appear also in the manuscript Paris, Bibliothèque Nationale, *graecus* 2179 (see : A. Touwaide, *Farmacopea Araba Medievale. Codice Ayasofia 3703*, Milano, 1992-1993 (4 vols), ill. n° 78 & 80, for example) (on the manuscript, see, recently : A. Touwaide, " Le Traité de matière médicale de Dioscoride en Italie depuis la fin de l'Empire romain jusqu'aux débuts de l'école de Salerne. Essai de synthèse ", in A. Krug (ed.), *From Epidaurus to Salerno. Symposium held at the European University Centre for Cultural Heritage, Ravello, April, 1990.* PACT, 34 [1992], Rixensart, 1994, 275-305, 288-295).

70. No study has been devoted to this point, which would deserve, instead, to be fully researched. For a full page colour reproduction of all the extant representations of human figures in the manuscript known in the bibliography as *Ayasofia 3703* (currently : Istanbul, Süleymaniÿe Kütüphanesi, *Ayasofia 3703*) (Baghdâd [?], 1224), see : A. Touwaide, *Farmacopea Araba Medievale. Codice Ayasofia 3703*, Milano, 1992-1993 (4 vols), passim.

71. Reproduction in colour in : A. Touwaide, *Farmacopea Araba Medievale. Codice Ayasofia 3703*, vol. 1, *op. cit.*, illustrations n° 81-88.

as significant as one would think at first glance), the presence of human figures could, instead, allow us to conclude that there was a possible influence of the Arabic system of plant representation on the manuscripts of Dioscorides' Greek text.

Towards a theoretical conceptualisation of the phenomenon of transfer of knowledge between Medieval cultures

From our examination, it emerges that the assimilation of Arabic data into Byzantine Pharmacology during the 11th century A.D. was probably by far more complex than what expected from the analysis currently made of astronomical treatises to which we referred at the beginning of our discussion, in order to clarify the case of Pharmacology. From the case of Dioscorides' *De materia medica* in the Athos manuscript, we have identified, indeed, at least the following phases (the sequence does not necessarily reflect the historical one) :

- drugs of Oriental origin were probably available on the market in Byzantium and, consequently, they were used in practical therapeutical treatment by common physicians ;
- Greek textual data of earlier epoch dealing with Oriental drugs (fallen into disuse, probably due to the absence of the drugs on the market) were recuperated ;
- Arabic illustrated books on Pharmacology were at disposal in Byzantium ;
- Arabic treatises on Pharmacology were probably translated into Greek ;
- Arabic illustrations were reproduced in Greek manuscripts containing pharmacological texts, and new illustrations were created on the model of the Arabic ones ;
- translated Arabic data were introduced into new Greek treatises.

Before further discussion, these possible phases rise an important question, that of the relation between textual and real data : did the written treatises reflect the concrete practice, or were they copied from other books (Arabic, in our case), without contact of any type with the reality they refer to ? In other words : which was the main historical sequence of these phases ?

Lacking an explicit information on this point, we may look at another phenomenon of transfer of data and knowledge to clarify the question. In order to avoid anachronism, we shall confront the data of our analysis with what is known of such phenomena of the Medieval World. In fact, a comparison with the best known phenomenon of importation of drugs from a culture to another one, that of the drugs of the New World introduced in Europe during the Renaissance and the Early Modern Times, could provoke wrong inferences, because of the deep gap between the societies, trade system and organisation, means of transport and others, in Byzantium and the West.

Among the phenomena of the Medieval World which are known enough to be compared with the case we are studying[72], a first one is probably the Pre-Salernitan West[73]. It leads to evidence both the presence of Oriental drugs in Western recipe books and their availability on the market[74]. The absence of information about commercial activity in this sector has been attributed to the fact that pharmaceutical products pertain to folk-practice, which escaped notice ; consequently, there would have been a gap between practice and written record, *i.e.* : the former preceded the latter. Since this analysis has been done, it has been confirmed by recent fresh study of material previously known, such as the so-called Lorscher Arzneibuch, written around 800 A.D. in the Royal Monastery of Lorsch, in Germany[75].

The case of Al-Andalus, *i.e.* of Southern Spain, is not dissimilar : it shows, indeed, that oriental vegetals quoted in later treatises written in the Western Arabic World were imported into Muslim Spain as early as 756-788 A.D.[76], *i.e.* no more than half a century after the first penetration of Omeyyad troupes into the Iberian Peninsula (711 A.D.). Later on, a local school of Pharmacology developed, supposed to be directly linked with the centre into which these plants were introduced, and which associated the local Latin and Wisigothic traditions with these newly arrived Oriental material[77].

The cultural history of Baghdâd in the 9th century A.D. will probably seem to contradict the conclusion to which these two phenomena lead, *i.e.* the antecedence of practice on the written record. In Baghdâd, in fact, textual data have been imported before practice of any type, by means of an enterprise of translation duly organized[78]. As a proof of this, one could allege the lack of Arabic

72. We do not take into account the problems of the translation itself, which is different.

73. For a general overview of Western Medieval herbalism, see recently : L. Ehrsam Voigts, " Herbals, Western European ", in J.R. Strayer (ed.), *Dictionary of the Middle Ages*, vol. 6, New York, 1985, 180-182.

74. See : J.M. Riddle, " The Introduction and Use of Eastern Drugs in the Early Middle Ages ", *Sudhoffs Archiv*, 49 (1965), 185-198 (reprint in : J.M. Riddle, *Quid pro quo : Studies in the history of drugs,* n° II, *op. cit.*).

75. See mainly : U. Stoll, " Das Lorscher Arzneibuch. Ein medizinisches Kompendium des 8. Jahrhunderts (codex Bambergensis Medicinalis 1) ", *Text, Übersetzung und Fachglossar, Sudhoffs Archiv*, Beiheft 28, Stuttgart, 1992.

76. See : J. Samsó, " Las ciencias de los antiguos en Al-Andalus ", *Colecciones MAPFRE 1492*, XVIII : *Colección Al-Andalus*, 7, Madrid, 1992, 20-22.

77. See : *Idem*, 40-42, 110-116. On this school, see also : S.K. Hamarneh, Sonnedecker, " A Pharmaceutical View of Abulcasis Al-Zahrâwî in Moorish Spain with Special Reference to the " Adhân " ", *Janus*, Suppléments, V, Leiden, 1963, 1-12 ; J. Vernet, *La cultura hispanoárabe en Oriente y Occidente*, Barcelona, Caracas, México, 1978 (revised edition [in French translation] : " Ce que la culture doit aux Arabes d'Espagne ", *La Bibliothèque Arabe, Collection l'Histoire décolonisée*, Paris, 1985 [quoted here], 81-89.

78. For a recent study of the introduction of Greek Pharmacology into the Arabic World, see : A. Touwaide, *L'intégration de la pharmacologie grecque dans le monde arabe. Une vue d'ensemble*, vol. 2 (1995), in L.R. Angeletti, A. Touwaide (eds), (1994-1995), 159-189 ; L.R. Angeletti, A. Touwaide, " Medieval Arabic Medicine ", in L.R. Angeletti, A. Touwaide (eds), *Medicina nei Secoli,* Roma, 1994-1995, 6, 2 [1994] & 7, 1 [1995], (2 vols).

names stated by the scientists of the 9th century to translate the Greek plant names. But the case is probably exceptional and, from a closer examination, Arabic Pharmacology reveals also a phenomenon of transfer of knowledge[79] : that dealing with Far East drugs, introduced into the body of data born within the Arabic World by means of the translation of Greek treatises[80]. These products were probably present on the market and in daily life before being assimilated into the textual body of knowledge artificially introduced into Arabic scientific culture, thanks to the translation. Consequently, the case of 9th century Baghdâd and the constitution of a new corpus of scientific data directly from earlier texts is an exception and does not contradict the data of the other phenomena of assimilation we have analysed.

When we come back to the case of Byzantium during the 11th century, it seems, on the basis of these comparisons, that the presence on the market of the drugs we thought to have traced, could have been a reality, and that it could have very well preceded the mention of these drugs in written works[81]. Consequently, the presence of Oriental drugs in the Athos text has probably not to be considered as the first sign of a transformation (or rather : its starting point), but as a first achievement of a previous phenomenon. In other words : the written treatise records practice, rather than generate it.

It is probably possible to go a little further and to identify some chronological data of this possible sequence of facts : if the Athos manuscript reproduced, as we suppose it did, an earlier text, the recuperation of Dioscorides' chapters dealing with Oriental drugs could be dated from the first half of the 11th century. This implies that these products were available previously on the market in Constantinople, so as to have lead to their re-introduction in Dioscorides' text ; *i.e.* a presence at least before the 11th century.

79. This point is not clearly pointed out by M. Levey, *Early Arabic Pharmacology. An Introduction Based on Ancient and Medieval Sources*, Leiden, 1973, although the work identifies the components (among others of the Far East) of Classical Arabic Pharmacology.

80. On this point, see e. g. : S.K. Hamarneh, " Sources and Development of Islamic Medical Therapy and Pharmacology ", *Sudhoffs Archiv*, 54 (1970), 30-48 (reprint in : S.K. Hamarneh, *Health Sciences in Early Islam*, vol. 2, *op. cit.*, 217-228), S.K. Hamarneh, " Development of Islamic Medical Therapy in the 4th/10th century ", *Journal of History of Medicine and Allied Sciences*, 27 (1972), 65-79 (reprint in : S.K. Hamarneh, *Health Sciences in Early Islam*, *op. cit.*, 229-241), and S.K. Hamarneh, " The Pharmacy and Materia Medica of Al-Biruni and Al-Ghafiqi. A comparison ", *Pharmacy in History*, 18 (1976), 3-12 (reprint in : S.K. Hamarneh, *Health Sciences in Early Islam*, *op. cit.*, 99-109) ; S.K. Hamarneh, *Health Sciences in Early Islam*, *op. cit.*

81. Given that the Oriental drugs were known in Classical Antiquity (see : M.G. Raschke, " New Studies in Roman Commerce with the East ", in H. Temporini (ed.), *Aufstieg und Niedergang der römischer Welt*, vol. II, 9, Berlin, New York, 1978, and J.I. Miller, *The Spice Trade of the Roman Empire. 29 B.C. to A.D. 641*, Oxford, 1969), it does not necessarily imply that their trade was totally interrupted from the end of Classical Antiquity to the 11th century. Surely reduced because of the wars between the Byzantine and the Persian and, later, the Arabic Empires, it probably did never cease completely, above all in frontier regions, but changed greatly. At the epoch under consideration, it developed so as to reach a level by far higher than in previous time, and, consequently, to have an impact on written records.

On the other hand, if the Athos' illustrations are original, it would denote a second phase of integration of Oriental drugs into Greek books, with the addition of new figures born from a direct observation of the drugs themselves. This would have happened at the epoch of the copy of the manuscript, it is to say about mid-11th century.

Finally, textual data resulting from the translation into Greek of Arabic information were introduced into new Byzantine treatises such as that of Symeon Seth.

As always, this tentative sequence of facts and chronology has not to be considered as absolute, but only indicative, as historical events are never simple and constitute a main stream, compound, on its turn, of a variety of tendencies in diverse senses.

Conclusions

The presence of Eastern drugs in Byzantine treatises of Pharmacology dating back to the 11th century seems thus to be a recent phase of a larger phenomenon began earlier. Expressed in terms of a general paradigm aimed to conceptualize the general phenomenon of transfer of knowledge between cultures, this would mean that practice preceded the registration in written records.

This paradigm has probably a limited validity and could not be transferred as such to other sectors, being specific of Pharmacology. In the case of Astronomy, for example, and probably also for theoretical Medicine[82], the relation between practice and presence of external data in written works could be inverted : practice would have been generated by written treatises.

As a consequence, the transfer of scientific data from a culture to another in the Medieval World could be analysed according to two opposite paradigms : the first, which could be qualified *ascendant*, in which practice preceded written records, and the second, which could be considered *descendant*, in which, instead, practice came from written data[83].

The presence of one or the other paradigm was probably determined by the field in which phenomena of transfer happened : the *ascendant* one is probably more typical of Pharmacology and the second one of other sectors.

82. For a study of the introduction into Arabic Science of a specific field of Greek theoretical Medicine (*psycho-pathology*), see : M.W. Dols, *" Majnûn " : the Madman in Medieval Islamic Society*, in D.E. Immisch (ed.), Oxford, 1992.

83. These adjectives do not refer to a concept of higher vs. lower science (viz. erudite vs. popular, or : codified vs. oral). They are adopted here for the clarity and correspond to practice of a scientific activity and formal science.

In other words : written Pharmacology reflects and records real practice[84], while, in other scientific sectors, textual data precede and, consequently, guide practical activity.

ABBREVIATIONS :

DKP = *Der Kleine Pauly* ;

NP = *Der Neue Pauly* ;

RE = *Realencyclopädie der classischen Altertumswissenschaft.*

84. For a similar theoretical analysis of another period of Pharmacology (Classical Antiquity), see : J. Scarborough, " Adaptation of Folk Medicines in the Formal Materia Medica of Classical Antiquity ", in J. Scarborough (ed.), *Folklore and Folk Medicines*, Madison, 1987, 22-32. For the example of a field of medical practice (contraception in Antiquity and in the West), see : J.M. Riddle, *Contraception and Abortion from the Ancient World to the Renaissance*, Cambridge (Mass.), London, 1992 ; J.M. Riddle, *Eve's Herbs. A History of Contraception and Abortion in the West*, Cambrige (Mass.), London, 1997. For a general analysis of Therapeutics in this sense, see : A. Touwaide, Une histoire du médicament en Occident, Paris (forthcoming).

It must be stressed that this concept does not appear in current bibliography on the History of Therapeutics and Pharmacology, although some works analyse precisely the birth of formal Pharmacology; see, among recent production : H.M. Koelbing, " Die ärztliche Therapie. Grundzüge ihrer Geschichte ", *Grundzüge*, 58, Darmstadt, 1985 ; G. Lorenz, " Antike Krankenbehandlung in historisch-vergleichender Sicht. Studien zum konkret-anschaulichen Denken ", *Bibliothek der klassischen Altertumswissenschaft*, Neue Folge, 2. Reihe, Band 81, Heidelberg, 1990 ; J. Mann, *Murder, Magic, and Medicine,* Oxford, 1992 ; G. Stille, *Der Weg der Arznei von der Materia Medica zur Pharmakologie. Der Weg von Arzneitmittelforschung und Arzneitherapie*, Mit Beiträgen von M.H. Bickel, H. Göing, Karlsruhe, 1994 ; C. Rätsch, " Heilkräuter der Antike in Ägypten, Griechenland und Rom. Mythologie und Anwendung einst und heute ", *Diederichs Gelbe Reihe*, 115, Alte Welt, München, 1995 ; W. Müller-Jahnce, C. Friedrich, unter Mitarbeit von J. Paulus, *Geschichte der Arzneimitteltherapie*, Stuttgart, 1996.

This question is fundamental, as it justifies the research procedure of current Ethnopharmacology (see : P.A. Cox, " Ethnopharmacology and the search for new drugs ", in D.J. Chadwick, J. Marsh (eds), *Symposium on Bioactive Compounds from Plants, held in collaboration with the Chulabhorn Research Institute at the Royal Orchid Sheraton Hotel, Bangkok, Thailand, Feb. 20-22, 1990*, Chichester, 1990, 40-47 (Ciba Foundation symposium ; 154) ; and N.R. Farnsworth, " Ethnopharmacology and drug development ", in D.J. Chadwick, J. Marsh (eds), *Symposium on Ethnobotany and the search for new drugs, held at The Hotal Praia Centro, Fortaleza, Brazil, 30 November - 2 December 1993*, Chichester, 1994, 42-51 (Ciba Foundation symposium ; 185).

ILLUSTRATIONS

Plate 1. Athous, Megistis Lauras, Ω 75, f. 6^{recto}.

Plate 2. Leiden, Universiteitsbibliothek, Or. 289, f. 8^{recto}.

Plate 3. Leiden, Iniversiteitsbibliothek, Or. 289, f. 6recto.

Contributors

Alparslan Açikgenç
International Institute of Islamic Thought and Civilization
Kuala Lumpur (Malaysia)

Cemil Akdogan
International Institute of Islamic Thought and Civilization
Kuala Lumpur (Malaysia)

Mashallah Ali-Ahyaie
Tehran (Iran)

Hélène Bellosta
Institut Français d'Etudes Arabes
Damas (Syrie)

Emilia Calvo
Universidad de Barcelona
Barcelona (Spain)

Mercè Comes
Universidad de Barcelona
Barcelona (Spain)

Gregg De Young
American University in Cairo
Cairo (Egypt)

Carmen Escribano Ródenas
Universidad de Madrid
Madrid (Spain)

Mehrnaz Katouzian-Safadi
Centre d'Histoire des Sciences et des Philosophies Arabes et Médiévales
Villejuif (France)

E.S. Kennedy
Doylestown, PA (USA)

Abdulhai Komilov
Tajik Academy of Sciences
Dushanbe (Tajikistan)

Paul Lettinck
ISTAC
Kuala Lumpur (Malaysia)

Irina Luther
Institute for the History of Science and Technology
Moscow (Russia)

Juan Martos Quesada
Universidad de Madrid
Madrid (Spain)

Roshdi Rashed
Centre d'Histoire des Sciences et des Philosophies Arabes et Médiévales
Villejuif (France)

Mònica Rius
Universidad de Barcelona
Barcelona (Spain)

Mariam M. Rozhanskaya
Institut d'Histoire des Sciences et des Techniques
Moscou (Russie)

Petra Schmidl
Johann Wolfgang Goethe University
Frankfurt am Main (Germany)

Nouha STÉPHAN
Université Paris 7
Paris (France)

Alain TOUWAIDE
Consejo Superior de Investigaciones Científicas
Madrid (Spain)

PROCEEDINGS OF THE XXth INTERNATIONAL CONGRESS

VOLUMES' LIST

VOLUMES' LIST

Vol. I : Dieter HOFFMANN, Benoît SEVERYNS, Raymond G. STOKES (eds), *Science, technology and political change* (De Diversis Artibus, Tome 41, N.S. 4).

Vol. II : Gérard EMPTOZ, Patricia Elena ACEVES PASTRANA (eds), *Between the natural and the artificial. Dyestuffs and medicine* (De Diversis Artibus, Tome 42, N.S. 5).

Vol. III : Charles GALPERIN, Scott F. GILBERT, Brigitte HOPPE (eds), *Fundamental changes in cellular biology in the XXth century* (De Diversis Artibus, Tome 43, N.S. 6).

Vol. IV : Paul BENOIT, Catherine VERNA (eds), *Le charbon de terre en Europe Occidentale avant l'usage industriel du coke* (De Diversis Artibus, Tome 44, N.S. 7).

Vol. V : Celina A. LÉRTORA-MENDOZA, Efthymios NICOLAÏDIS, Jan VANDERSMISSEN (eds), *The spread of the scientific revolution in the European periphery, Latin America and East Asia* (De Diversis Artibus, Tome 45, N.S. 8).

Vol. VI : Ekmeleddin IHSANOGLU, Ahmed DJEBBAR, Feza GÜNERGUN (eds), *Science, technology and industry in the Ottoman world* (De Diversis Artibus, Tome 46, N.S. 9).

Vol. VII : Michel LETTE, Michel ORIS (eds), *Technology and engineering* (De Diversis Artibus, Tome 49, N.S. 12).

Vol. VIII : Pier Daniele NAPOLITANI, Pierre SOUFFRIN (eds), *Medieval and classical traditions and the Renaissance of physico-mathematical sciences in the 16th century* (De Diversis Artibus, Tome 50, N.S. 13).

Vol. IX : Alain ARRAULT, Catherine JAMI (eds), *Science and technology in East Asia* (De Diversis Artibus, Tome 51, N.S. 14).

Vol. X : Goulven LAURENT (ed.), *Earth sciences, geography and cartography* (De Diversis Artibus, Tome 53, N.S. 16).

Vol. XI : Denis BUICAN, Denis THIEFFRY (eds), *Biological and medical sciences* (De Diversis Artibus, Tome 54, N.S. 17).

Vol. XII : Suzanne DÉBARBAT, Gérard SIMON (eds), *Optics and astronomy* (De Diversis Artibus, Tome 55, N.S. 18).

Vol. XIII : Eberhard KNOBLOCH, Jean MAWHIN, Serguei S. DEMIDOV (eds), *Studies in history of mathematics dedicated to A.P. Youschevitch* (De Diversis Artibus, Tome 56, N.S. 19).

Vol. XIV : Helge KRAGH, Geert VANPAEMEL, Pierre MARAGE (eds), *History of modern physics* (De Diversis Artibus, Tome 57, N.S. 20).

Vol. XV : Hans-Joachim BRAUN, Alexandre HERLEA (eds), *Materials : research, development and applications* (De Diversis Artibus, Tome 58, N.S. 21).

Vol. XVI : Maurice DORIKENS (ed.), *Scientific instruments and museums* (De Diversis Artibus, Tome 59, N.S. 22).

Vol. XVII : Michael Ciaran DUFFY (ed.), *Engineering and Engineers* (De Diversis Artibus, Tome 60, N.S. 23).

Vol. XVIII : Michel BOUGARD (ed.), *Alchemy, chemistry and pharmacy* (De Diversis Artibus, Tome 61, N.S. 24).

Vol. XIX : Andrée DESPY-MEYER (ed.), *Institutions and societies for teaching, research and popularisation* (De Diversis Artibus, Tome 62, N.S. 25).

Vol. XX : Erwin NEUENSCHWANDER, Laurence BOUQUIAUX (eds), *Science, philosophy and music* (De Diversis Artibus, Tome 63, N.S. 26).

Vol. XXI : S.M. Razaullah ANSARI (ed.), *Science and technology in the Islamic world* (De Diversis Artibus, Tome 64, N.S. 27).

Proceedings of the XXth International Congress

Author Index

AUTHOR INDEX

Note to the reader : the present index indicates the authors of the papers included in the proceedings of the XXth International Congress of History of Science (21 volumes), Liège, 20-26 July 1997. The first column gives the authors' name and first name, the second, the volumes' number, the third, its number in the collection " De Diversis Artibus ", and the fourth the papers' first page.

NOM	VOLUME	TOME	PAGE
ABALLAGH Mohamed	VI	46	75
ABDOUNUR Oscar João	XIII	56	173
ABIKO Seiya	XIV	57	205
ACEVES PASTRANA Patricia	II	42	119, 165
AÇIKGENÇ Alparslan	XXI	64	7
AKDOGAN Cemil	XXI	64	23
AKHUNDOVA Svetlana	X	53	81
ALFONSO-GOLDFARB Ana Maria	II	42	119, 125
ALI-AHYAIE Mashallah	XXI	64	155
ALLAIRE Bernard	VII	49	125
ALMAÇA Carlos	X	53	9
ÁLVAREZ LIRES Mari	XIX	62	271
ALVES PORTO Paulo	XVIII	61	51
ALYABIEVA G. Valentina	XIII	56	259
AMARAL I.	XI	54	251
D'AMBROSIO Ubiratan	XIII	56	59
AMORIM DA COSTA M. Antonio	XVIII	61	169
ANCEL Bruno	IV	44	153
ANDERSON R.G.W.	XVI	59	145
ANDRIEU Bernard	XI	54	287
ANSARI S.M. Razaullah	XXI	64	
ARAMAKI Seiya	XIV	57	323

ARANDA Andrés	II	42	131
ARBOLEDA Luis Carlos	V	45	137
ARRAULT Alain	IX	51	125
AUGER Réginald	VII	49	125
AVRAMOV Iordan	XIX	62	265
AYED Naceur	II	42	103
AZMANOV Iskren	XI	54	303
BAKASOVA B. Zaryl	I	41	169
BALDWIN Martha	II	42	157
BALFOUR-PAUL Jenny	II	42	29
BANGE Christian	XI	54	263
BARJOT Dominique	VII	49	185
BARKER Peter	XII	55	43
BARTHA Lajos	XVI	59	319
BASTABLE J. Marshall	VII	49	309
BATES L. David	XIX	62	143
BAUDET C. Jean	XX	63	165
BEAUJOUAN Guy	XIII	56	19
BEKTAS Yakup	VI	46	139
BELHOSTE Jean-François	IV	44	187
BELLOSTA Hélène	XXI	64	71
BENNETT Jim	XVI	59	165
BENOIT Paul	IV	44	7, 49, 217
BENOIT Serge	XIX	62	175
BERGASA LIBERAL Javier	XII	55	177
BERNHARDT Hannelore	XX	63	105
BERTOL DOMINGUES Heloisa Maria	VII	49	207
BESSOT Didier	XVI	59	55
BETTAHAR Yamina	XIX	62	207
BEYLER H. Richard	III	43	39
BHATTACHARYA Asitesh	XV	58	165
BIANCHI Assunta	IV	44	199
BLANCHARD Ian	IV	44	61
BLAY Michel	XIV	57	17
BLET M.	XV	58	211
BLOUIN Daniel	XIX	62	175
BLUE Gregory	IX	51	11

BOLOTOWSKY Boris Michailovich	I	41	155
BOUGARD Michel	XVIII	61	7, 77
BOUQUIAUX Laurence	XX	63	
BOUTHIER Alain	IV	44	99
BRAUN Hans-Joachim	XV	58	7, 185
BRENNI Paolo	XVI	59	191
BRET Patrice	VI	46	101
BRIGAGLIA Aldo	VIII	50	47
BRUNET Sébastien	XX	63	181
BRUWIER Marinette	IV	44	199
DE BRZEZINSKI PRESTES Maria Elice	XI	54	73
BUICAN Denis	XI	54	9, 333
BURIAN M. Richard	III	43	121
BÜTTNER Johannes	III	43	11
BÜTTNER Manfred	X	53	139
BYÉ Pascal	XIX	62	89
CABALLER VIVES María Cinta	VII	49	291
CADDEN Joan	XVIII	61	11
CALAPÉS Maria Elvira	XV	58	139
CALVO Emilia	XXI	64	109
CAMEROTTA Michela	VIII	50	141
CANO PAVÓN M. José	XVII	60	107
CANTELAUBE Jean	IV	44	177
CARDONE Francesco	XVIII	61	217
CARDOSO DE MATOS Ana Maria	XIX	62	161
CARNEIRO Ana	V	45	67
CARNEIRO L. Fernando	VII	49	27
CAROLINO Luís Miguel	XII	55	167
CARRÉ Anne-Laure	XV	58	81
CARRION ARREGUI M. Ignacio	VII	49	265
CASTAÑEDA Luzia Aurelia	XI	54	229
CAUVIN Aldo	XIII	56	217
CECCHINI Michela	XIII	56	165
CERRETA Pietro	XIV	57	249
CERVERA JIMÉNEZ José Antonio	V	45	183
CHAMPEAU Virginie	XV	58	97
CHANG Hao	XVIII	61	209

CHEMLA Karine	XIII	56	25
CHEN Meidong	IX	51	139
CHEN-MORRIS Raz Dov	XII	55	83
CHEZEAU Nicole	XV	58	31
CHINENOVA Vera	XIII	56	201
CHRISTIANIDIS Jean	XIII	56	153
CLIFTON Gloria Christine	XVI	59	179
COHEN H. Floris	IX	51	21
COLAN Horia	XV	58	177
COLIN Stéphane	XIV	57	77
COMES Mercè	XXI	64	121
CORNELIS C. Gustaaf	XII	55	229
COTTE Michel	XIV	57	125
COUNIHAN Martin	XX	63	173
COWBURN Ian	IV	44	125
CREMO A. Michael	X	53	39
DA CONCEIÇÃO BURGUETE Maria	XVIII	61	207
DAMEROW Peter	VIII	50	115
DA ROCHA BARROS Alberto Luiz †	XIV	57	261
DAVOIGNEAU Jean	XVI	59	211
DÉBARBAT Suzanne	XII	55	137, 195
DEBRU Claude	XI	54	271
DE JONG Ella	XI	54	157
DEMBREVILLE Monique	II	42	17
DE MEY Mark	XII	55	51
DEMIDOV S. Serguei	XIII	56	33
DE NIL Erwin	XII	55	51
DÉPREZ-MASSON Marie-Claude	VII	49	115
DÉRÉ Anne-Claire	II	42	17, 69
DESPY-MEYER Andrée	XIX	62	9
DE VRIES J. Marc	VII	49	95
DE YOUNG Gregg	XXI	64	83
DHOMBRES Jean	VI	46	91
DI GIROLAMO Giulia	XIV	57	33
DIJKSTERHUIS Fokko Jan	XIV	57	57
DIOGO Paula	V	45	67
DJEBBAR Ahmed	VI	46	7, 49

DOATI Roberto	XX	63	217
DONNELLY F. James	XVIII	61	177
DORIKENS Maurice	XVI	59	9, 153
DORIKENS-VANPRAET Liliane	XVI	59	153
DOROFEEVA-LICHTMANN Véra	IX	51	111
DRIEVER L. Steven	X	53	149
DUDA Roman	XX	63	139
DUFFY Michael Ciaran	XVII	60	7
DUGAC Pierre †	XIII	56	11
DUMONT Simone	XII	55	195
DUPONT Jean-Claude	III	43	137
	XI	54	293
EGÍDIO REIS Fernando José	XIX	62	295
EMPTOZ Gérard	II	42	9
	XV	58	97
ENDREI Walter	VII	49	133
ERMOLAEVA S. Natalja	XIII	56	43
ESCRIBANO RÓDENAS Carmen	XXI	64	43
EVANS B. Rand	XVI	59	171
FAIDIT Jean-Michel	XVI	59	109
FALK Raphael	III	43	77
FARAGGIANA Giorgio	XVI	59	287
FAUST Don	XX	63	15
FERNANDES THOMAZ Manuel	XIV	57	105
FERRAZ H.M. Márcia	II	42	173
FIGUEIRÔA Silvia	XVIII	61	129
FIOCCA Alessandra	VIII	50	131
FISCHER Jean-Louis	III	43	117
FORD E. Charles	XIII	56	345
FOUGÈRES Dany	VII	49	201
FOURNIER Marian	XVI	59	253
FOX KELLER Evelyn	III	43	173
FREDETTE Raymond	VIII	50	125
FREIRE Olival Jr.	XIV	57	261
FRÍAS NÚÑEZ Marcelo	XI	54	83
FUCHS Eckhardt	XIX	62	337
FUNG Kam-Wing	V	45	147

GADISSEUR Jean	XX	63	119
GAGNON Robert	VII	49	201
GALERA Andrés	X	53	25
GALLES D. Carlos	V	45	117
GALPERIN Charles	III	43	75, 103
GAPAILLARD Jacques	XIV	57	41
GARÇON Anne-Françoise	XV	58	11
GARIBALDI A.C.	VIII	50	93
GARMA Santiago	XIII	56	269
GATTO Romano	VIII	50	83
GAUKROGER Stephen	XX	63	75
GAYON Jean	III	43	151
GILAIN Christian	XIII	56	207
GILBERT F. Scott	III	43	75, 163
GIL SAURÍ Miguel Ángel	XIII	56	311
GINGRAS Yves	XX	63	157
GIUSTI Enrico	VIII	50	33
GODARD Roger	XIII	56	237
GOICOLEA ZALA Javier	XIX	62	23
GOLAS J. Peter	IX	51	43
GOLDFARB José Luiz	II	42	137
GOMIS Alberto	XI	54	95
GONZÁLEZ BUENO Antonio	XI	54	115
DE GORTARI RABIELA Rebeca	X	53	57
GRATUZE Bernard	XV	58	211
GRMEK D. Mirko †	XI	54	11
GROULT Martine	XII	55	185
GUBIN I. Oleg	XX	63	83
GUERGOUR Youcef	VI	46	67
GUIJARRO MORA Víctor	XVI	59	309
GUILLAUME Anne	XIII	56	187
GÜNERGUN Feza	VI	46	7, 127
GUNTAU Martin	X	53	19
GUTINA Vera	III	43	63
HAMOU Philippe	XII	55	111
HASHIMOTO Keizo	V	45	159
	IX	51	85

HAYASHI Haruo	XIV	57	133
HEIN Katharina	I	41	57
HEIZER Alda	X	53	177
HELBING M.O.	VIII	50	141
HELEREA Elena	XVII	60	171
HERCOCK Marion	X	53	159
HERLEA Alexandre	XV	58	7
HESSENBRUCH Arne	V	45	11
HIDALGO CÁMARA Encarnación	VII	49	281
HILLS Richard Leslie	VII	49	37
HOCHADEL Oliver	XIX	62	289
HOFFMANN Dieter	I	41	13
HOLLO Szilvia Andrea	XVI	59	319
HOPPE Brigitte	III	43	9, 29
HORIO MONTEIRO Rosana	XVIII	61	129
HORNIX J. Willem	XVIII	61	137, 149
HOUZIAUX Léo	XII	55	217
HUERTA JARAMILLO Ana María	II	42	181
IAKOVLEV I. Vadime	XIV	57	65
IBÁÑEZ Itsaso	XIX	62	61
IHSANOGLU Ekmeleddin	VI	46	7, 11
ITAGAKI Ryoichi	XIV	57	243
JAMI Catherine	IX	51	7
JANKO Jan	III	43	21
JEANFILS Edward	XI	54	193
JOLY Bernard	XVIII	61	67
JONGMANS F.	XIII	56	251
KABZIŃSKA Krystyna	XIX	62	135
KAÇAR Mustafa	VI	46	81
KAISER Walter	XIV	57	327
KARPENKO Vladimír	XVIII	61	17
KATOUZIAN-SAFADI Mehrnaz	XXI	64	217
KATZ Claudio	VII	49	245
KELLER G. Alexander	XIV	57	349
KENNEDY E.S.	XXI	64	65
KERBE Friedmar	XV	58	155
KHABELASHVILI V. Albert	XIII	56	125

KHEIRANDISH Elaheh	XII	55	17
KHELFAOUI Hocine	I	41	191
KIM Yung Sik	IX	51	75
KININI Angélique	II	42	93
KIPNIS Nahum	XIV	57	189
KISTEMAKER Jacob	IX	51	59
KLAIRMONT LINGO Alison	XI	54	33
KNIGHT David	II	42	49
KNOBLOCH Eberhard	XIII	56	
	XVII	60	23
KÖHLER Piotr	XI	54	129
KOJIMA Chieko	XIV	57	275
KOLLERSTROM Nicholas	XII	55	145
KOMILOV Abdulhai	XXI	64	183
KONOVETS Olexander	I	41	107
KOSTINSKIY Grigoriy	X	53	125
KOZENKO V. Alexander	XII	55	213
KRAGH Helge	XIV	57	9, 175
KRANZ Horst	IV	44	21
KRIIS-ILVES Leili	XVI	59	261
KRIKSTOPAITIS A. Juozas	I	41	129
KUBBINGA Henk	XI	54	241
KURS Ott	XIX	62	41
LAFUENTE Antonio	V	45	99
LANGTON John	IV	44	141
LARDIN Philippe	IV	44	41
LA SALVIA Vasco	VII	49	13
LAURENT Goulven	X	53	7, 31
LA VERGATA Antonello	XX	63	97
LE GUET TULLY Françoise	XVI	59	211
LE GUYADER Hervé	III	43	159
LEIKOLA Anto	XIX	62	75
LENZNER Claudia	III	43	47
LEONE Matteo	XIV	57	197
LE ROUX Muriel	XV	58	107
LÉRTORA MENDOZA A. Celina	V	45	89, 123
LETTE Michel	VII	49	9, 61

LETTINCK Paul	XXI	64	189
LINDSAY FAULL Margaret	IV	44	13
LINS DE BARROS Henrique	VII	49	81
LINS DE BARROS Mauro	VII	49	81
LLOMBART José	XIX	62	307
LOCHER Kurt	XX	63	21
LOPES Maria Margaret	XVI	59	221
LÓPEZ-OCÓN CABRERA Leoncio	XIX	62	329
LÓPEZ VILLEGAS Virginia	XIX	62	349
LU Dalong	V	45	169
LUTHER Irina	XXI	64	91
LÜTHY Christoph	XVIII	61	29
MACHLINE Vera Cecília	XI	54	39
MACKIE Robin	XIX	62	125
MACRAKIS Kristie	I	41	15
MAFFIOLI S. Cesare	XVII	60	67
MAIER Helmut	XV	58	117
MALANIN Vladimir	XIV	57	65
MALAQUIAS Isabel	XVI	59	299
MALET Antoni	XII	55	127
MARAGE Pierre	XIV	57	
MARTINSON Helle	I	41	135
MARTINSON Karl	I	41	135
MARTOS QUESADA Juan	XXI	64	43
MASIULIS Kestutis	I	41	119
MATAGNE Patrick	XVII	60	185
MATHIJS Ernest	XX	63	57
MATSUMOTO Miwao	VII	49	317
MATSUMURA Noriaki	IX	51	85
MATVIICHINE Iaroslav	X	53	93
MAVEL S.	XVIII	61	117
MAWHIN Jean	XIII	56	
MAZZOTTI Massimo	XIII	56	107
MEHEUS Joke	XIV	57	143
MENGE Wemo	XIX	62	105
MÉTAILIÉ Georges	IX	51	33
MEYER Bruno	XVII	60	91

MGALOBLISHVILI Levan	I	41	163
MILLS A. Allan	XVI	59	29
MINELLI Carla	XX	63	237
MIRKOVIĆ Miroslav	XIV	57	89
MOHAMMAD A.H. Helmy	VI	46	115
MOM Gijs	XV	58	147
MONROY-NASR Zuraya	XI	54	59
MOREIRA I.C.	XIV	57	229
MORIMOTO Eiichi	XIII	56	319
MORIN Denis	IV	44	85
MOSCHEO Rosario	VIII	50	15
MOSSELMANS Bert	XX	63	57
MOSSMAN Susan	XV	58	127
MOUNTRIZA Ioanna	XIII	56	115
MOVSUMZADE M. Eldar	XV	58	239
MULDER Mimi	XI	54	157
MÜÜRSEPP Peeter	XX	63	45
NAKAJIMA Hideto	XII	55	161
NAKANE Michiyo	XIII	56	277
NAPOLITANI Pier Daniele	VIII	50	7, 9
NAVARRO LOIDI Juan	XIII	56	95
NEUBAUER Alfred	III	43	47
NEUENSCHWANDER Erwin	XX	63	
NICKEL Gisela	XIV	57	113
NICOLAÏDIS Efthymios	V	45	7, 33
NIKOLANTONAKIS Konstantinos	XIII	56	141
NISIO Sigeko	VII	49	225
NOBRE Sergio	XIII	56	89
NUNES Maria de Fátima	XIX	62	153
NUNES DOS SANTOS A.M.	XI	54	251
OESTMANN Günther	XVI	59	291
OGAWA Mariko	XI	54	219
O'HARA G. James	XVII	60	77
OKADA Tomoji	XIV	57	287
ONOPRIENKO Valentin	I	41	111
ORIS Michel	VII	49	9, 139
OSTROVSKY M.A.	VII	49	107

OTERO H. Mario	XX	63	113
OUESLATI Issam	II	42	103
PALLADINO Franco	XVI	59	271
PALLÓ Gábor	I	41	89
PATY Michel	XIV	57	261
PECHENKIN A.A.	V	45	59
PELLÓN GONZÁLEZ Inés	XVIII	61	231
PEREIRA Amélia	I	41	181
PÉREZ-GARRIDO María de Lurdes	XVIII	61	117
PERFETTI Amalia	XX	63	67
PEYRIERAS Nadine	III	43	89
PIASECKI Ian	IV	44	77
PIASKOWSKI Jerzy	XV	58	195
PIERSEAUX Yves	XIV	57	217
PIGHETTI Clelia	XIX	62	229
PIMENTEL Juan	V	45	99
PINEL Pierre	VIII	50	59
PINTO R.E.	XI	54	251
POSTNIKOV V. Alexei	X	53	105
POURPRIX Bernard	XIV	57	155
RAINA Dhruv	IX	51	95
RAMBALDI MORCHIO Gabriella	XVI	59	237
RAMOS Mariblanca	II	42	131
RAMOS LARA María de la Paz	V	45	91, 111
RASHED Roshdi	XXI	64	101
RATCLIFF M.J.	XI	54	67
RAWLINS Dennis	XII	55	139
RENN Jürgen	VIII	50	115
RHEINBERGER Hans-Jörg	III	43	133
RIBEIRO DE ANDRADE Ana Maria	XIV	57	313
RIEGER Simone	VIII	50	115
RIERA Juan	XI	54	141
RIUS Mònica	XXI	64	143
RIVADULLA Andrés	XX	63	149
ROBERT-HAUGLUSTAINE Anne-Catherine	XV	58	65
ROBERTS K. Gerrylynn	XVIII	61	189
ROBIC Marie-Claire	X	53	183

ROBOTTI Nadia	XIV	57	197
RODRÍGUEZ Martha Eugenia	XI	54	173
RODRÍGUEZ DE ROMO Ana Cecilia	XI	54	213
RODRÍGUEZ NOZAL Raúl	II	42	81
ROMÉRO Ricardo	XIV	57	165
ROSELL-GONZALEZ Pablo	XIX	62	257
ROSNER W. Robert	XVIII	61	161
ROSSI Arcangelo	XIV	57	183
ROTH Michael	XIX	62	81
ROXO BELTRAN Maria Helena	II	42	145
ROZHANSKAYA M. Mariam	XIII	56	51
	XXI	64	177
RUDAYA Svetlana	XIX	62	199
RUIZ-SOTO Galo	XIX	62	257
SAITO Ken	VIII	50	41
SALDAÑA Juan José	V	45	91, 111
SALZ Hanuš	XVII	60	161
SÁNCHEZ DÍAZ Gerardo	II	42	41
SANTOS María Josefa	X	53	69
SAPRYKIN D.L.	V	45	59
SARAIVA M.R. Luis	XIII	56	325
SATO Ken'ichi	IX	51	145
SATOFUKA Fumihiko	I	41	203
SAVCHUK Warfolomey	I	41	97
SCALDAFERRI Nicola	XX	63	207
SCHIMEK Michael	VII	49	171
SCHIRRMACHER Arne	XIV	57	295
SCHLOTE Karl-Heinz	XIII	56	285
SCHMIDL G. Petra	XXI	64	195
SCHMIDT Wilson	XIX	62	89
SCHMITT Stéphane	III	43	81
SCHOPMAN Joop	VII	49	231
SCHRÖDER Eberhard	XIII	56	81
SCHUPPENER Georg	XIII	56	73
SEGAL Jérôme	I	41	67
SEILS Markus	I	41	23
SELIGARDI Raffaella	XVIII	61	105

SEPPÄNEN Jouko	XI	54	339
SERRA Isabel	I	41	181
SETTLE B. Thomas	VIII	50	99
SÉVÉRINOV Konstantin	XVII	60	123
SÉVÉRINOVA Véra	XVII	60	123
SEVERYNS Benoît	I	41	9
SHAMIN N. Alexei	XI	54	279
SHEA R. William	XII	55	93
SICHAU Christian	VII	49	49
SIMÕES Ana	V	45	67
SIMON Gérard	XII	55	9, 119
SINDING Christiane	III	43	57
SMITH A. Mark	XII	55	11
SOBOLEVA V. Elena	XIX	62	193
SOFONEA Liviu	XVII	60	143, 171
SOLDATOV V. Alexander	XIX	62	193
SOUFFRIN Pierre	VIII	50	7, 107
SOULIER Isabelle	XV	58	211
SOULLARD Eric	IV	44	111
ŚREDNIAWA Bronisław	XIX	62	35
STAGNITTO Giuseppe	XIII	56	217
STÉPHAN Nouha	XXI	64	209
STIDWILL Robert	III	43	95
STOKES G. Raymond	I	41	13, 39
STOTCHIK M. Andrei	XI	54	165
von STROMER Wolfgang	XX	63	27
SUÁREZ Y LÓPEZ GUAZO Laura	XI	54	223
SUGIYAMA Shigeo	XVII	60	101
SUN Xiaochun	IX	51	59
SUTTO Jean-Pierre	VIII	50	73
SYMONS A. Sarah	XVI	59	13
SZABÓ N. József	I	41	83
TAHA Abdel-Kaddous	VIII	50	59
TAKAHASHI Ken'ichi	XII	55	29
TAMMIKSAAR Erki	XIX	62	41
TANKLER Hain	I	41	145
TAPDRUP Jan	XVI	59	199

TARDENT Pierre †	III	43	97
TASSORA Roberta	VIII	50	23
TAUB Liba	XVI	59	21
TCHESNOV V.M.	VII	49	107
TEN E. Antonio	V	45	77
THEILE Dominique	VII	49	193
THÉODORIDÈS Jean †	XI	54	207
THIEFFRY Denis	III	43	111
	XI	54	363
THIJSEN-VISSER Elisabeth	XVIII	61	137, 149
TOCA Ángel	XVIII	61	199
TOLMASQUIM A.T.	XIV	57	229
TOMSIN Philippe	VII	49	163
TOSI Lucía	XVIII	61	99
TOUSSAINT Friedrich	XV	58	49
TOUWAIDE Alain	XXI	64	223
TRÉPANIER Michel	VII	49	201
TRIADOU Patrick	III	43	143
TRIARICO Carlo	XVI	59	45
TSUKAHARA Togo	IX	51	85
UEMATSU Eisui	VII	49	225
ULLRICH Peter	XIII	56	297
UNGURU Sabetai	XII	55	83
VAMOS Éva Katalin	XIX	62	237
VAN BALEN Koenraad	XV	58	229
VANDERSMISSEN Jan	V	45	
VAN DE VIJVER Dirk	XV	58	229
VANPAEMEL Geert	XIV	57	
VAN TIGGELEN Brigitte	II	42	59
VASANTHA Arsampalai	XI	54	179
VAVILOV Yury Nicolaevich	I	41	155
VELHO Léa	XVIII	61	129
VERNA Catherine	IV	44	7, 31, 217
VIDOLIN Alvise	XX	63	227
VIESCA T. Carlos	II	42	131
VILADRICH GRAU Maria Mercè	XVI	59	73
VIZGUINE Victor	XVIII	61	61

VLAHAKIS N. George	V	45	45
VOGT Annette	XIX	62	111
VORONINA M. Margarita	XVII	60	117
WARDENGA Ute	X	53	117
WEISS Burghard	I	41	47
WIESENFELDT Gerhard	XIX	62	13
WILCZYŃSKI J. Witold	XX	63	197
WOHLMAN Avital	III	43	167
WÓJCIK Wiesław	XIII	56	339
WORONOFF Denis	IV	44	169
XÉNAKIS Christos	V	45	53
YAGI Eri	XIV	57	133
YAMAZAKI Masakatsu	I	41	223
YAROCHEVSKY M.G.	VII	49	107
YOSHIDA Haruyo	I	41	211
ZATERKA Luciana	II	42	137
ZAUN Jörg	XVI	59	325
ZEITHAMMER Karel	XVII	60	155
ZEMANEK Alicja	XI	54	47
ZERNER Martin	XX	63	9
ZHANG Baichun	XVII	60	133
ZOLOTOVITSKAYA A. Tamara	X	53	89
ZOTT Regine	XIX	62	317